Ceramic Change and the Maya Collapse

A study of pottery technology, manufacture and consumption at Lamanai, Belize

Linda Howie

BAR International Series 2373
2012

Published in 2016 by
BAR Publishing, Oxford

BAR International Series 2373

Ceramic Change and the Maya Collapse

ISBN 9781407309637 paperback
ISBN 9781407339429 e-format
DOI https://doi.org/10.30861/9781407309637
A catalogue record for this book is available from the British Library

BAR Publishing is the trading name of British Archaeological Reports (Oxford) Ltd. British Archaeological Reports was first incorporated in 1974 to publish the BAR Series, International and British. In 1992 Hadrian Books Ltd became part of the BAR group. This volume was originally published by Archaeopress in conjunction with British Archaeological Reports (Oxford) Ltd / Hadrian Books Ltd, the Series principal publisher, in 2012. This present volume is published by BAR Publishing, 2016.

BAR
PUBLISHING

BAR titles are available from:

BAR Publishing
122 Banbury Rd, Oxford, OX2 7BP, UK
EMAIL info@barpublishing.com
PHONE +44 (0)1865 310431
FAX +44 (0)1865 316916
www.barpublishing.com

ACKNOWLEDGEMENTS

This study would not have been possible without the generous financial support of the Social Sciences and Humanities Research Council of Canada and the Association of Commonwealth Universities. I would also like to extend my sincere gratitude to the many individuals and institutions that contributed to its successful completion, particularly the Institute of Archaeology, N.I.C.H., Belize. Your support and assistance with various aspects of the research were greatly appreciated.

I am particularly grateful to Dr. Elizabeth Graham for taking me under her wing and giving me access to such a wonderful set of material. Her insights, guidance and never-ending enthusiasm contributed greatly to this work, and its completion, in countless ways. I would also like to thank her husband, Dr. David Pendergast, for numerous conversations that clarified my thinking on the archaeological contexts of the pottery, as well as different aspects of the stylistic data.

I am also grateful to Dr. Peter Day, who supervised the PhD project from which this study has emerged, and to Dr. Dean Arnold and Dr. Roger Doonan, for their thoughtful critical feedback and encouragement of this publication.

There are several individuals who contributed to the laboratory analysis of the Lamanai pottery and geological specimens, by offering assistance and guidance: Dr. Charles Frederick, Erica Petersen, Rocky Hyacinth, Dr. Miguel Angel Cau Ontiveros, Dr. Patrick Quinn, Rosario and Millie Ku, Darcy Wiewall and Jorge Ramos. Their time and efforts were much appreciated. I would also like to thank Dr. Vassilis Kilikoglou of the Archaeometry Laboratory at the National Centre for Scientific Research in Athens, Greece, for facilitating the analysis of the pottery by NAA and SEM. His guidance at the microscope, as well as with the statistical analysis and interpretation of the chemical data were very much appreciated. In this regard, I am also grateful to Dr. Anno Hein, who provided guidance with sample preparation and with the preliminary analysis of the NAA data set.

The geological survey would not have been possible without the assistance of Jorge Vasquez and Raul Arevalo. I would like to extend my sincere gratitude to them for sharing their knowledge of the Lamanai area and for spending numerous hot days trekking around on foot and by canoe to collect clay and rock samples. I am also grateful to the caretakers of the Lamanai Archaeological Reserve, particularly Nasario Ku and 'Lupe', for their assistance with navigating the on-site collections.

The pottery illustrations that appear throughout this study were produced by Rebecca Curran and Louise Belanger. I am extremely grateful for their talents, skill and professionalism. I would also like to extend my heartfelt thanks to the many others who provided thoughtful commentary and assistance with various editorial tasks: Maurizio Guadagnini, Susy Kirk, Emma Wager, Harriet White, Emma Barker, Anna Jorge, Caroline Jackson and Patrick Flarety.

TABLE OF CONTENTS

CD ROM

Please note that the CD referred to above and throughout the text has now been replaced with a download available at www.barpublishing.com/additional-downloads.html

List of Figures

List of Maps

LIST OF TABLES

CHAPTER 1
INTRODUCTION

Archaeology is just another form of detection; the artifacts are the clues and the history they reveal is the solution. (Dr. Larry Fleinhardt, 2005)

This study examines patterns of pottery production and consumption at the Maya city centre of Lamanai during the Terminal Classic to Early Postclassic period (A.D.800-A.D.1250). The central focus is on pottery that was deposited in the central precinct as part of ritual and ceremonial activities and events. Through an analytical framework that involves the detailed examination of stylistic and technological variation within the ceramic assemblage, as well as ceramic depositional patterns and the wider cultural and environmental setting within which pottery was manufactured, used and deposited, this study investigates the kinds of factors that contributed to ceramic change during this period of cultural transition.

1.1. BACKGROUND TO THE TIME PERIOD UNDER INVESTIGATION

The Terminal Classic period (A.D.750-A.D.1050) is well recognized as a time of significant cultural change in the Maya lowland region. It was an era that brought about a dramatic shift in archaeological and material manifestations of Maya culture and society and has been regarded variously as a period of 'collapse', 'decline', 'transition' and 'transformation'(e.g. Thompson 1954; Culbert 1973a; Sabloff and Andrews 1986; Demarest et al. 2004a). A central focus of research on this time period has been investigation of the cultural and environmental factors and processes that contributed to the decline and eventual abandonment of city centres situated throughout the southern area of the lowland region. This widespread trajectory towards decline, commonly referred to as the *Maya collapse,* is viewed as having had far ranging effects on cultural-historical developments, on both the local and regional levels, marking the failure of Classic period elite culture and socio-political institutions, causing disruptions in networks of exchange, communication and political affiliation, and stimulating population migrations. Parallel developments in the northern area of the lowland region during this period involve the successive expansion and contraction of hegemonies tied to different city centres situated in the north of the Yucatan Peninsula. Coupled with the collapse of major city centres in the southern lowlands, the rise and fall of these realms of political and economic influence are considered to have fostered military conflicts and population shifts, as well as changes in regional economic infrastructures as part of the emergence of new trade networks. A central focus of the present work is investigating the ways in which the local community at Lamanai was affected by these regional level developments and pressures.

It is generally agreed that the effects and/or repercussions of these external disruptive forces were felt, to a greater or lesser degree, by communities and city centres across the lowland region. Nonetheless, as the state of knowledge of this time period has expanded, it has become increasingly apparent that local cultural-historical sequences document an array of highly variable and complex responses and experiences. Conventional models of culture change that focus on the disruptive influence of environmental or external factors and pressures often cannot account for these divergent local trajectories, since they are built around a specific set of events and circumstances that simply do not apply to the material evidence at some sites. Such is the case at Lamanai, where community patterns during the Terminal Classic to Early Postclassic period often present a striking contrast to those observed at other sites in northern Belize, as well as in other areas of the lowland region. Recognition that conventional explanatory models of the Classic to Postclassic transition often serve to subsume important differences among local and sub-regional patterns, therefore masking the complex and inter-related cultural processes at work, has led to a recent call, among archaeologists working in the field, for a return to a 'descriptive' phase of work (e.g. Demarest et al 2004a, 2004b). The ultimate objective of this new phase of study is to generate a body of interpretative theory that is consistent with and informed by the variable and complex nature of the Terminal Classic to Postclassic record. Studies designed specifically to investigate variability and patterning in the local material record, as well as the factors and processes involved in shaping it, form a foundational component of this endeavour. The present work represents a step in this general direction.

1.2. A BRIEF INTRODUCTION TO THE SITE AND THE POTTERY

Lamanai is situated on the western bank of the New River Lagoon, in the interior of northern Belize. Excavations at the site, which have been ongoing since 1974, have concentrated on architectural structures and adjacent public areas of the central precinct, the sector of the settlement dominated by ceremonial and administrative structures and residences of the elite, and adjacent residential areas (Pendergast 1975, 1978, 1980, 1981a, 1981b, 1981c, 1981d, 1984; Graham 2004). The ceramics that form the basis of this study derive from offerings and burials recovered from within different buildings and large refuse deposits (middens), primarily those associated with two of the major ceremonial structures within the central precinct. These contexts are described and discussed in Chapter 6. As with most other sites in the lowland region, the Terminal Classic to Early Postclassic ceramic sequence at Lamanai is characterized by striking changes in pottery styles (Pendergast 1981a; Graham 1987). However, continuity in certain elements of vessel design within the Lamanai ceramic sequence suggests that these changes evolved gradually (Graham 1987). As will be seen in Chapter 2, this dynamic of continuity and change also characterizes other aspects of archaeological record of the Terminal Classic to Early Postclassic period of occupation at the site. A central focus of this study is uncovering the material and human factors and processes contributing to this dynamic, as concerns the ceramic record, through the reconstruction of patterns of ceramic production and consumption.

1.3. Methodological Framework

This study advocates a particular way of conducting ceramic research targeted ultimately at illuminating the human behaviours and motivations that underlie patterns and variability in the material record. Accordingly, the conceptual and analytical framework upon which it is based, represents a significant departure from prevailing methodologies employed in Maya ceramic studies. It combines traditional methods of macroscopic analysis with scientific techniques, including thin section petrography, neutron activation analysis (NAA) and scanning electron microscopy (SEM), in order to examine in detail the stylistic and technological characteristics and provenance of the Lamanai pottery. This examination of physical aspects of the ceramics is situated alongside a detailed study of ceramic depositional patterns within the archaeological contexts examined, as well as the environmental and cultural setting within which the ceramics were manufactured and used and deposited as part of religious and ceremonial activities that took place within the community at Lamanai. This information provides an environmental and cultural framework of material-human interactions within which to situate and interpret the physical ceramic evidence, in order to reconstruct and characterize community-level patterns of ceramic production and consumption. The process of building this framework begins in Chapter 2, which presents an overview of Terminal Classic to Early Postclassic patterns of occupation, as well as the character of site assemblages, for the northern Belize region. The focus is then narrowed to the community level, documenting community patterns at Lamanai relating to construction activities, the use and organization of community living space, burial and offertory patterns, and patterns of ceramic deposition within these different contexts. The process continues in Chapter 5, which provides an overview of regional weather and climate patterns and the geography and geology of Belize, as well as the environmental and geological setting of local pot-making activities.

1.4. Structure of the Study

The study begins, with a summary of previous research on the Terminal Classic to Postclassic period, focussing on the ways in which perceptions have changed as the level of knowledge of Maya cultural patterns during this time period has expanded (Chapter 2). Previous investigations and interpretations of local cultural-historical sequences in northern Belize are documented and discussed, and current interpretative models that attempt to delineate cultural processes during the Classic to Postclassic transition are presented. It is argued that current perspectives of culture change during this time period ignore the intra-community dynamics and interactions that contribute to changes in the material record. An approach to the study of local-level patterns is advocated whereby the processes by which individual community histories *and* variability in the local material record are created together form the focal points of investigation. Chapter 2 concludes with a description of the cultural setting within which community activities at Lamanai involving pottery took place. It documents continuity and change in different aspects of the archaeological record of the Terminal Classic to Early Postclassic period of occupation and discusses the factors that appear to have contributed to shaping these wider community patterns.

In Chapter 3, the focus narrows to current characterizations of ceramic economic patterns during the Classic and Postclassic periods. It presents and discusses the pervading economic model proposed to explain changes in ceramic styles and distributional patterns during the Classic to Postclassic transition, as well as the evidence and assumptions upon which this model is based. Through this critical review it is demonstrated that current conceptualizations and characterizations of patterns of ceramic production and consumption are based on an incomplete body of evidence and knowledge, as well as questionable assumptions. It is argued that a different approach to ceramic analysis and interpretation would help to fill this knowledge void and outlines what this approach should ideally involve.
Chapter 4 presents a methodology for the analysis and interpretation of Maya ceramics that attempts to overcome the limitations of conventional approaches by:

1. expanding the range of physical evidence that is considered
2. focussing on both stylistic *and* technological characteristics of pottery
3. situating ceramic economic activities within a wider cultural and environmental setting.

The conceptual and analytical framework that forms the basis of the ceramic analysis is presented and the analytical methods employed in the analysis are described and discussed, with particular attention paid to their use in previous studies of Maya pottery.

Chapter 5 presents the geological and environmental setting of ceramic production on both the local and regional levels. This review of the geological literature reveals that the pervading assumption held by many Maya archaeologists and ceramic analysts that the homogenous limestone geology of northern Belize and adjacent areas precludes the discrimination of ceramic fabrics based on their raw materials (especially through petrographic analysis), as well as the ability to ascribe provenances, represents a historical bias that is unjustified. A thorough review of the geological literature in publication reveals that considerable variation exists in the compositional characteristics of the different geological formations and deposits that occur in this region. The results of the geological survey and compositional analyses of local raw material resources available for pottery manufacture at Lamanai document the character and extent of this variation on the local level.

Chapter 6 presents the archaeological contexts of the whole and reconstructed vessels and sherd assemblages that were included in this study. The discussion then expands to consider the broader cultural associations and significance of the different types of ceramic deposits analysed, as well as the nature and depositional patterns (placement and fragmentation) of the ceramics they contain. The chapter concludes with a discussion of the dating of the Terminal Classic to Early Postclassic ceramic sequence at Lamanai.

Chapters 7 to 10 present and discuss the results of the physical analyses of the pottery, highlighting temporal trends in their stylistic and technological characteristics. Chapter 7 presents the results of the macroscopic study of vessel forms, morphology and surface treatment. Stylistic groups are identified and their temporal relationships are examined. The focus of Chapter 8 is compositional variability within the assemblage on the microscopic level. This chapter presents the results of the petrographic analysis and describes the geological characteristics and paste technology of the different classes and groups of fabrics that were differentiated. It identifies fabric groups consistent with the local geology and ascribes provenances to non-local fabrics. Chapter 9 examines and characterizes compositional variability within the ceramic assemblage on the chemical level, as revealed through the statistical analysis of the NAA results. The focus of Chapter 10 is surface-finishing and decorative technologies and firing methods. This chapter investigates and characterizes these aspects of vessel technology, as revealed through SEM analysis, for selected styles of fine ware pottery, and examines similarities and differences within and among different styles of pottery.

In Chapter 11, all of the lines of evidence presented and discussed in the preceding chapters are brought together and community level patterns of ceramic production and consumption at Lamanai are reconstructed for the time period under investigation. Changes in pottery demand, manufacturing priorities, as well as aspects of consumption patterns and ritual practice, are identified and discussed. The patterns that emerge at Lamanai are considered in light of conventional models and characterizations of ceramic economic activities and ceramic change, revealing a significant departure from predicted patterns. The contributing role of external factors and pressures in shaping local patterns of ceramic production and consumption at Lamanai are examined in light of the physical evidence, and contributing factors and dynamics internal to the community are characterized and discussed.

1.5. Objectives of the Study

The objectives of this study are as follows:

1. To examine, in detail, the stylistic, technological and provenance characteristics of the pottery comprising the Terminal Classic to Early Postclassic assemblage.
2. To examine and characterize stylistic and technological variation within the assemblage in terms of the morphology and visual appearance of vessels, their petrographic and chemical characteristics, and decorative technology and firing methods.
3. To identify technological similarities and differences within and among different stylistic and functional categories of pottery.
4. To investigate and characterize techniques of manufacture relating to paste and decorative technology and firing methods for different stylistic, compositional and functional categories of vessels.
5. To investigate continuity and change in ceramic depositional patterns for the different depositional contexts examined.
6. To establish a geological and environmental framework for the interpretation of the petrographic results, in order to ascribe provenances to different fabric groups and to characterize local paste technologies.
7. To establish a cultural framework, on the local and regional levels, for the interpretation of the physical evidence.
8. To reconstruct patterns of ceramic production and consumption over the course of the Terminal Classic to Early Postclassic period and to identify aspects of continuity and change in these patterns.
9. To compare local patterns to existing characterizations of pottery production and ceramic economic trajectories during the Classic to Postclassic transition.
10. To identify factors, internal and external to the community, that contributed to shaping patterns of ceramic production and consumption at Lamanai during the Terminal Classic to Early Postclassic period.

CHAPTER 2
THE TERMINAL CLASSIC TO EARLY POSTCLASSIC PERIOD: REGIONAL AND LOCAL PERSPECTIVES

It seems to me as true of the Postclassic as it is of the Classic that, especially at our present level of knowledge, we can derive more profit from focussing on the variety in the archaeological record than we can from seeking some means of subsuming that variety within a single unifying theory (Pendergast 1986:248).

Historical events in the Maya lowlands at the close of the first millennium A.D., culminating with the apparent collapse or decline of city centres and polities in many parts of the southern lowlands and cultural florescence in areas of the northern lowlands, constitute a period of cultural transformation that remains one of the most intriguing and intensively investigated chapters of Maya prehistory. More than a century of research has produced numerous detailed reconstructions of local and regional chronologies and cultural-historical sequences (a main focus of the 'state of the art' reviews edited by Culbert [1973a] and Demarest et al. [2004a]) and a variety of explanatory models of cultural collapse, ascent and culture change, postulating economic (e.g. Rathje 1973; Rathje and Sabloff 1973; Sabloff et al. 1974; Culbert 1977), socio-political (e.g. Thompson 1954; Hamblin and Pitcher 1980; Friedel 1986), ideological (e.g. Ringle et al. 1998; Houston et al. 2001) ecological (e.g. Sanders 1973; Webb 1973; Harrison 1977; Culbert 1988), climatological (e.g. Gunn and Folan 1981, 2000; Gill 2000) or a combination of factors (e.g. Sharer 1977; Hosler et al. 1977; Ringle et al. 2004; Carmean et al. 2004) as the driving forces of culture process. This chapter presents a general summary of research on the Terminal Classic to Postclassic period and the changing perspectives that have accompanied the steady influx of new data over the past three decades. It provides a critical review of previous investigations of Terminal Classic to Postclassic occupations in Northern Belize, as well as of various interpretative models that have been proposed to explain cultural-historical developments in this particular area of the lowland region, and presents the evidence of the Classic to Postclassic transition at the site of Lamanai. It outlines an alternative way of viewing cultural manifestations of this time period, in which individual community histories and variability in the local material record form the focal points of investigation.

2.1. THE TERMINAL CLASSIC TO EARLY POSTCLASSIC TRANSITION

A central focus of research on the Terminal Classic period has been investigating the 'Maya collapse' and defining its nature, timing and causes on the local and regional levels. Manifested in the decline and abandonment of major city centres that appear to have flourished in the preceding Late Classic period, the collapse is broadly perceived as the terminal, degenerative period of development of Classic Maya culture and society. The first systematic consideration of the collapse and its underlying causes was presented in *The Classic Maya Collapse* (Culbert 1973a), which sought to define these changes in temporal and cultural terms, based on the data available at that time. The collapse was characterized as a widespread phenomenon that occurred over a seemingly brief period of time during the ninth to early tenth centuries A.D., and which coincided with the Tepeu 3 ceramic horizon (Culbert 1973b:16-17; Adams 1973a:22). Comparative study of the existing ceramic data showed that this period was marked by a dramatic decline in the amount and quality of polychrome pottery and the concurrent appearance and increasing prevalence of fine paste wares, particularly Fine Orange and Fine Gray Wares (Rands 1973:56-61). The introduction of the fine paste tradition was thought to reflect the intrusion of foreign peoples and ideas into the Maya area. In addition to its relatively short duration (being confined to a single ceramic period), other defining characteristics of the collapse that were identified include the rapid depopulation of ceremonial centres and the surrounding countryside and the apparent failure of elite-class culture, as indicated by:

1. The abandonment of administrative and residential structures (palaces).
2. Cessation of erection and refurbishment of funerary monuments and foci of ritual activities (temples).
3. Cessation of manufacture of sculptured historical monuments and records (stelae).
4. Cessation of the manufacture of luxury items such as the finest polychrome pottery, fine stonework, and jade carving for the use of the elite class.
5. Cessation of the use of calendrical and writing systems, at least in Classic Period forms.
6. Cessation of nearly all behavioural patterns associated with the above and other elite-class-directed activity, for example, the ball game played in formal courts. The processionals, rituals, visits and conferences characteristic of Maya elite-class life lapsed (Adams 1973a:22).

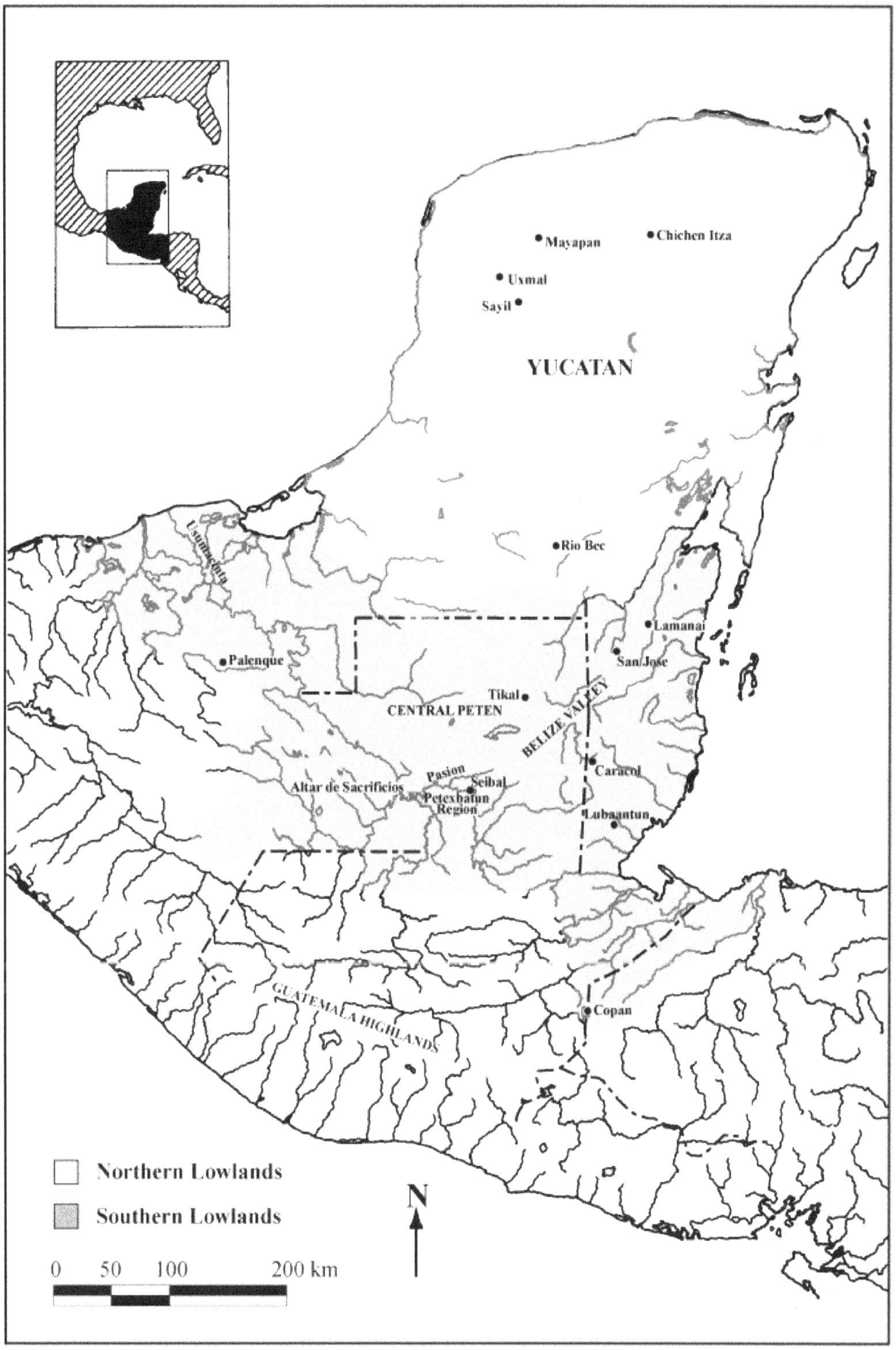

Map 2.1: Map of the lowland region showing sites and areas mentioned in the text.

Central to this characterization of the collapse was the notion that the activities, management capabilities and very existence of the elite-class was of such fundamental importance to maintaining the workings of Maya society as a whole, that the disappearance of the Classic period manifestations of elite-class culture from the archaeological record, invariably meant that the society itself had ceased to exist. From this perspective, the collapse was envisioned as a general failure of the Maya socio-cultural system, from which there was no subsequent recovery (Willey and Shimkin 1973:490-491). Recognizing that the specific causes and circumstances relating to the decline and abandonment of the city centres and areas considered in the study appeared to vary, both regionally and locally, it was concluded that this general 'systemic failure' was caused by the coalescence of a combination of existing internal and external 'stresses' that made Late Classic society vulnerable to dissolution (Willey and Shrimkin 1973; Willey 1974). Internal factors identified as having contributed to local and regional crises included environmental limitations; overpopulation; a widening social gulf between elites and commoners causing social unrest; and escalating, and increasingly costly, competition between centres attempting to expand their wealth, prestige and spheres of influence (Willey and Shimkin 1973:484-487). In terms of external pressures, the disruption of trade networks was identified as having played a particularly significant role in the collapse, on all levels, and foreign invasion (physical and intellectual) and military pressures were suggested as having had an important impact on the course of events on the regional level, and especially at centres situated along the western and northwestern frontier of the southern lowlands (Willey and Shrimkin 1973:488-490; Webb 1973; Adams 1973b; Sabloff 1973).

The portrait of the Terminal Classic to Early Postclassic period presented in *The Classic Maya Collapse* relied heavily upon data obtained from excavations undertaken at major city centres situated in the Central Peten region of Guatemala and adjacent areas to the northwest, south and southwest (Map 2.1) – i.e. the area of the Maya lowlands in which major investigations had been traditionally centred. Consequently, the resulting characterization of the time period can be considered as primarily reflecting events and circumstances in that particular area of the southern lowlands, rather than the lowland region as a whole. A major issue that could not be resolved at that time was whether the conclusions of the study could be extended to include the northern lowlands. Owing to the fact that comparatively few major investigations had been undertaken in the north, the cultural-historical sequence was still poorly understood. In addition, there was considerable debate among archaeologists as to how events in the north such as the decline and fall of the Puuc centres, Uxmal, Sayil and Rio Bec, and the rise and decline of Toltec Chichen Itza correlated with the systemic collapse seen in the southern lowlands (Culbert 1973b:17-18; Andrews 1973). This unresolved problem of the alignment of the southern and northern sequences severely limited understanding not only of the nature and effects of the collapse within the broader context of the Maya area, but also the more pressing issue of how the rise of centres of power in the north related to the collapse of the southern polities or territories.

Another problem was that the general perception of the collapse as having had devastating and irreparable consequences was based largely on negative evidence and an incomplete understanding of the Postclassic record (Graham 1985). Since the Postclassic period had rarely been a focus of investigations in the southern lowland region, the nature of society and culture after the Terminal Classic period remained poorly understood. Interpretations relied on a limited range of evidence, primarily from the Belize Valley and the Central Peten Lakes District (e.g. Bullard 1973). The obvious contrast between the smaller, nucleated, riverine/lacustrine oriented Postclassic settlements and the city centres and dispersed settlement pattern typical of the Classic period led to an emphasis on cultural discontinuity over time and a perceived dichotomy between the nature and socio-cultural complexity of Classic vs. Postclassic society. Unlike their culturally advanced predecessors, the Postclassic Maya of the southern lowlands were characterized as having lived an isolated existence as "peasant-level groups without strong social class differences or strong political controls" and having had no appreciable influence on or participation in cultural-historical events taking place elsewhere in the lowland region (Bullard 1973:241). An important implication of this characterization is that it promoted a general belief that everything of cultural-historical significance in the lowland region after the Terminal classic period had transpired in the northern lowlands. The rise and decline of hegemonies centred at Chichen Itza and subsequently, at Mayapan were perceived as central to these developments.

Since the time of its publication, *The Classic Maya Collapse* has remained one of the most influential studies of the Terminal Classic to Early Postclassic period, having had a significant impact on subsequent research trajectories, the interpretation of new data and the ways in which the time period has been generally perceived in cultural-developmental terms. Over the past three decades, however, a proliferation of new excavation projects and an expanded geographical focus of investigation have produced a wealth of new information, shedding light on this time period in previously unstudied areas of the lowlands, and enabling more detailed reconstructions of local and regional cultural-historical sequences The focus of archaeological investigation has also expanded beyond larger city centres and their central precincts to include a wider range of settlement types and greater coverage of individual settlements and their surrounding hinterlands. This broadened perspective has enabled researchers to add flesh to the bones of previous local and regional archaeological frameworks, which were built around a more restricted range of evidence. On a more specific level, advances in glyphic decipherment have led to a deeper understanding of the political inter-

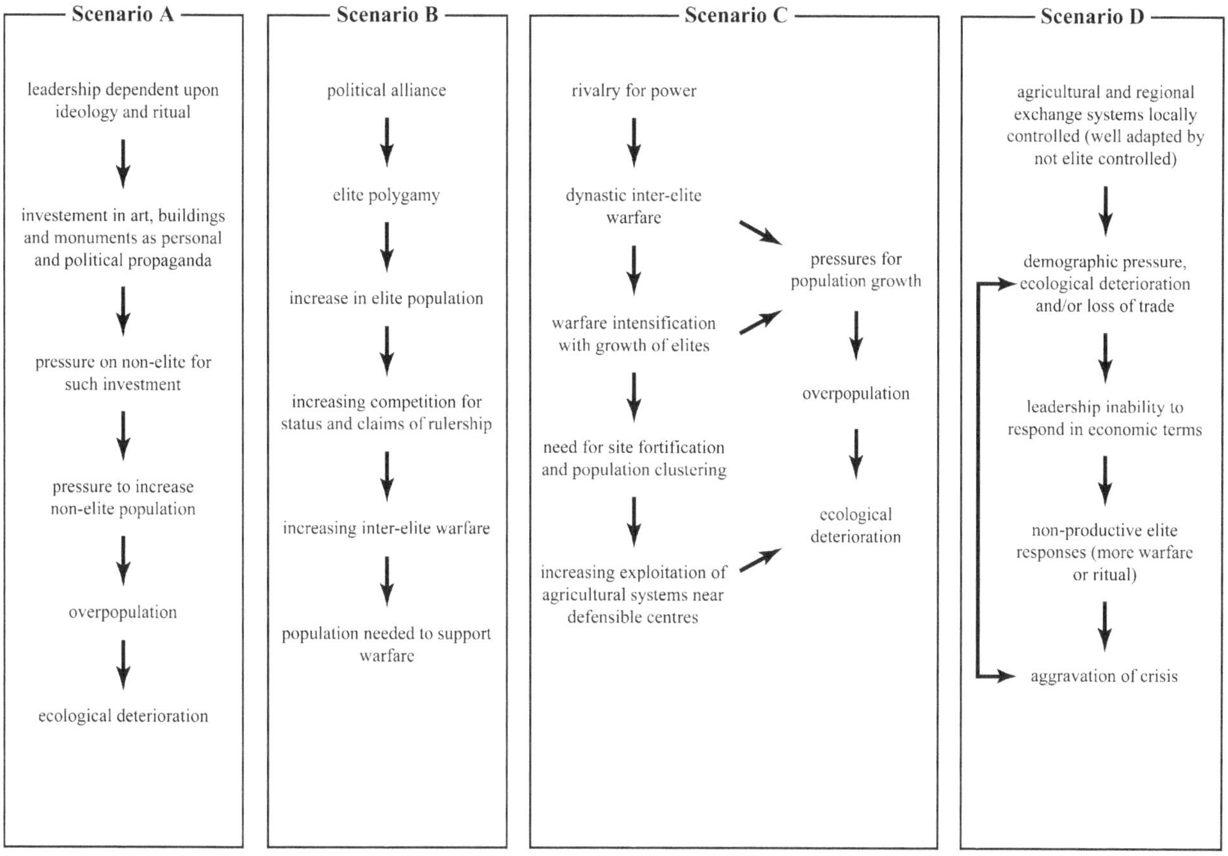

Figure 2.1: Factors that led to instability during the Late Classic period (after Demarest et al. 2004b: 365, Table 23.1).

relationships among sites during the Late to Terminal Classic period, as well as the history of rulership, alliance formation and inter-site warfare among the better-documented 'kingdoms' or polities leading up to the point at which divine kingship ended (e.g. Stuart 1993; Martin and Grube 2000). These studies have revealed that some basic characteristics of the Maya political system, such as a flexible system of royal succession, a heavy ideological basis for power and elite competition in the form of status-rivalry on the inter- and intra-site levels, led to escalating conflict and instability, thereby contributing greatly to the rapid decline and collapse of city centres and polities (Figure 2.1) (Webster 1998; Pohl and Pohl 1994). Meanwhile, studies correlating epigraphic, ethnohistorical and archaeological data have offered new insight into some intrinsic features of Maya socio-political organization and world view that would have been capable of generating expansions and contractions of realms of political and ideological influence and the associated fluctuations over time in population levels, construction activity and patterns in the production and consumption of material items. Such features include: the pre-programmed transfer of ritual authority among centres associated with the *may* politico-calendrical cycle (Rice and Rice 2004; Demarest et al. 2004b); a segmentary lineage-based system of social and political organization (Fox 1987); and the Maya conception of human and human-environmental interactions as an emulation of the epic battle between life and death, and regeneration and pestilence, portrayed in the creation myth, which fostered a shamanistic style of leadership centred around ritual performance, ceremonialism, alliance formation and war (Freidel and Shaw 2000).

Perhaps most significantly, an abundance of information has emerged about Postclassic Maya occupations and culture, leading to new perspectives on the Maya collapse and the Terminal Classic period in general (e.g. Sabloff and Andrews 1986; Chase and Rice 1985). Mounting evidence of the persistence, and in some cases flourishing, of populations and centres in various areas of the lowlands during and after the collapse has led to the conclusion that the collapse was not as disastrous and complete an event as once thought. Instead, it has become increasingly evident that the collapse was just "part of an array of responses to crisis in the wider lowlands which includes florescence and persistence as well as decline" (Freidel 1985:309). As new evidence of the nature of Terminal Classic to Early Postclassic occupations has come to light, it has also become apparent that the record of this period is highly complex and variable, being characterized, often simultaneously, by instances of disjunction, continuity and 'syncretism'. In addition, some studies have demonstrated that many aspects of Postclassic institutions have traceable Classic roots

(Chase and Chase 2004). This situation has led some researchers to adjust their thinking on the Terminal Classic period and to consider it as representing *a transitional* rather than *the terminal* phase of Maya cultural development (e.g. Freidel 1985; Andrews and Sabloff 1986; Demarest et al. 2004a).

The most recent considerations of the Terminal Classic to Early Postclassic record (Demarest et al. 2004a) demonstrate the ways in which general thinking and the state of knowledge of this period have evolved in response to the steady influx of a range of new data and evidence. Greater control over local and regional chronologies and cultural-historical sequences has led to the recognition that, as a period of cultural development, the transition between the Classic and Postclassic is not reducible to an identifiable moment of profound culture change that transpired over a mere hundred years. It is now clear that the timing and duration of processes and events relating to the divergent, archaeologically-visible, developmental phenomena observed across the lowland region – i.e. the cessation and changes in the intensity of elite-sponsored activities, population growth and decline, site abandonment and expansion, fluctuations in the geographical distribution of material culture – vary widely and occur over a more protracted, 300 year period (A.D. 750-1050). For example, processes leading to depopulation and the cessation of construction activities at sites in the Petexbatun region of Guatemala begin comparatively early, prior to A.D. 760, with nearly complete abandonment of the area occurring by A.D. 830 (O'Mansky and Dunning 2004). During the same period, Puuc cities in the northern lowlands were experiencing a cultural florescence, marked by population growth, settlement expansion and an increased investment in public architecture. By A.D. 950, however, these same cities had also fallen into a rapid decline, whereas the nearby site of Chichen Itza, under Toltec control, was expanding its sphere of influence, only to fall into a similar decline a century or so later (Carmean et al. 2004; Suhler et al. 2004; Cobos Palma 2004). The wider chronological framework that has been adopted for investigation of the Terminal Classic period represents a significant advancement in thinking as it acknowledges that sub-regional and local trajectories do not reflect isolated and completely unrelated developments and will never be fully understood if envisioned as such. Nor do they represent a unified passive response to a specific set of external pressures or stimuli. Accordingly, it permits an important shift in focus towards broader patterning and the examination of the complex interplay of "intra- and inter-site dynamics and broader, interregional interactions between the north and south", which are equally fundamental to the transition between the Late Classic and Postclassic archaeological manifestations of Maya culture and society (Rice et al. 2004:8).

The steady influx of new data from local and regional excavations and surveys, and the sometimes conflicting interpretations of this data (e.g. the interpretations forwarded by Webster et al. 2004 vs. Fash et al 2004 regarding the Terminal Classic period at Copan) have produced a more detailed knowledge of the Classic to Postclassic record. However, with greater detail has come an increasingly diverse and variable body of data which is ill-fitted to broad characterizations. Due to this fundamental complexity, the nature of the Terminal Classic to Early Postclassic record calls into question the usefulness (and appropriateness) of conventional model-based approaches to the reconstruction of regional and pan-regional culture-histories. For example, diversity and variability in local archaeological manifestations of the time period suggest that the collapse and/or decline witnessed at many Classic centres, cannot be considered as reflecting a pan-lowland phenomenon that was brought about by a specific set of events and/or circumstances. It is now clear that some sites remained occupied throughout the period, experiencing continued or heightened prosperity, or alternatively, varying degrees of decline. Even when considering the sites where the evidence for abandonment, and thus collapse, is fairly clear it is evident that such developments range from a gradual process to an abrupt or sudden event, and are alternatively violent or peaceful in character. In addition, as recent reviews of regional level patterns demonstrate (e.g. Demarest et al. 2004b), the individual paths followed by particular regions or local communities embody highly variable responses to often differing sets of regional and local influences, events, processes, changes and continuities. Consequently, it has become increasingly apparent that monolithic models that attempt to account for culture change at the end of the Classic period for the lowland area as a whole, only serve to subsume important differences among local, sub-regional and regional patterns, and to mask the complex and inter-related cultural processes at work.

In the conventional approach to the construction of explanatory models of culture change, the nature of and processes attributable to local and regional patterns are projected onto a pan-lowland scale as a universal template for Maya culture-history. A specific example of this tendency are recent models that argue for the catastrophic effects (demographic, economic and ecological collapse) of either climatological (e.g. Gill 2000; Gunn and Folan 2000) or anthropogenic environmental changes (e.g. Shaw 2003) as the root cause of culture change at the end of the Classic period, based on the evidence obtained from limited studies of particular localities or sub-regions. As Demarest et al. (2004b:548) observe, models constructed in this way only serve to perpetuate "a fundamental misconception of the unity and nature of the Classic to Postclassic transition", since they ignore that the evidence for the existence and contributing role of such factors varies both locally and regionally (e.g. Gunn and Folan 1996; Ringle et al. 2004:505-506; Braswell et al. 2004:185-188). As a result, such models fail in their basic aim – to achieve a more comprehensive understanding of the cultural-historical record – since they do not offer any explanation for why certain sites or areas appear to have been affected differently by such factors, and more specifically, why particular communities survive or prosper in spite of them. Explanatory frameworks that emphasize socio-political rather than environmental factors, such as inter-

and intra-site competition, conflict and warfare among elites are equally problematic for the same reasons (e.g. Pohl and Pohl 1994; Webster 1998; Freidel and Shaw 2000).

Recognition that the state of knowledge concerning the Classic to Postclassic record has generally out-grown the existing body interpretive theory, has recently led some Maya archaeologists (e.g. Demarest et al. 2004b) to call for a new phase of study in which variability and the delineation of areal and temporal trends in the archaeological data are the focal points of investigation. In this new phase of analysis and interpretation, which might be considered as a return to a 'descriptive' phase of work, systematic comparative studies undertaken at successively higher levels of integration, form the basis of cultural-historical reconstruction and are a prerequisite for theory building and synthesis:

> We must systematically compare and link site sequences to understand subregional processes and variability, compare these to adjacent zones to construct regional patterns, and then compare regions to begin reconstructing pan-lowland histories and correlating these to other zones in Mesoamerica. The subsequent interpretation of these culture-histories must take into account the economic shifts, changing ideologies, political dynamics, population movements, and other complexities of the variable Late Classic to Postclassic landscape. (Demarest et al. 2004b:549, 572)

A main objective of this approach is to generate interpretations and theory that are both consistent with and informed by the variable and complex nature of the Classic to Postclassic record. In doing so, it provides a means of moving beyond site-focussed isolationism and ill-suited global projections of local events and circumstances, which loom largely in current scholarship on the time period (Demarest et al. 2004b:545-549).

The present work argues that this new phase of research should begin on the *local* rather than *inter-site* level, with studies that are designed specifically to investigate the nature and meaning of variability and patterning in the local record, as well as the processes involved in shaping it. Since local-level studies are the building blocks of regional studies, coming to terms with the complexity of the Classic to Postclassic record must necessarily rely on a firm understanding of variability in the local record. It is generally the case that interpretations of the local record are skewed by perspectives that emphasize the primacy of either environmental, economic, socio-political or ideological factors to the exclusion of other, equally plausible, alternative explanations that consider, for example, the contributing role of a variety of factors. The focus is on discovering patterns that are most relevant to the analyst's preferred perspective, and not on investigating the nature and meaning, in cultural terms, of both emergent patterns and variation. When considering the material record, few would deny that patterning, or lack thereof, arises out of the combined influence of a multitude of environmental and cultural factors. However, studies aimed at developing method and theory for examining the nature of this integration of environmental, economic, socio-political and ideological factors, as well as the cultural processes involved (within archaeological contexts), are surprisingly rare. As recent considerations of Maya political economy have shown, general patterning in the material record is as much determined by economic factors, such as the use value of particular objects, as it is by ideological factors, such as the cosmological significance of the material from which the object is made (Freidel et al. 2002). On a more fundamental level, however, patterning in the nature and distribution of material culture is also a reflection of *people's choices*, as well as *responses to demand*, which, in turn, are informed by an array of cultural, material and environmental factors and considerations (Graham 2002). It is only through evaluating the contexts of these choices and responses that the nature and meaning of variability and patterning can be discovered. Local-level studies offer a good opportunity to investigate these contexts since they provide a means of situating the material evidence within specific environmental and cultural frameworks of *material-human interaction* (Day 2004). At the same time, however, local manifestations of cultural behaviour are affected by both internal and regional dynamics of human interaction (Graham 1987). Consequently, these local contexts must be understood not only on their own terms, but also within the broader context of external factors and regional developments that might have contributed to continuity and change in local patterns of economic, socio-political and ideological activity and behaviour.

2.2. Archaeological Investigation at the Northern-Southern Lowland Interface

Situated at the geographical crossroad between declining spheres of power and influence centred in the southern lowlands and ascending hegemonies in the north, sites in northern Belize offer a unique insight into the ways in which local communities were affected by larger regional-level developments that transpired during the Terminal Classic to Early Postclassic period (Map 2.2). These developments include:

1. The movement of immigrant populations from collapsing southern lowland polities and areas of ongoing conflict in the northern lowlands – groups seeking to re-establish themselves in more stable surroundings.

2. The disruption of networks of exchange, communication and political affiliation as a result of the decline and collapse of southern lowland centres, and the expansion and contraction of competing northern lowland hegemonies.

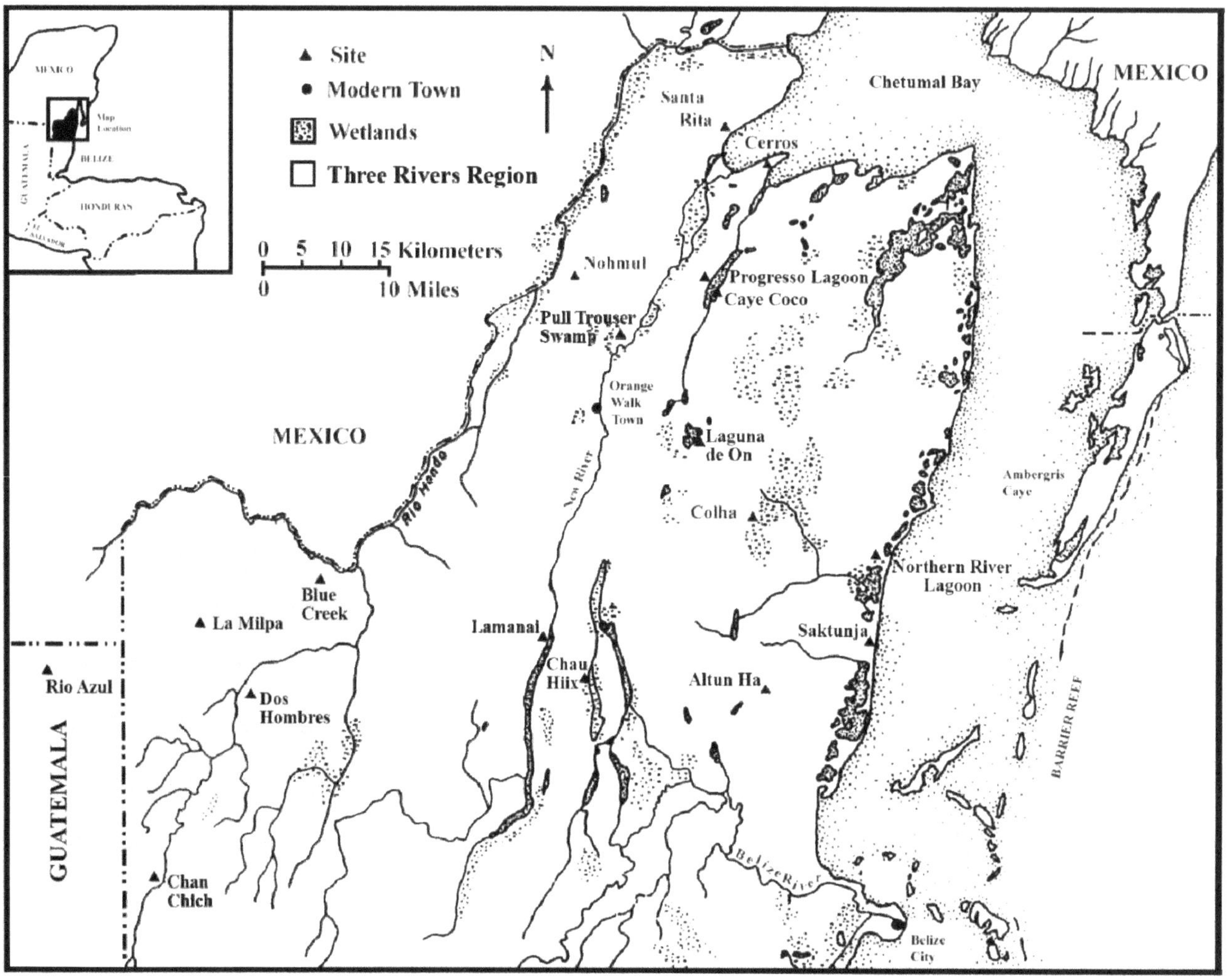

Map 2.2: Map of northern Belize showing sites mentioned in the text.

3. Ideological shifts
4. The predatory expansion of northern lowland states.

Due to a range of theoretical and practical problems, however, the reconstruction and interpretation of local and regional cultural-historical sequences for Terminal Classic and Postclassic occupations in northern Belize has proved a challenge.

Since few large excavation projects were undertaken in northern Belize prior to the mid-seventies, Terminal Classic and Postclassic occupations have borne the brunt of traditionally held assumptions about the nature and role of northern Belize centres within broader cultural and historical constructs for the lowland region. Central to these assumptions is the physical distance of northern Belize from the perceived 'heartland' of Maya culture during the Classic and Postclassic periods. As pointed out by Pendergast (1990), and more recently by Chase and Chase (2004), the early focus of Maya archaeology on sites located in the Central Peten and adjacent areas has led to the conceptualization of this region as the cultural 'heartland' of the Maya world, with Late Classic manifestations representing the pinnacle of Maya cultural development. This paradigm has affected not only the way in which Postclassic society and culture is viewed, but also the interpretation of Maya occupations in areas, such as northern Belize, that are situated at some distance from the heartland area. The absence of the material hallmarks of Classic period culture in the Postclassic record has been viewed as evidence of cultural decadence and devolution. Likewise, since centres situated outside the Peten core are generally smaller and have fewer sculptured monuments, inscriptions and lesser amounts of figure-painted polychrome pottery, they have been generally perceived as having a secondary and peripheral status within lowland socio-political and religious infrastructures. In addition, it is generally assumed that areas surrounding the heartland played a supporting role in the Classic period economy, being envisioned as the 'bread basket' of the southern lowlands.

A central facet of pervading perspectives on lowland Maya cultural development over the course of the Classic to Postclassic transition is the notion that, with the political collapse of southern lowland polities, the cultural heartland of the Maya world shifted to the

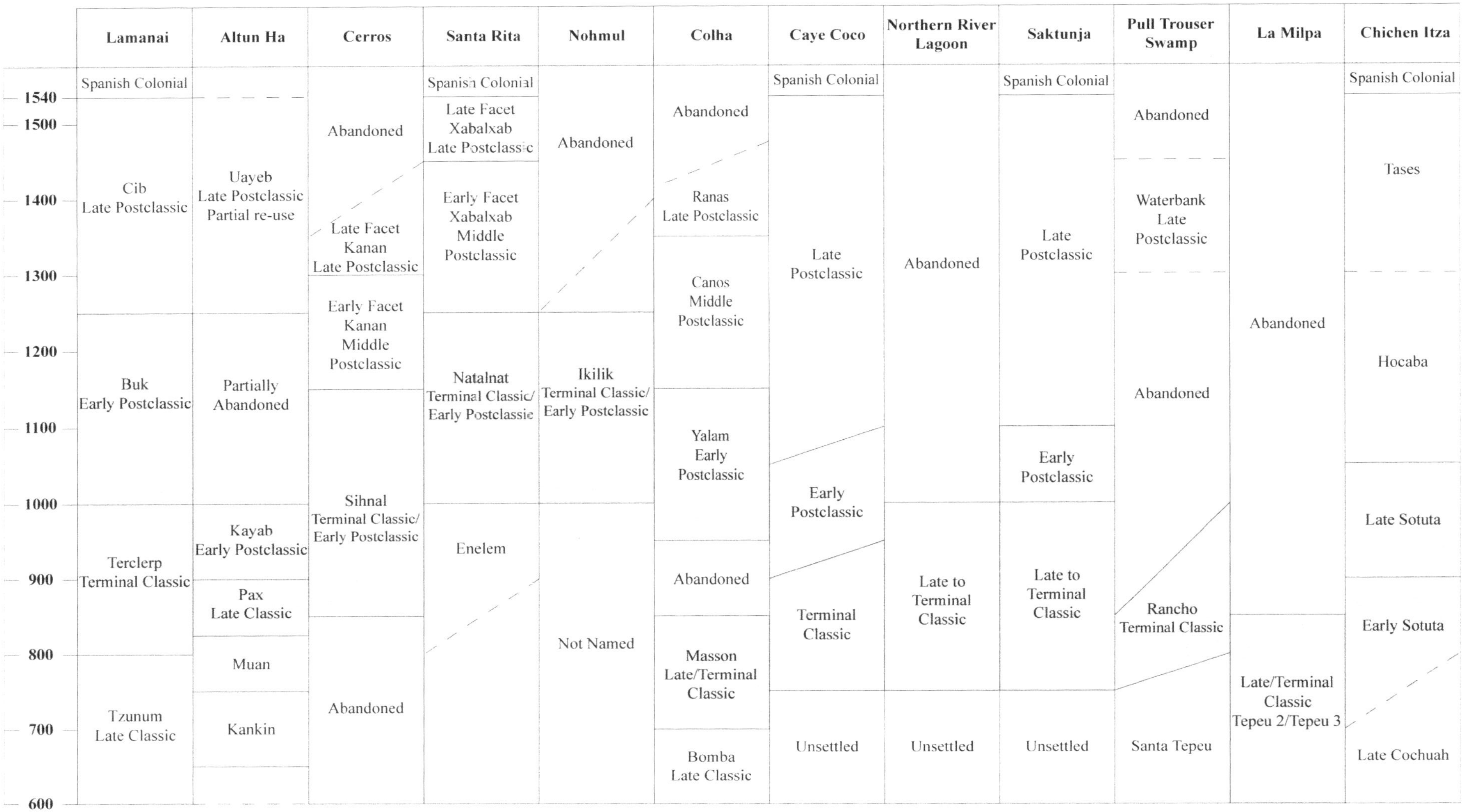

Figure 2.2: Late Classic and Postclassic chronological sequences (and ceramic complexes) published for northern Belize sites (Lamanai, Graham 2004; Altun Ha, Pendergast 1982: 2, Table 1; Cerros, Walker 1990; Santa Rita and Nohmul, A. Chase 1986: 115-122; Colha, Valdez 1994: 10, figure 1; Caye Coco, Northern River Lagoon and Saktunja, Masson and Mock 2004; Pull Trouser Swamp, Fry-1989: 95, Table 1 and 104; La Milpa, Hammond and Tourtellot 2004; Chichen Itza, Cobos Palma 2004).

northern lowlands. This conceptualization, in a very real sense, echoes the earlier views of Morley (1956) and others who perceived the expansion and flourishing of sites in northern Yucatan as evidence of the physical movement of a flourishing culture across the landscape. Since northern Belize is situated at the northern-southern lowland interface, an important implication of this general perception is that sites in this area are often assumed to have retained their peripheral position within the human landscape[1] during the Postclassic period by way of their geographical location. It is most often the case that cultural-historical interpretations of occupations in northern Belize that span and/or continue beyond the collapse period tend to be deeply embedded within this broader frame of reference. That is, there is a tendency to regard local culture-histories, up until the point of the collapse, as invariably linked to events and processes ongoing in the Peten core, and Postclassic trajectories as linked to developments in the northern lowlands.[2] This tendency is perhaps best exemplified by scholars (e.g. Masson 2000, 2002; Masson and Rosenswig n.d.; Masson and Mock 2004) who invoke Rathje's *core area-buffer zone* model (Rathje 1972, 1973; Molloy and Rathje 1974) in their interpretations of Terminal Classic and Postclassic occupations, which are perceived as an adaptive response to a restructuring in trade networks that was initiated by ascending centres of power in the north (discussed below). A major implication of this tendency is that local-level trajectories are presented and described as passive responses to regional-level developments.

Comparison and correlation of the data pertaining to Terminal Classic to Postclassic occupations in northern Belize is complicated by a range of practical problems, the most important one being the significant disparities that exist among site chronologies. As illustrated in Figure 2.2, some analysts identify a single Terminal Classic/Early Postclassic period, while others recognize two separate periods – i.e. a Terminal Classic period and an Early Postclassic period. Similarly, the Late Classic and Terminal Classic periods are conjoined in some site chronologies and split in others. There are also significant differences among sites in terms of the start and end dates assigned to different periods and their duration. It is well recognized that these chronological disparities are at least partly due to the nature and stratigraphic context of the ceramic materials that form the basis of local chronological sequences. As Walker (1990:20-21) observes, Terminal Classic and Postclassic ceramics often occur in "deflated erosional contexts lacking visible stratigraphy" and sometimes intrude into earlier deposits (also see Masson and Mock 2004:378; Pendergast 1990; Graham 1985). The interpretation of these stratigraphic contexts varies; and they are considered to reflect varying degrees of cultural and ethnic continuity/discontinuity, or alternatively postdepositional disturbance, depending on the personal opinion of the analyst (e.g. Masson and Mock 2004:378; Walker 1990:21-22). Other interpretive problems relate to the nature of ceramic materials. Since site assemblages are often dominated by poorly preserved and fragmentary material, typological classification according to the type:variety scheme, which emphasizes surface treatment attributes, is often difficult. This general problem is further compounded by the fact that Terminal Classic ceramic deposits in northern Belize often have mixed chronological indicators and, consequently, do not fit easily into a conventional, uniform, regional and chronological typology (Masson and Mock 2004:381). Ceramic analysts have handled these problems differently, and as a result, interpretations of local ceramic assemblages, which in reality share many commonalities, vary considerably.

[1] By 'human landscape' I mean the cultural geography of the Maya area, which has political, social, economic, and ideological dimensions that are closely intertwined.

[2] Pendergast's work on Lamanai, which adopts a contextual approach to interpretation, is an exception (e.g. 1981a, 1985, 1986, 1990).

General Socio-Political and Socio-Economic Paradigms for Northern Belize

In cultural-historical syntheses of the greater lowland region, sites located in northern Belize are most often grouped together for the sake of convenience, such that the area represents a geographical sub-region within the Belize zone of the southern lowlands (e.g. Culbert 1973b:5), or most recently within the eastern lowlands (Demarest et al. 2004b:556-559). The socio-political dimensions of this geographical construct remain poorly understood for all time periods, since little is known about the political interrelationships among sites, settlement hierarchies that may have existed on various levels, and the ways in which these relationships might have changed over time. In regard to Terminal Classic and Early Postclassic occupations, archaeologists working in the area surrounding Chetumal Bay have suggested that settlement patterns indicate a hierarchy involving at least two tiers of differentially ranked settlements (Sidrys 1983:393-400). According to this model, the relative importance and position of settlements in the area is manifested by differences in settlement size, architecture and other conspicuous indicators of wealth and power. Following the same line of argument, Masson (2000:31-38) has proposed a provincial model for occupation in the Chetumal area. Working backwards from ethnohistoric accounts, which document the Chetumal 'province' as a cohesive political unit headed by a capital city (tentatively identified as the site of Santa Rita Corozal [D. Chase 1986]), she envisages a Terminal Classic/Early Postclassic political infrastructure consisting of three tiers: a paramount centre that held ultimate political power and authority; lower-ranking secondary centres, which might have been previous seats of government, and a third tier of supporting villages and hamlets. She notes, however, that contemporaneous Terminal Classic/Early Postclassic centres might have been equivalently-ranked, in which case a two-tiered hierarchy of settlements would apply. Similarly, obvious differences in size and monumentality among contemporaneously occupied settlements in northern Belize are often considered to reflect differences in wealth, power and relative status on the regional level.

(e.g. Chase and Chase 1982; Masson 2000:31-32; Adams et al. 2004; Hammond 1974). These interpretations derive from conventional models of the political geography of the lowland region, which envision the socio-political landscape as comprising a patchwork of territories and polities with an identifiable hierarchy of settlements (e.g. Culbert 1977:512-513). However, whether these perceived status differences among sites can be directly translated into political relationships has more recently become a matter of debate. As Freidel and Shaw (2000) have shown, it might not be entirely appropriate, or accurate, to view these observed differences in the material record as having a purely political significance, since ideological, economic and political dimensions of Maya cultural behaviour appear to be largely inseparable. Accordingly, just as the political nature of the relationships among sites require further investigation, current interpretations regarding the existence of political units in northern Belize, and their internal organization, require corroborating archaeological evidence to demonstrate that these frameworks represent a valid conceptualization of the socio-political landscape.

Viewed on a more general level, Terminal Classic and Early Postclassic occupations have been described as reflecting dispersed, decentralized and semi-autonomous communities (Masson and Mock 2004:400). Nucleated settlements containing civic and ceremonial architecture, as well as residential structures, have been the traditional focus of archaeological investigation. These sites are thought to have consisted of two functional components: a central precinct, which was the focal point of religious, ceremonial and administrative activities and the main area of elite residence; and a surrounding sustaining area, or hinterland, primarily inhabited by the non-elite classes, where economic and subsistence production took place. There is considerable debate surrounding whether these sites functioned primarily as *administrative* centres or *regal-ritual* centres, although it is well recognized that the central precincts of these sites potentially played an important role as central places in regard to a variety of religious and economic activities (Fox 1977, see also Ball 1993:247-249 for a discussion). It is generally held that individual centres maintained a basic level of self-sufficiency, in terms of the production of subsistence items and fulfilling the labour requirements for a range of community activities. (e.g. Adams et al. 2004:329-330; Eaton 1982:17-19). Intra-regional socio-economic interaction among sites (i.e. within northern Belize) is assumed, based on the commonalities observed among local ceramic assemblages and the widespread occurrence of material items deriving from raw material resources, such as chert and marine shell, that only occur naturally in particular areas of northern Belize (e.g. Masson and Mock 2004; Masson 2000:105-193; Masson and Rosenswig n.d.; Andres and Pyburn 2004; Valdez 1994; Fry 1989:103-106; Ball 1983). Local access to goods and/or materials deriving from more distant locations is demonstrated by the occurrence of artifacts made of exotic raw materials such as precious stones, obsidian, gold and copper, which originate from deposits located in Mexico and highland Guatemala and Honduras (e.g. Pendergast 1986; Sidrys 1983; Masson 2000:153-164 and 173-179). The occurrence of these exotic artifacts is often taken as evidence that the inhabitants of the site participated in long distance exchange networks. However, the comparative rarity of these items would seem to imply that either access to or the distribution of these items was fairly restricted, or alternatively, that these commercial connections operated on a relatively limited scale.

Terminal Classic to Postclassic Occupations in Northern Belize

Terminal Classic to Postclassic occupations in northern Belize are often described as embodying a "variable mosaic of historical events and processes" (Demarest et al. 2004b:557-558) and individual site histories attest to an array of differing and complex developmental trajectories (Table 2.1). At Altun Ha and sites in the Three Rivers Region, the emergent pattern is similar to that observed at many other southern lowland centres, where a period of decline during the Terminal Classic is followed by partial or complete site abandonment (Pendergast 1979:190-199, 1982:260-263; Adams et al. 2004). In the Three Rivers Region, the abandonment of sites is thought to have occurred at around A.D. 850 (Adams et al. 2004:336-337), thereby coinciding with the collapse of polities in other areas of the southern lowlands, whereas the partial abandonment of Altun Ha appears to have occurred much later, closer to A.D. 1050 (Pendergast 1979:34, Table 5; Pendergast 1982:2, Table 1). In contrast, at many other sites in northern Belize the final centuries of the Classic period saw heightened construction activity, settlement expansion and increasing population levels. Sites such as Northern River Lagoon, Saktunja and Caye Coco (Masson and Mock 2004:368-375) appear to have been initially settled or founded during this period, and the previously abandoned Preclassic centre at Cerros (Walker 1990:8-10) was extensively reoccupied at this time. Developments at these sites are thought to reflect the movement of immigrant populations into these areas, perhaps from neighbouring collapsing centres and polities or those situated in the Maya heartland (Masson and Mock 2004; Andres and Pyburn 2004; Adams et al. 2004). At sites such as Colha (Hester et. al. 1982; Eaton 1982; Valdez 1987, 1994; Mock 1994), La Milpa (Hammond and Tourtellot 2004), Pull Trouser Swamp (Fry 1989) and Northern River Lagoon (Masson and Mock 2004), this period of heightened activity and rising population levels appears to be relatively short lived, being interrupted or immediately followed by the permanent or temporary abandonment of these sites by A.D. 850 - A.D. 1000. In the case of Colha, the Classic period occupation appears to have ended rather violently, with the execution of a group of elite inhabitants and the destruction of buildings in the site's central precinct (Eaton 1982:14). Some of these abandoned centres appear to have been subsequently reoccupied on a more limited and ephemeral scale during the Postclassic period. Radiocarbon dates obtained for Early Postclassic deposits at Colha suggest that reoccupation had occurred by A.D.

Table 2.1: Cultural-historical developments at various sites in northern Belize during the Terminal Classic and Postclassic periods.

Site	Settlement Type	Terminal Classic to Postclassic Developments	Source
Altun Ha	Ceremonial/administrative center	•Decline in construction activities, especially in the central precinct, during the Terminal Classic •At least partial abandonment by A.D. 1050 •Limited reuse of certain areas during the Late Postclassic	Pendergast 1979, 1982, 1990
Colha	Ceremonial/administrative center - industrial level production of chert implements	•Abandoned ca. A.D. 850 after execution of an elite group (possible Yucatec Maya involvement) •Small scale reoccupation A.D. 900-A.D. 1000 with complete abandonment by A.D. 1250	Hester et. al. 1982; Eaton 1982; Valdez 1987, 1994; Mock 1994
Santa Rita Corozal	Ceremonial/administrative center regional capital at the time of European contact	•Continuity in occupation through to the contact period	D. Chase 1985, 1986; A. Chase 1986
Northern River Lagoon	Coastal settlement - salt production - cadet settlement/trading node	•Initially settled in the Late Classic (ca. eighth century A.D.) •Abandoned after A.D. 900, with no substantial reoccupation	Masson and Mock 2004
Saktunja	Coastal settlement - salt production -cadet settlement/trading node	•Initially settled in the Late Classic (ca. eighth century A.D.) •Continuity in occupation through to the contact period	Masson and Mock 2004
Progresso Lagoon	Inland lagoon settlement - - administrative node	•Continuity in occupation through to the contact period	Masson and Mock 2004
Caye Coco	Inland lagoon settlement - elite shrine	•Founded in the Terminal Classic with continuity in occupation through to the contact period	Masson and Mock 2004; Masson and Rosenswig n.d.
Chau Hiix	Inland lagoon settlement - trade node and defensible refuge during the Postclassic	•Continuity in occupation through to the contact period	Andres and Pyburn 2004

Table 2.1: Cultural-historical developments at various sites in northern Belize during the Terminal Classic and Postclassic periods (continued).

Site	Settlement Type	Terminal Classic to Postclassic Developments	Source
Lamanai	Ceremonial/administrative center	•Extensive construction activities and settlement expansion during the Terminal Classic •Continuity in occupation through to the contact period	Pendergast 1981a, 1985, 1986
Nohmul	Ceremonial/administrative center	•Continuity in occupation through to Late Postclassic •Strong connections to Chichen Itza starting in the Terminal Classic	Chase and Chase 1982; A. Chase 1986
Cerros	Ceremonial/administrative center	•Extensive reoccupation of abandoned Preclassic mounded architecture in the Terminal Classic •Population decline and limited ephemeral occupation after A.D. 1150 •Complete abandonment after A.D. 1300, prior to the contact period	Walker 1990
Pull Trouser Swamp	Settlement associated with a raised field system	•Population increase during the Terminal Classic •No evidence of Early Postclassic occupation •Limited localized occupation during the Late Postclassic (after A.D. 1300)	Fry 1989
La Milpa	Ceremonial/administrative center	•Sudden abandonment ca. A.D. 850. amidst extensive expansion and refurbishment activities in the central precinct	Hammond and Tourtellot 2004
Three Rivers Region (sites include Dos Hombres, Chan Chich, Blue Creek and Rio Azul)	Settlements/centres of varying size	•Population decline during the Terminal Classic •Abandonment of centers ca. A.D. 850 •Limited reoccupation of some areas by 'new' peoples 50 to 100 years after abandonment •No substantial population after A.D. 1250	Adams et al. 2004

900 - A.D. 1000 (Hester et al. 1982:8). Sites in the Three Rivers Region are thought to have been reoccupied at around the same time (A.D. 900 – A.D. 950) (Adams et al. 2004:337), while the evidence obtained from Pull Trouser Swamp suggests that resettlement of this area did not take place until after A.D. 1300 (Fry 1989:106-107). The Classic to Postclassic record at many other sites exhibits greater continuity. Sites such as Nohmul (Chase and Chase 1982; A Chase 1986) and Cerros (Walker 1990) are continuously occupied well into the Postclassic period, whereas at Lamanai (Pendergast 1981a, 1986), Santa Rita (D. Chase 1985, 1986; A. Chase 1986), Chau Hiix (Andres and Pyburn 2004), Saktunja (Masson and Mock 2004) and Caye Coco (Masson and Mock 2004; Masson and Rosenswig n.d.) occupation extends even further, into the Spanish Colonial period. Continuity in occupation at these sites over the bridging period between the Classic and Postclassic is considered to reflect the long term capacity of these communities to adapt to a changing socio-politico-economic environment (e.g. Pendergast 1986; Masson and Mock 2004; Chase and Chase 1982). Nonetheless, long term stability during the Postclassic appears to vary from site to site. For instance, population levels at Cerros decline substantially by A.D. 1150, and the site is completely abandoned some time after A.D. 1300 (Walker 1990:9-10). Nohmul also appears to have been abandoned prior to Late Postclassic times (A. Chase 1986:121-124). Such a decline is less apparent at other sites such as Santa Rita, Lamanai, Caye Coco, Saktunja, where there is abundant evidence of thriving communities throughout the Postclassic period.

Explanatory Models of Culture Change and Process

The 'Delayed' Collapse Model - The disparate local manifestations of the Classic to Postclassic transition in northern Belize have given rise to a range of explanatory models that attempt to delineate culture process at this juncture in local and regional cultural-historical sequences. In many instances, observed discontinuities in local occupation are seen as linked to the political collapse of the Maya heartland. Sites that experience a period of decline, followed by permanent or temporary abandonment by the end of the Classic period, are generally thought to have succumbed to a similar combination of external and internal disturbing factors that brought about the collapse of other southern lowland centres and polities (e.g. Adams et al. 2004). Developments at these centres are seen as part of a general pan-lowland trajectory towards collapse, culminating in population dispersal and the termination of Classic elite culture and socio-political institutions. Alternatively, at sites where abandonment is preceded by a period of heightened construction activity and population growth, local patterns are seen as reflecting direct repercussions of the collapse (e.g. Hammond and Tourtellot 2004; Sullivan 2002), perhaps reflecting the heavy influx of immigrants from areas of the southern lowlands subject to earlier disruptions. The rapid increase in local populations, as a result of immigration, is argued to have led to the same crisis situation as had evolved previously at other southern lowland centres – i.e. population levels eventually surpassed the carrying capacity of the local environment, and local rulers were unable to put into effect an adequate response to a worsening situation. Another contributing factor may have been the loss of traditional trading partners, as economic interaction with areas first affected by the collapse dramatically declined. As has been suggested for other areas of the southern lowlands, the end result was the eventual breakdown of social, political and economic infrastructures and the forced abandonment of these centres. Frequently characterized as a delayed collapse, developments at these sites are often viewed as a case of history repeating itself.

The Mercantile Model - The fact that some sites in northern Belize maintained substantial populations throughout the Collapse period, some being settled for the first time, has been viewed by some scholars as evidence of a broader pan-lowland politico-economic transformation that started in the Terminal Classic period. Following the model developed by Rathje (1975), Masson and others (Masson 2002; Masson and Rosenswig n.d.; Masson and Mock 2004), have suggested that the founding of new centres in northern Belize and the persistence of some existing centres can be best explained by a reorientation of local economies in response to the rise of mercantilism. According to Rathje's model (1975:431-433; also Rathje and Sabloff 1973; Sabloff and Rathje 1975) the Terminal Classic to Postclassic period marks the transition of a regional trade system based on overland transport into an external, large-scale commercial system utilizing the seagoing canoe[3]. This shift in the technology of distribution made feasible the transport of items such as pottery, which were difficult to transfer successfully using previous means, and created a new 'mass market potential' for a wide range of trade items. It also led to the mass production of ceramics and other items as well as the greater dispersion of material items over social and geographical space. Over time, these changes would have promoted a certain level of community specialization in economic activities, with specific products being produced at fewer centres and distributed over a wider geographical area, thereby fostering economic interdependence among local populations. Rathje (1975:439-440) suggests that a key ingredient to the development of this large-scale commercial network was the emergence of material culture with a shared ideological value and significance. These material components of the system not only symbolized the integration and unity of the system itself, but also the shared beliefs and values of network participants.

By applying Rathje's model to Terminal Classic occupations in northern Belize, Masson and colleagues have argued that the occupational continuity observed at

[3] As Walker (1990:16) points out, a major flaw in Rathje's model is the lack of evidence that the seagoing canoe was in fact invented at this particular time.

some sites was largely due to the strategic location of these communities within a 'buffer zone' area. (following Rathje 1972, 1973). Located in a resource rich environment and in an intermediate position between the southern lowland interior and the Yucatan area of the northern lowlands, these communities were in an advantageous position to participate in emerging maritime trade networks of the type that Rathje describes. These scholars point out that many of the sites that were initially settled or were able to maintain substantial populations during the Terminal Classic to Early Postclassic period are situated strategically either in coastal areas or along waterways. This, they argue, attests to both the importance of inter-site communication and trade, and the central role economic activities played in the establishment and/or survival of these communities. They envision the inhabitants of these sites as innovators, entrepreneurs and opportunists, who were able to capitalize on their advantageous position by re-orienting their local economies towards intra-regional and maritime trade. Masson and her colleagues suggest that the maritime trade network that emerged during the Terminal Classic to Early Postclassic period was initiated and driven by the economic and political interests of Chichen Itza. Hence, their model departs slightly from Rathje's, in that they see the interaction between Chichen Itza and communities in northern Belize as having an overtly political dimension as well as an economic one. They argue that observed stylistic and technological similarities between Terminal Classic ceramic assemblages in northern Belize and contemporaneous complexes of the Chichen Itza sphere provide evidence, not only that sites in northern Belize participated in exchange networks driven by this distant city, but also of Chichen Itza's direct influence on some aspects of local manufacturing traditions. They suggest that these similarities reflect an international ideological emulation that accompanied the development of close trading ties, and point to a re-orientation of local political and economic ideology. Masson and Rosenswig (2003; Masson 2002) argue that additional support for Rathje's model lies in the trend towards standardization observed in regard to Postclassic pottery at sites such as Caye Coco, since this trend may point to the implementation of mass production techniques. Furthermore, they suggest that widespread stylistic similarities in pottery items might imply fewer production centres and extensive distribution. They admit, however, that other production-related mechanisms such as the involvement of fewer ceramic specialists, more tightly regulated manufacturing processes or similar production techniques could also account for these patterns.

A fundamental aspect of the mercantile model is that it presumes that Postclassic society was fundamentally distinct from Classic society. Since it is proposed that the transformation of one into the other was the direct result of the development of a merchant class that controlled government as well as trade, a main underlying premise is that Postclassic occupations constitute a significant reconfiguration of previous social, political and economic structures. A direct implication is that the Terminal Classic to Early Postclassic period was a time of fairly radical culture change. This interpretation has been called into question by others working at sites with occupations spanning the Terminal Classic period, who observe that many of the purported Postclassic 'innovations' have antecedents in the Classic period or earlier. For example, Chase and Chase (2004:24) have argued that internally complex economic institutions involving centralized distribution and a middle social level engaged in the manufacture of items for trade existed in the Late Classic period at Caracol. In addition, it is well recognized that stackable vessels and standardized forms, often considered as indicative of mass production, also occur during the Late Preclassic and Classic periods (c.f. Fry 1981). Moreover, the Classic-Postclassic dichotomy that is suggested by the mercantile model would appear to be at odds with the abundant evidence of cultural continuity at some northern Belize sites, where many Postclassic patterns appear to be a logical outgrowth or modification of previous conventions (e.g. Cerros, Walker 1990; Lamanai, Pendergast 1985; Graham 1987; Nohmul and Santa Rita, Chase and Chase 2004).

The Regional Cult Model - In contrast to explanatory models that focus specifically on the Maya lowland region and identify economic or politico-economic factors as the driving forces of change, Ringle et al. (1998) see a broader level of cultural transformation within Mesoamerica, which they suggest is driven by a cross-cultural shift in ideology. According to this model, cultural-historical developments in the Basin of Mexico, the Yucatan Peninsula of Mexico and areas of the Central Peten and Guatemala highlands during the Terminal Classic and Postclassic periods are linked directly to the spread of a 'regional cult'[4] dedicated to the worship of the god Quetzalcoatl. This cult focussed on the god's aspects as creator, as wind god and as patron of merchants and leaders, and his associations with rebirth, fertility, war and portals of emergence and transformation (Ringle et al. 1998:223-225). Central features of the cult that led to its proliferation were military proselytism and the fact that it offered a common belief system of 'universal truths', communicated through a uniform set of visual symbols conveyed in sculpture, murals and other forms of prestige art, which transcended ethnic and political boundaries. The authors contend that the cult was based on the establishment of regional shrine centres for pilgrimage and worship and that it spread through military conquest and the active founding of new centres and dynasties by conquering elites. Moreover, they suggest that dissemination of religious beliefs and practices of the cult would have been further facilitated through political alliances, trade and pilgrimage (Ringle et al. 1998:213-214). Ringle et al. see the existence of the

[4] Ringle et al. (1998:185, footnote 1) define a 'regional cult' as "the adoption of a belief system by several politically and ethnically independent polities throughout a highly developed culture area". They equate the term loosely with 'world religion'.

cult manifested in an 'international style'[5] in architecture and iconography, during the Terminal Classic and Postclassic periods. This 'international style' had a distinctive emphasis on warfare, human sacrifice and calendrical symbolism. They argue that additional evidence of cult activities lay in the increased incidence of defensive structures in association with Postclassic occupations as well as in the widespread occurrence of cult paraphernalia such as figural jades, which they interpret as religious tokens, and a subcomplex of ceremonial vessels used in cult rituals and worship – i.e. fine-paste wares (particularly Silo Fine Orange), plumbate pottery and an *incensario complex* consisting of ladle-style incense burners, open-work censers, spiked hourglass incensarios and Tlaloc pots (Ringle et al. 1998:207, 214-218).

According to Ringle et al.'s model, conflicts between centres in the northern and southern lowlands during Terminal Classic and Postclassic periods, and the rise of various centres of power in the north from Terminal Classic times onwards, are direct repercussions of the spread of the cult of Queztalcoatl. Thus, the emergence of this new religious institution is considered to be the prime mover of cultural process during this period of cultural development (Ringle et al. 1998:218-222). The authors argue that cult activities best explain both the rise of international styles in art, architecture and iconography and the escalation of conflicts among sites and elite factions that is observed. They envision cult membership as a strategy employed by local elites to further their personal political ambitions, since cult allies could be solicited for military support. In addition, they suggest that the enhanced trade opportunities, and enormous material wealth and symbolic prestige that likely accompanied cult membership and status as a major shrine centre, might have been important motivating factors behind the conflicts waged over the control of particular northern lowland centres (e.g. Chichen Itza). Ringle at al. also postulate that the spread of the cult network had an important role in the collapse of the southern lowlands. In considering Late Classic southern lowland iconography, they see the appearance of certain 'Mexican' symbols and glyph forms, the increase in sacrificial and military themes and the regularization of warrior costumes, as evidence of the emergence of the Quetzalcoatl cult in this area (Ringle et al. 1998:226). Viewed in light of this evidence, the intensification of conflict and competition among elites during the centuries leading up to the collapse can be considered as arising, at least in part, out of cult activities and the spread of this new religion. This interpretation closely follows the views of Thompson (1954:87) and others, who have suggested that the collapse was precipitated by an ideological invasion, with origins in Mexico, that had a heavy militaristic emphasis.

Since northern Belize borders the geographic area considered to be the 'hot spot' of cult activity in Ringle et al.'s model, one might expect to find evidence that the cult spread into this region. Whereas the military invasion of Colha and the collapse of centres in the east of the sub-region might be construed as related to the cult's spread, the rapid decline and abandonment of these centres suggests that they became neither shrine centres nor members of the cult network. Even at sites where occupation continues into the Postclassic period, the evidence of local participation in the cult network is less than convincing. Many of the proposed indicators of cult activities such as defensive structures appear to be absent at most centres, and fine paste and plumbate pottery, identified as forming an integral part of the ceramic subcomplex, are comparatively rare at most sites, often being represented by only a few sherds. Nonetheless, at least some components of the *incensario complex* would appear to occur at some sites (e.g. Laguna De On, Masson 2000; Lamanai, see Chapter 7). Late Postclassic deposits of censer fragments at previously abandoned centres are often thought to relate to pilgrimage visitation (e.g. Hammond 1982:70; Walker 1990; Pendergast 1979:199). The connection between these deposits and cult activities, however, is less than clear, since these sites were unoccupied and, hence, not functioning as shrine centres at the time these offerings were made (at least not of the sort that Ringle et al. describe).

Site Specific Interpretations - Some scholars working in northern Belize have favoured a more 'contextual' approach in their interpretations of Terminal Classic to Postclassic occupations. In most instances, these site-specific accounts focus on evidence of ethnic and occupational discontinuity and/or foreign influences, which almost invariably are considered to reflect the physical invasion or intrusion of foreign groups from the northern lowlands. Such is the case with Colha, where there is abundant evidence of a militaristic invasion at the end of the Classic period. Invasion is indicated by the occurrence of large deposits of disarticulated human skeletal material, the destruction of buildings in the central precinct, and the shift in lithic workshop production during the Terminal Classic towards implements used in warfare (Hester et al. 1982; Eaton 1982). It is suggested that Colha was completely abandoned for a brief period (50-100 years) after this invasion, and was subsequently reoccupied to a more limited extent by a 'new' group, possibly from northern Yucatan. This formulation of the cultural-historical sequence is largely based on Valdez's interpretation of the Colha ceramic data (Valdez and Adams 1982; Adams and Valdez 1980; Valdez 1987, 1994). Valdez (1987:270) argues that there is a temporal gap in the ceramic sequence at the end of the Terminal Classic period based on the absence at Colha of an intermediary ceramic complex in which vessels exhibit a combination of earlier (Terminal Classic) and later (Early Postclassic) stylistic

[5] Ringle et al.'s (1998:185 and footnote 4) usage of 'international style' connotes specific ethnic and cultural linkages. Their 'Early International' style is typical of the Terminal Classic period and refers to iconographic conventions and imagery that derive from a heavy Maya-Toltec influence. Their 'Late International' style, which is characteristic of the Postclassic period, has Mixteca-Puebla influences and is associated with the painting style of the Mixteca codices.

attributes. Since such continuities are recognized at sites such as Lamanai (Pendergast 1981a; Graham 1987), Nohmul (D. Chase 1982; Chase and Chase 2004) and Santa Rita (Chase and Chase 1988; Chase and Chase 2004), Valdez argues that their absence at Colha represents a clear break in local ceramic traditions. Consequently, the succeeding Early Postclassic Yalam complex is viewed as a typological and functional departure from previous complexes. Valdez suggests that this change in the ceramic inventory reflects the reoccupation of the site by a new ethnic group, possibly the same group that had invaded the site. That this group came from northern Yucatan is inferred by stylistic similarities between Yalam ceramics and northern Yucatecan wares (Valdez 1987, 1994; Mock 1994).

A similar line of argument has been followed in the construction of a cultural-historical framework for the Three Rivers Region (Table 2.1). Although there is less evidence in this area of an actual military invasion (Rio Azul is a possible exception, Sullivan 2004), Adams et al. (2004) link the abandonment of urban centres and the rapid depopulation of the countryside to an intrusion of Puuc Maya peoples from the north. They argue that after a period of 50-100 years, certain parts of the Three River region were reoccupied by Puuc Maya or other groups from northern Yucatan. As at Colha, a gap in the occupational sequence is inferred based on the absence of 'intermediary' ceramics that combine Terminal Classic and Early Postclassic stylistic traits.

The presence of a foreign influence has also been suggested for Nohmul, where architectural and ceramic assemblages bear similarities to that at Chichen Itza (Chase and Chase 1982). At Nohmul, however, there is no evidence of a militaristic invasion, and occupational and ceramic continuity are readily apparent. As a result, the emergence of foreign architectural and ceramic traits during the Terminal Classic period has been viewed as evidence that the site came under Chichen dominance, and perhaps was even an outpost of this northern city (Chase and Chase 1982; A. Chase 1986). Unlike at Colha and sites in the Three Rivers Region, where discontinuities in material culture are considered to correlate with an ethnic break in occupation, patterns at Nohmul are viewed as reflecting the presence of foreign elite group. Hence, at least some semblance of continuity in the local population is assumed for Nohmul. The resulting interpretation, therefore, is that the local population of Nohmul came under the control of a group of foreign leaders during the Terminal Classic period. Chase and Chase (1982) see these developments at Nohmul as related directly to the territorial expansion of the Toltec hegemony based at Chichen Itza. In the case of Cohla and the Three Rivers Region, however, the link between local trajectories and regional-level developments associated with the expansion and contraction of predatory states in northern lowlands is stated less explicitly.

The Inadequacies of Single Factor Models

At first glance, it would appear that the various models that have been developed or invoked to explain Terminal Classic and Postclassic occupations in northern Belize, present highly contrasting accounts of culture change and process during this period. This is especially evident in regard to interpretations of the Terminal Classic to Early Postclassic ceramic record, which have led to wildly different formulations of local cultural-historical sequences, despite the acknowledged similarity of local assemblages. A central feature that nearly all of these models share, however, is that they identify external factors and pressures as the driving forces of culture change and process during the Terminal Classic to Postclassic transition (the exception is the conventional collapse model, Willey and Shimkin 1973). Such factors include: population migrations and disruptions in economic activities caused by the collapse of southern lowland centres and polities, the emergence of maritime trade networks initiated and driven by Chichen Itza, the spread of a new world religion with a heavy emphasis on militarism and physical invasion by foreign groups, military or otherwise. What the foregoing discussion also demonstrates is that the contributing role of these factors in shaping local-level trajectories varies from site to site. For example, physical evidence of military invasion is completely absent at most sites, whereas factors such as the disruption or reconfiguration of trade-networks and/or the immigration or intrusion of non-local groups appear to have had differential effects, contributing to the rapid decline of some centres, and the continued or heightened prosperity of others. The persistence of local communities over the bridging period between the Classic and the Postclassic is most often put down to the ability vs. inability of local communities and their leaders to respond adequately and adapt to external pressures and a changing socio-politico-economic landscape, both locally and regionally. This view would seem to imply that either social compliance or an instinctive willingness to follow a particular developmental path lay at the heart of culture process during this period. Accordingly, it would appear to ignore the possibility that all levels of society had a vested interest in both their situation and continued prosperity of the community, and made decisions that bore on various aspects of local-level development – for the better or for the worse. Surely the fact that some communities appear to have been less vulnerable to external pressures suggests that factors and dynamics internal to the community played an equally significant role in shaping local trajectories. The nature and role of these internal factors are largely overlooked in current explanatory models, since they tend to focus on the 'who's, where's and when's' of historical development and not so much the 'how's and why's' of the intra-community interaction that is fundamental to this development. As a result, while these models may address the effects of regional-level developments on local communities, they provide only limited insight into the ways in which local communities were affected by these developments.

Despite obvious differences in interpretive and theoretical perspectives, there are two issues that the advocates of the different explanatory models would seem to agree upon: firstly that stylistic similarities in the material record reflect shared beliefs and values, regardless of whether they are identified by the analyst as being specifically ideological, political or economic in nature; and secondly that political, economic and ideological aspects of Maya behaviour are closely intertwined. If these two assumptions can be judged as being reasonably valid, it is not surprising that models that argue for the primacy of one of these components of culture over the others often appear to ignore or be at odds with certain facets of the evidence at hand. At the same time, however, different formulations of cultural change and process need not necessarily be considered as representing competing explanations of the same developmental phenomena. Rather, they can be viewed as reflecting diverse and complex manifestations of contemporaneous cultural behaviour, as well as the variable experiences, actions and reactions of local communities during a time of changing world conditions. As the forgoing discussion demonstrates, local manifestations of the Terminal Classic to Postclassic transition in northern Belize do not conform to a single universal pattern. Yet, regardless of differences in local events and circumstances, communities share a common, distinctly Maya material identity. What is intriguing is that local communities maintained these shared beliefs and values through time, while adapting to changing local and regional conditions and circumstances in distinctly individual ways. A central premise of the present study is that these individual community histories arose out of human interaction embedded in intra-community dynamics that are a reflection of both local experiences and circumstances and extra-local developments and pressures. These human interactions can be studied as they are manifested in the material remains of community based activities and behaviours – i.e. material-human interactions. The next section looks at cultural-historical developments over the course of the Classic to Postclassic transition at the site of Lamanai, focusing on behavioural patterns associated with contexts and settings of ritual and ceremonial activity and material-human interactions relating to the use and deposition of pottery items.

2.3. Lamanai during the Terminal Classic to Early Postclassic Transition

Lamanai lies on the northwest bank of the New River Lagoon in the interior of northern Belize, and is one of the few Maya sites for which the ancient name is recorded, appearing on early maps and in early Spanish documents regarding the church that was established there in the sixteenth century (Pendergast 1988, 1981a:31). The site is perhaps best well known for its lengthy and continuous history of occupation, which spans the Middle Preclassic through to the Spanish Colonial period, and also for the fact that it boasts one of the most robust and well preserved records of the Terminal Classic to Postclassic transition. Material evidence of community life during the Terminal Classic and Early Postclassic periods is ubiquitous, being represented by extensive artifact and architectural assemblages and numerous burials and offerings. Since archaeological investigations at the site have involved the extensive excavation of architectural structures and features, the archaeological and cultural contexts of the material evidence are well understood in terms of the history of construction, use and maintenance of different structures and areas within the settlement. As a result, Lamanai presents a rare opportunity to examine continuity and change in community-level patterns of economic, socio-political and ideological activity and behaviour, as manifested in the material record.

Situated in a narrow strip along the lagoon shore, the site is characterized by an unusual settlement pattern, with a fairly compact central zone, or central precinct, comprising an irregular arrangement of large structures and building groups, surrounded by zones in which both the scale and density of structures markedly decreases (Pendergast 1981a:32; Loten 1985:86) (Map 2.3). As with other sites in northern Belize (e.g. Altun Ha, Pendergast 1982), mapping of the area revealed no obvious community boundaries with surrounding hinterlands characterized by an increasingly dispersed pattern of structural remains. The mapped area of the site includes 4.5km^2 of settlement and a recorded total of 718 structures (Loten 1985:86), the majority of which lie today within a protected archaeological reserve.

The first major excavations at the site were undertaken by Pendergast from 1974 to 1985, recognizing that the site potentially had a lengthy history of occupation, as suggested by the presence of both a Spanish church and large structural remains of earlier buildings, and thus could provide important insights into cultural developments in the Maya lowlands over a considerable period (Pendergast 1981a, 1985, 1986). More recent excavations, directed by Graham since 1997, have concentrated on investigating critical periods of cultural transition such as the Preclassic to Classic, the Terminal Classic to Early Postclassic and the Late Postclassic to Spanish Colonial Period (Graham 2004). The ongoing focus of investigations at Lamanai has been the densest area of settlement situated along the lagoon shore and extending 0.5km inland. This part of the settlement includes the central precinct, which comprises the main ceremonial and administrative structures as well as residences of the elite or ruling class, and adjacent residential areas. These investigations have revealed more about the Preclassic to Early Classic period and the Terminal Classic and Postclassic periods than the Middle to Late Classic period. The evidence suggests that this area of the site was continuously occupied from at least 600 B.C. to A.D. 1675, with an extensive and well-developed occupation prior to 300 B.C. (Pendergast 1981a:42, 1986:226). Architectural and other material evidence indicate that the settlement enjoyed trade and intellectual contacts with other centres in the Maya region and beyond throughout its history (Pendergast 1981a, 1985), while archaeobotanical studies have revealed a

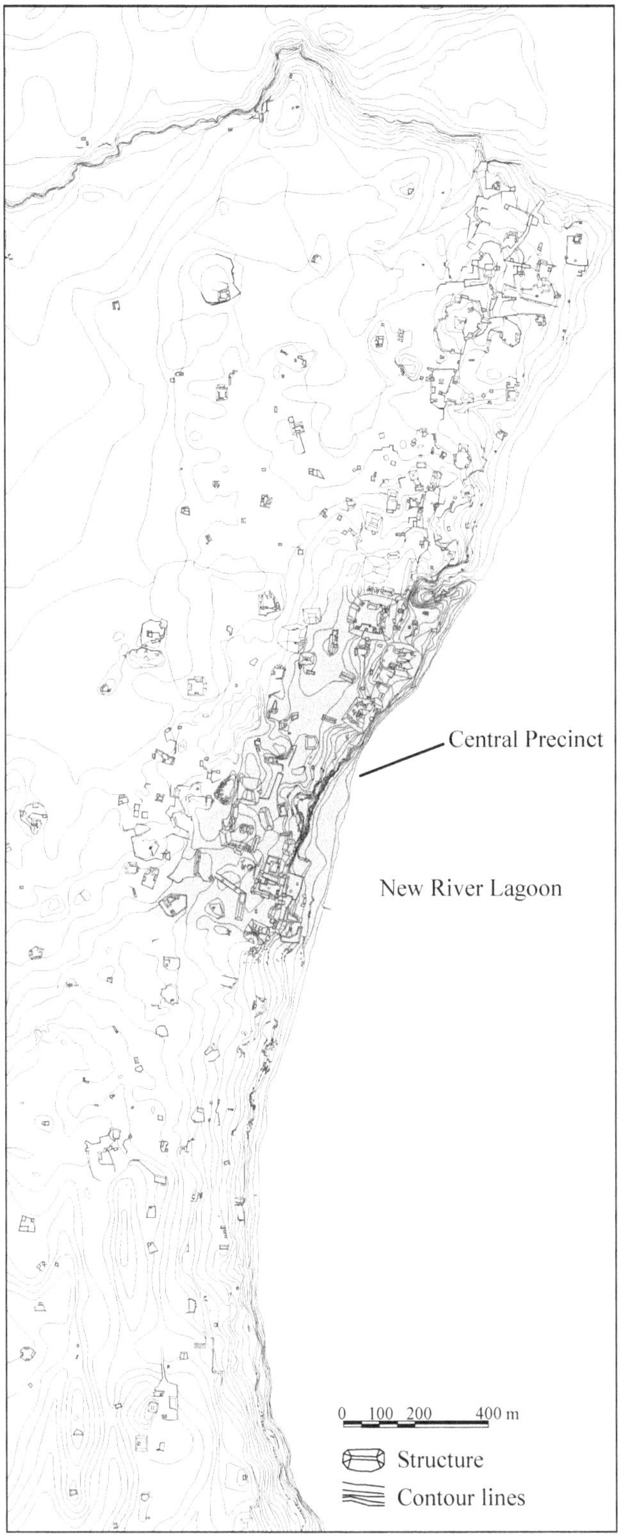

Map 2.3: Map of Lamanai.

strong economic base in maize agriculture (Pendergast 1990; Loten 1985).

The archaeological manifestations of what can be considered as Classic vs. Postclassic culture at Lamanai are as difficult to distinguish on typological grounds as they are to fix temporally as architectural, ceramic and other material traits "do not come to neat ends to be supplanted by another set" (Pendergast 1986:234). Instead, the evolution of one to the other, which took place roughly over the eighth to twelfth centuries, appears to form a complex and fluid history of development in which earlier traditions were retained, sometimes in a slightly modified form, alongside shifts in direction and obvious instances of innovation. Recent work on the Classic to Postclassic transition (e.g. Chase and Chase 2004) advocates an approach to enquiry centred on the identification of the Classic roots of Postclassic cultural phenomena as a means to shed previous biases surrounding the perceived dichotomy between Classic and Postclassic lifeways. What is clear at Lamanai, however, is that not all aspects of the Postclassic record have traceable Classic origins. Some developments are clearly based on new ideas (local innovations and concepts from other areas), while others appear to be a distinctive end result of the merging of the 'new' with the 'old'. It is this blending and merging of new ideas and approaches with established customs and conventions that is perhaps the most revealing aspect of the time period and, hence, a potentially illuminating focus of investigation.

The best-documented aspects of the Terminal Classic to Postclassic transition at Lamanai derive from Pendergast's detailed study of monumental architecture and construction activities (Pendergast 1980, 1981a, 1981b, 1981c, 1981d, 1982, 1984, 1985, 1986, 1988, 1990, 1992, 1998). These investigations have yielded a ubiquity of information regarding various aspects of cultural development during this time period and have produced an extensive and diverse data set, which is rarely paralleled elsewhere in the southern lowland region. Through detailed examination of the architectural and settlement data, and by integrating these lines of evidence within a broader framework of cultural activity that considers a range of other evidence of ceremonial, ritual and day-to-day life within the community, Pendergast has been able to shed light on various socio-political, socio-economic and ideological developments that are definitive of the transition from Classic to Postclassic culture at Lamanai. He generally sees this evolution as

> ...an essentially indigenous development out of Classic and earlier roots, into which ideas and materials from outside sources were introduced as necessary, with the degree of alteration required to make them blend into the local pattern. (1986:245)

Graham's more recent investigations (Graham 2004) , which have targeted areas of the central precinct that offer the most robust record of the bridging period between the Classic and the Postclassic, have produced additional insights into the cultural processes at work during this transitional phase of the community's development. She observes:

> ...there is little doubt at Lamanai that the Classic to Postclassic transition was distinguished by a socio-cultural dynamic of continuity and change. Aspects of change in material culture [architecture and material items]...suggest that those pulling the strings of power in the Early Postclassic had cultural, political, religious and perhaps economic priorities that were different from those of Classic Period rulers, although key aspects of the way society was organized seem to have remained the same (2004:4).

Of particular relevance to the present study is the detailed information Pendergast's and Graham's work has yielded regarding the cultural and archaeological contexts of structure-associated ceramic deposits. This information provides a frame of reference within which the cultural significance and meaning of the patterns that emerge through the detailed analysis of the ceramics themselves can be evaluated. For example, Classic to Postclassic period construction activities provide evidence of continuity and change in patterns of use of different structures and areas within the settlement (the associated depositional contexts of the ceramics) as well as the attitudes and priorities reflected in archaeological manifestations of ceremonial and ritual activity and aspects of day-to-day life. Along these same lines, the depositional patterns of ceramics within offertory, burial and midden contexts offer insight into continuity and change in the associated behaviours, practices and ceremonies that give rise to these deposits. Finally, the circumstances that appear to have contributed to the community's continued prosperity provide some clues as to the factors influencing patterns of ceramic production and consumption during the Terminal Classic to Early Postclassic period.

Construction Activities and Patterns of Use of Ceremonial and Residential Areas within the Community

Generally speaking, as with most sites located within the Peten or 'core area' of Classic Maya culture, the Classic period at Lamanai is characterized by standardized architectural forms and offertory practices, suggesting a community under the rigid control of ruling elites, and architectural features that have recognizable affinities with the Peten (Pendergast 1986:226). Most of the large ceremonial structures within the central precinct, as well as their associated plaza groups, (e.g. structures N10-43, N9-56, N10-27 and N10-9) underwent large scale modifications during the Classic period and there is evidence for their continued maintenance and use into the Terminal Classic and, in some cases (N10-9), into succeeding periods (Map 2.4). A significant aspect of building renewal during the Classic period was the major frontal reshaping of communally built ceremonial structures to include a chambered room athwart the central stair. This building type (the Lamanai Building Type), which according to Pendergast is distinctive of Lamanai and Altun Ha, becomes the standard edifice form at Lamanai by Late Classic times (1981a:36-41). The basic form of these ceremonial structures remained essentially unaltered in succeeding time periods, suggesting strong links to conventions established in the Classic period in terms of the physical setting of ritual and ceremonial activities as well as some basic aspects of ritual practice.

During the Terminal Classic, while the central precincts at many other Maya centres were experiencing a diminishing amount of maintenance attention and use, and often abandonment, construction activities at Lamanai continued on a scale at least equal to that of earlier time periods. The maintenance and periodic renewal of ceremonial structures continued (for example structures N10-27 and N10-9 and Plaza Group N10[3]), although the focus of these efforts shifted to the central portion of the structure surrounding the primary axis and major reshaping projects were no longer undertaken. For example, frontal modification of structure N10-9, the principal ceremonial structure within the N10-9 plaza group, included renewal of the central stair, while the terraces of the main platform were left to exposure and decay (Pendergast 1981a:43). Other large scale construction efforts within the N10-9 plaza group include the building of a new ceremonial structure (N10-7) and a major reshaping of the elite residential and administrative complex (the Ottawa Group N10[3]) situated directly across the plaza from structure N10-9. Structure N10-7 is a large ceremonial structure that was built on the site of an earlier platform atop the burial of a high status individual. The styles of the pottery included in the burial securely date the interment to the Terminal Classic period, indicating not only that new large-scale ceremonial constructions were being undertaken during the Terminal Classic, but also that new sites for ritual activity were being established. Structure N10-7 continued as a site for ritual activities and as a burial repository for high status individuals into Early Postclassic times, with several interments dating to this period intruded into its upper core (Pendergast 1981a:42).

The elite residential and administrative complex of the N10-9 plaza group (N10[3]) was perhaps the site of the most intensive construction activities undertaken during the Terminal Classic period, undergoing a successive series of alterations and ultimately a radical transformation. Pendergast (1985:93-94, 1986:231-233) notes that, up until the late ninth or early tenth centuries, the architecture that characterized this building complex was largely a continuation of earlier traditions, perhaps first established in the Middle Classic, although with innovations in form and in façade treatment. The first extensive alterations to the complex included the addition of a colonnaded entryway to one structure (N10-15) – a feature previously unseen at the site that is more typical in northern Yucatan – and the erection of a building with an elaborately painted stuccoed façade (N10-28). Since the iconographic content of the façade is largely religious in nature, with 'rulership' as a central theme (Shelby 1999), this would seem to mark the entry of strong visual

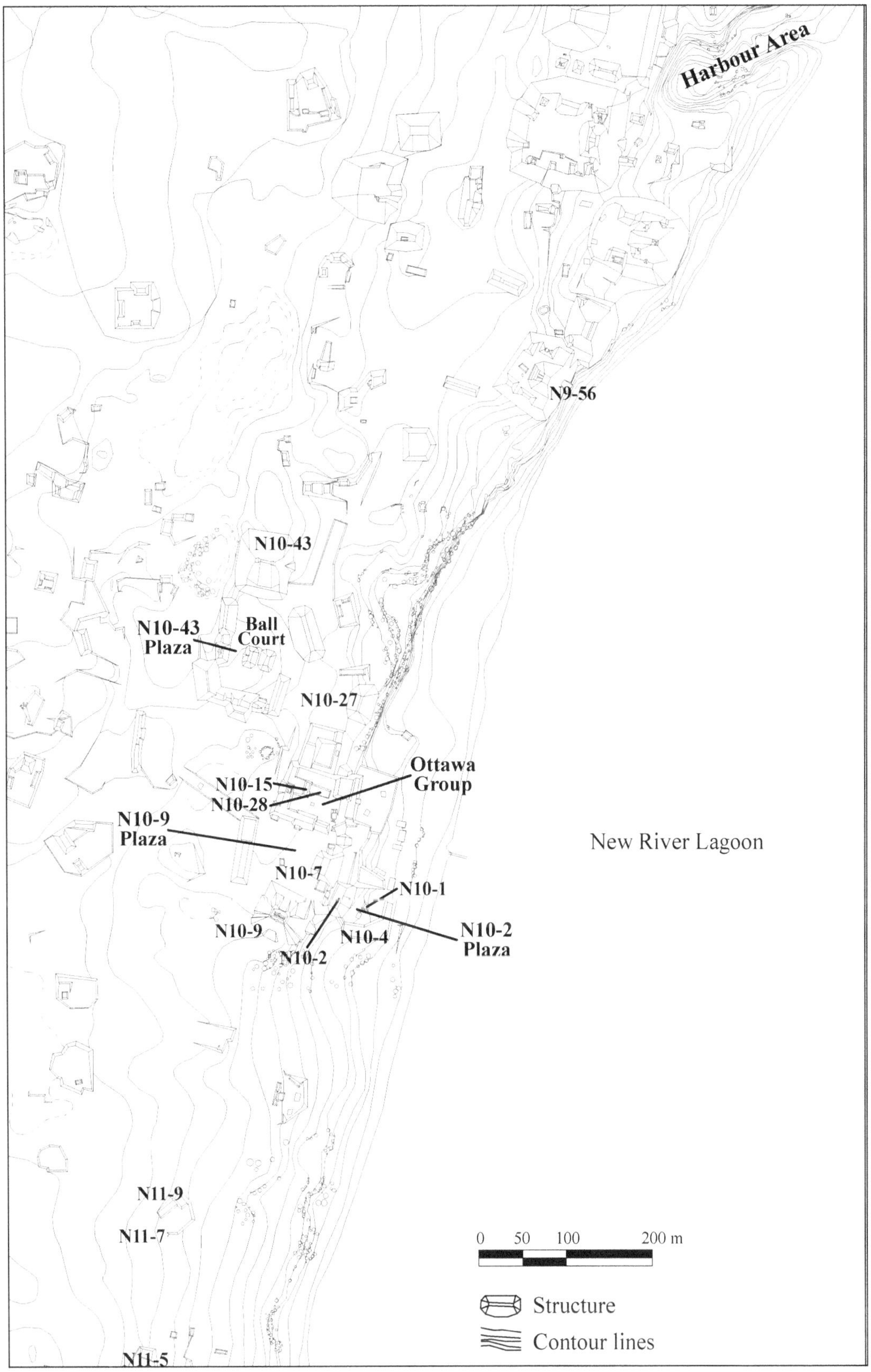

Map 2.4: Map of the central precinct at Lamanai showing structures and areas mentioned in the text.

expressions of religious values and notions of authority, which previously were restricted to ceremonial structures, into the domestic realm (in this case elite) (Pendergast 1985:94, 1986:231). The increased importance of ceremonial and religious activities within elite residential contexts is also suggested by the fact that many of the Terminal Classic alternations to the Ottawa Group buildings, such as the repositioning of doorways and benches, appear to have been dictated by ceremonial rather than practical concerns (Pendergast 1985:95).

Following the modifications that took place at the end of the Late Classic or early in the Terminal Classic, the entire complex underwent a major reshaping. The upper portions of many of the existing buildings were razed and their lower rooms and the courtyard onto which the buildings fronted were in-filled to a depth of approximately 2.5m with more than 21,000 metric tonnes of uncut boulders and fill (Pendergast 1986:232-233). At this point, the entire area was capped to create a single large platform and a new complex of residential structures bearing completely different architectural features and elements of design was built on top of it. Unlike the previous masonry buildings, the new structures had largely perishable superstructures - of wood, of masonry and wood, or of wood but built on low stone platforms (Pendergast 1985:95; Graham 2004:21). The peripheral buildings of the group also experienced some alterations and structure N10-15 underwent numerous internal and external modifications (Pendergast 1986:232-233; Graham 2004:18-21).

As with other structures within the N10-9 plaza group (e.g. N10-9 and N10-7), those comprising the Ottawa Group underwent subsequent modifications and alterations well into the Postclassic period (Late Postclassic). Some of these construction efforts represent clear departures from previous architectural conventions (for example, the form of the stairside outsets constructed on N10-9 during the Early Postclassic), while others appear to follow earlier patterns (for example, residences within the Ottawa Group built during the Early Postclassic also had perishable superstructures). It is significant, however, that as the structures within the plaza group changed in appearance and form, their respective functions appear to have remained essentially the same. This continuum of use would seem to suggest the continuation, from at least Late Classic through to Late Postclassic times, of at least some activities and practices for which the structures served as back drops or settings.

Whilst the N10-9 plaza group was a clearly a hub of construction activity during the Terminal Classic period, other areas of the settlement also underwent significant changes. New residential structures were built on the periphery of central precinct and in the vicinity of the N10-43 plaza group (Pendergast 1981a:40). Since these structures appear to be non-elite residences (Pendergast 1986:229), their intrusion into this area of the settlement indicates a significant change in the pattern of use of areas that were previously reserved for ritual and ceremonial activities. In addition, Pendergast (1986:227) notes that, at this time, the maintenance and use of most of the large ceremonial structures situated in the northern part of the central precinct, especially in the vicinity of the 'Harbour' area, had generally ceased. It would appear, therefore, that the establishment of the new residential structures coincided with a general withdrawal from some areas of the northern part of the central precinct and a southward shift in ceremonial activities to the area dominated by the N10-43 and N10-9 plaza groups. The increasing importance of the southern area of the central precinct as a locus for ceremonial activities is further emphasized by the ball court that was built in the N10-43 plaza. The pottery styles recovered from the dedicatory offering interred below the central marker stone of the ball court, date its construction to the early part of the Terminal Classic (ninth to tenth centuries) (Pendergast 1982, 1986:229). The undertaking is significant not only because it occurs at time when such activities (the construction of the facilities and the ball game) had generally ceased at other lowland centres, but also because it represents the introduction of a form of public ceremony and ritual that is previously unseen at Lamanai, although well documented at other sites (Sharer 1994). In addition, since the ball court appears to have been built at a time when maintenance and other construction activities within the N10-43 plaza group had diminished, such that many of the buildings were "in various states of decay and neglect" (Pendergast 1986:229), its creation can be considered as a very public ceremonial statement, perhaps reflecting a reaffirmation or re-entrenchment of religious beliefs (Pendergast 1985:94).

By contrast, in the residential areas to the north of the central precinct, the emergent pattern during the Terminal Classic is more one of continuity. Excavations in this area of the settlement, which involved several different plaza groups, based on the pottery styles recovered from construction fill and associated garbage dumps and other material evidence, showed that some of the groups were occupied in the Early Postclassic period (tenth century or later) (Pendergast 1981b, 1985:99, 1986:229). So, while the focus of ceremonial activities was shifting to the southern part of the ceremonial precinct, the pattern of occupation of the northern part of the settlement remained fairly consistent.

The late Terminal Classic to Early Postclassic period marks the expansion of the settlement southward. A group of new ceremonial structures was established immediately east of the N10-9 plaza a group (the N10-2 plaza group), extending the central precinct, and new residential structures such as N11-5, N11-7 and N11-9, were built in the area adjacent to the lagoon just to the south. The principal structure within the N10-2 plaza group was structure N10-2, which, based on its architecture, likely functioned as an administrative building and a site for ceremonial activities, and perhaps partly as a residence. The earliest construction of N10-2 dates to the Early Classic, although both the building and the plaza a group as a whole, appear to have been of little importance during the Classic period (Pendergast

1981a:43). The Classic period structure was replaced by a series of four buildings, "all a distinct departure from earlier architectural traditions" (Pendergast 1981a:44). At least one of the structures (the second of the four) had a colonaded entryway and a masonry altar, which was set in the rear centre of the building. The columned portico, a feature that also occurs within the Ottawa Group, gives the structure affinities with building styles more typical in northern Yucatan. However, the building materials used in the construction of N10-2 – thick wattle and daub walls and the extensive use of timber – represent a significant departure from Yucatecan conventions (Pendergast 1981a:44). Another significant aspect of N10-2 is that it contained 50 burials. The richness of the grave accoutrements accompanying many of the interments suggests these individuals were of high rank or status.

Other structures of ceremonial importance within the N10-2 plaza group include N10-1, a small platform built in the centre of the plaza and N10-4, a large ceremonial structure erected on the plaza's east side. As with N10-2, both of these structures contained numerous burials (e.g. N10-4 contained 46), many reflecting high status, and Pendergast (1986:236) suggests that the primary function of N10-4, in particular, might have been as a burial mound. The situation of the N10-2 plaza group adjacent to structure N10-9, and especially the placement of structure N10-2 directly beside N10-9, is also noteworthy as it would appear that the intention was to underscore a direct relationship between the newly built structures and the existing ceremonial structure, as well as associated beliefs and practices. Pendergast (1986:236) suggests that this physical expression of strong links with the past and established traditions would have "enhanced the visual and psychological effects of continuity accompanied by change" within the community.

With the expansion of the settlement southward, the focus of ceremonial and domestic activities shifts to the southern area of the site and to the lagoon shore. The N10-2 plaza group becomes the focal point of the community, with a substantial Postclassic residential area extending south along the lagoon. (Pendergast 1985, 1986; Graham 2004). Many of the residential complexes situated north of the central precinct appear to have been abandoned during the Early Postclassic (Pendergast [1985:99, 1986:229] suggests by the mid twelfth century), suggesting that the physical extent of the community was reducing in size.

Structure-Associated Offerings and Offertory Practices

The Terminal Classic to Early Postclassic period marks a significant change in offertory practices at Lamanai. The best evidence comes from offerings interred as part of construction activities associated with the renewal and modification of ceremonial and administrative/residential structures within the central precinct. The general pattern remains strikingly uniform from Preclassic times through to the Terminal Classic period, involving the placement of two or more pottery vessels, arranged as lid and container and containing other artifacts, on and around the central, rear-front axis of the structure. However, while the general form of these offerings remains consistent, the quantity and nature of the objects placed within the vessels changes somewhat over time. Early Classic offerings nearly always contain small quantities of marine material and jade, whereas by the Late Classic period, the contents of the offerings are more variable and diverse, most often consisting of varying combinations of shell and jade objects and obsidian blades, cores and chips (Pendergast 1981a:40-42). Offerings interred during the Terminal Classic period often contain carbonized remains of burnt materials or items, sometimes in addition to other objects, usually of jade or obsidian (Graham 2004:26). As with earlier periods, paired vessels are the most common form but single vessel interments also occur.

Offertory patterns of the Postclassic period, however, are a clear departure from Classic traditions. The practice of interring paired vessels containing other objects ceases and offerings involving the placement of whole vessels are less frequent in general. The most common offering form consists of one or more vessels that were intentionally broken prior to or upon interment or deposition. These offerings not only occur within architectural features and building foundations but also on the exterior of structures, often in the vicinity of the central stair. For example, at structure N10-9 a large offertory deposit found over the central stair consisted of numerous vessels - Early Postclassic in style - smashed *en masse* and left where they lay. Pendergast (1981a:44) suggests that the ceremony in which the vessels were broken likely occurred at or after the abandonment of the structure. A similar deposit comprising Late Postclassic censers was found at structure N9-56, a ceremonial structure situated in the northern part of the central precinct that had fallen into disuse by Terminal Classic times. In this case, however, the deposit, which consisted of numerous figurine censers smashed and scattered over the front, sides and back of the abandoned structure, appears to be associated with renewed construction activities in the area (Pendergast 1978, 1981a:51). While offertory deposits involving intentionally broken figurine censers are a common feature of the Late Postclassic record in general, occurring at sites throughout the Maya lowlands, it is clear that at Lamanai, the practice commenced as early as the Early Postclassic and perhaps even during the Terminal Classic period. Another important characteristic of the smash-and-scatter offerings at Lamanai is that attempts to reconstruct the vessels comprising these deposits revealed that, in each case, pieces were missing from all of the vessels (Pendergast 1981a:44). This would seem to suggest that an integral part of the practice entailed fragments of each vessel being either retained or deposited elsewhere by those participating in the ceremony.

Classic to Postclassic Burial Patterns

Although a detailed synthesis of burial patterns at Lamanai has yet to be completed, it is still possible to make some general observations regarding specific aspects of the patterns that characterize the Classic vs. the Terminal Classic and Postclassic periods. On a very general level, burials dating to the Terminal Classic and Early Postclassic periods are far more numerous than those dating to the Classic period (White 1997). Since Classic construction phases of the structures and building complexes that have been the focus of excavations have been as intensively investigated as later phases, it would seem that the interment of deceased members of the community within building cores and foundations became a more common practice in later time periods. For example, ceremonial structures built in the Terminal Classic and Early Postclassic periods such as N10-7, N10-2, N10-1 and N10-4 were found to contain large numbers of burials, while those built in the Classic period or earlier such as N9-56, N10-43, N10-27 and N10-9 contained comparatively few (Pendergast 1981a). The ubiquity of interments within later ceremonial structures also suggests a broadening of the function of some ceremonial buildings, which in effect, take on an additional role as burial repositories.

The Terminal Classic period also marks a significant change in the depositional pattern of ceramics within burial contexts. During the Classic period and earlier, pottery vessels are placed in the burial along with the corpse, as grave accoutrements and, at least in some cases, as containers for food, drink and other perishable and non-perishable substances and materials (Pendergast 1981a). Starting in the Terminal Classic, however, the ceramics within burials exhibit patterns of pre-interment breakage – i.e. whole vessels are usually entirely absent, and the fragments of broken vessels are placed alongside and scattered over the corpse. This pattern becomes the standard by Early Postclassic times. Since the vessels within these burials are largely restorable into complete forms, it would appear that they were intentionally smashed just prior to interment as part of the funerary ceremony. As with the vessels comprising contemporaneous smash-and-scatter offerings, in every instance, pieces of each of the vessels recovered from these burials are missing, suggesting that fragments of the smashed vessels that were to be included in the interment were retained by participants in the burial ceremony, perhaps as a memento of the occasion or for some other purpose. This change in the general treatment of vessels within funerary contexts has several implications. Not only does it reflect a significant change in burial practices, in terms of the nature and kinds of grave goods that were included in burials, but also in the established form of funerary ceremonies, which come to incorporate new forms of ritual behaviour. In addition, it represents a significant alteration in the patterns of use of pottery vessels within the context of both funerary rituals and the burials themselves. The underlying implication, therefore, is that beliefs surrounding the function and appropriate treatment of ceramic items within these contexts had changed.

Other characteristics of Early Postclassic burials that represent a clear departure from earlier patterns include the positioning of corpse within the burial and the types of objects included as grave goods. In over half of the burials within N10-4, the ceremonial structure on the east side of the N10-2 plaza group, the body was laid face down with the legs bent back or drawn up so that the feet rested on the pelvis area (Pendergast 1975). The same body positioning occurs in burials containing Early Postclassic pottery styles within structure N10-7 of the N10-9 plaza group (Pendergast 1975) and also in those associated with an Early Postclassic residential/ administrative structure (N10-12) recently excavated in the Ottawa Group (the dating of these burials is based on the presence of Early Postclassic pottery styles) (Graham 2004:22-23). In contrast, Classic period burials at Lamanai never occur in this position, and those within ceremonial structures are usually in a seated position. Specific examples are the Early Classic tomb burials within structure N9-56 (both in a seated position) (Pendergast 1980, 1981a) and the Late Classic multiple interment beneath the stela at structure N10-27 (the principal interment is seated) (Pendergast 1988, also see Chapter 6). Items that were included as grave goods in Classic period burials include pottery vessels, objects made of marine shell, jade, chipped stone artifacts of chert and obsidian, textiles, wooden objects, sometimes with jade inlays, and other perishable items such as mats and possibly nets (Pendergast 1978, 1981a, 1988). The accoutrements accompanying Terminal Classic and Early Postclassic burials comprise a similar range of items, including pottery vessels and objects of worked stone, marine shell and bone. Pendergast (1986:241) observes, however, that the occurrence of obsidian artifacts in burials greatly diminishes during this time period and that Postclassic burials are most often entirely devoid of the material. At the same time, an important addition to the repertoire of grave goods generally found in Early Postclassic and later burials is that of metal artifacts. These include an array of copper objects such as bells, strap workings and clothing ornaments and fasteners; gold sheet coverings from perishable objects, likely of wood; rings of an as yet unidentifiable alloy; and in one case, a pyrite mirror (Pendergast 1975, 1981a). The raw material resources from which the metal artifacts derive are undoubtedly of non-local origin, and in the case of the copper artifacts, Pendergast (1986:240) observes that they are typologically similar to comparable items that occur in the Oaxaca area of Mexico. However, the presence at Lamanai of copper artifacts that appear to represent items that were damaged during manufacture, provides some evidence for on-site metal working (Graham 2004:9-10). Whether manufactured locally using imported materials or imported as finished products, the metal artifacts are significant because they provide evidence of distant trade contacts at a time when the central precincts at most other Maya centres in the southern lowland region had been long since abandoned. Perhaps more importantly, they also reflect the emergence of new forms of material wealth and prestige within community.

Structure-Associated Refuse Deposits

Structure-associated middens provide a wealth of information about various aspects of community life. They shed light on dietary patterns, provide clues as to the kinds of activities that took place at particular locations and yield considerable insight into the consumption and use of material culture items within the associated cultural contexts. Likewise, continuity and change over time in patterns of refuse deposition or accumulation offer direct evidence of the history of use of the particular structures or areas within the community, as well as the general attitudes towards the maintenance of public areas and buildings and of private living space. At Lamanai, the Terminal Classic period marks a significant change in the patterns of use and maintenance of principal areas of ceremonial activity within the central precinct. Large accumulations of primary refuse, which based on ceramic stylistic data date to the Terminal Classic and Early Postclassic periods, occur in the immediate vicinity of ceremonial structures and administrative/residential buildings situated within the core area of ceremonial activity. In preceding periods, however, comparable areas within the central precinct are generally kept cleared of such deposits.

The largest of the Terminal Classic to Early Postclassic middens occurs at the front of N10-27 and between N10-9 and N10-2 (both are described in detail in Chapter 6), and less substantial accumulations occur in direct association with the Ottawa Group complex (Pendergast 1988, 1981; Graham 2004). In the case of the two largest middens, the deposits extend into the adjacent plazas – public areas that could have easily accommodated large gatherings. Pendergast (1992:70) suggests that it is likely that middens also occurred within the central precinct prior to the Terminal Classic, observing that there is ample evidence that such deposits were periodically cleared away and often incorporated into construction fill. He points out, however, that although we may never know how garbage removal was organized – for example, whether it was organized at the single family, neighbourhood, or citywide level – an important implication of Terminal Classic and Early Postclassic middens is that they may reflect "the disappearance of the organizing force", which in preceding periods ensured that public areas were kept clear of such accumulations (Pendergast 1992:70). In addition, since the large deposits at N10-27 and N10-9 appear to have at least started to amass when portions of these structures and the associated plazas were still being actively maintained and hence still in use, the fact that the deposits were allowed to accumulate clearly suggests a significant change in the general attitude towards the treatment of ceremonial structures. In the case of the midden abutting N10-9 and N10-2, it is clear that N10-2 and the associated plaza group was a site of particularly intensive ceremonial activity throughout the period of the midden's accumulation. An equally plausible explanation, therefore, is that deposits were left intentionally where they accumulated for some other reason, as yet unknown. Regardless of the underlying reasons for their occurrence, the important point is that the middens are clearly indicative of a significant change in the behavioural patterns associated with areas of the central precinct that served as important settings for ceremonial and ritual activity.

The contents of the middens associated with structures N10-9 and N10-27 are also informative. In both cases, the deposits consist mainly of pottery vessels and faunal remains with a comparatively smaller volume of stone and shell artifacts. In regard to the ceramics, vessel fragments are large and in many cases substantial portions of the vessels are represented. This pattern of fragmentation suggests that the deposits comprise primary refuse, and may possibly imply that the vessels were purposefully broken after use. Taking into account the disproportionately high volume of ceramics and faunal remains within these deposits, as well as the nature of the ceramic artifacts, there appears to be some suggestion that the deposits are connected to activities with an underlying ceremonial or ritual purpose, rather than the clearing or cleaning up of domestic areas.

The nature and kinds of faunal remains contained within the deposit at structure N10-9 provide some additional insight into the kinds of activity that can be linked to these deposits. Ongoing investigations undertaken by Stanchly (2001) indicate that the faunal component of this midden is distinct from other middens that have been examined at the site. Over 60% of the remains derive from reptiles, the vast majority of which are from turtles and crocodiles. Mammals and large bird species, which are well-documented as forming an important part of the Maya diet, are poorly represented, both within this particular deposit and in comparison to other middens at the site. The cut marks observed on the reptilian remains reveal an equally unusual pattern – those observed on turtle shells are consistent with meat extraction, while those on the crocodile remains (especially the skull) are more consistent with removal of the skin. In addition, many of the faunal specimens within the midden show evidence of burning. Given the location of the midden between a ceremonial structure and an elite residential/administrative building, an inescapable conclusion is that the deposit may be at least partly the result of feasting activities. However, the disproportionately large amount of reptile remains, together with the fact that the butchering pattern associated with the crocodile remains indicates that meat extraction was not a primary aim, clearly suggests that the deposit cannot be considered as standard commensal or domestic refuse, neither of elites nor the community in general. It is more likely that both of the deposits reflect a range of ceremonial and ritual activities that were centred on the N10-2 and N10-9 plaza groups. This interpretation is explored in greater detail in subsequent chapters in light of the ceramic evidence.

Factors Contributing to the Community's Resilience During the Terminal Classic to Early Postclassic Transition

Set against the backdrop of the socio-political and economic upheaval that seems to define the Terminal Classic to Postclassic transition, Lamanai stands out not only as a centre that appears to have survived this period relatively unscathed, but as a community that managed to maintain a certain level of administrative and socio-economic stability despite the upheaval that was going on around it. As the foregoing discussion demonstrates, the Terminal Classic to Early Postclassic record at Lamanai suggests a time of continued vitality in community life; a period of growth which saw the construction of new ceremonial and residential structures and the expansion of the settlement southward. While continuity in some fundamental aspects of community and social organization, administration and ceremonial and religious practice is clearly indicated, it is also apparent that this was a time of considerable innovation, in which the inception of new ideas and new ways of doing things was accompanied by a merging of new concepts with established conventions to form something unique. These complex developments are summarized in Table 2.2.

Pendergast (1981, 1992, 1990:171-172) identifies two underlying factors that appear to have promoted Lamanai's continued well-being: 1) the site's lake side location within a major river system and 2) the nature and qualities of community leadership. The New River Lagoon would have provided an abundant source of protein from fish and other aquatic species and an ample supply of water to fulfil a range of community needs, including the irrigation of the raised field system situated to the north of the site. Studies of paleodiet and dental pathology at Lamanai provide direct evidence of the good health of the population during the Terminal Classic and Postclassic periods and indicate a uniform mixed diet that becomes more dominantly maize-based during the Postclassic period (White 1997). Extending from interior regions just to the north of the Maya Mountains, to Chetamal Bay on the Caribbean coast, the New River would have enabled trade and communication with sites in both the southern and northern lowlands. The Peten and northern Yucatecan influences observed in Classic and Terminal Classic to Early Postclassic architecture, respectively, provide evidence of the flow of ideas from these areas into the local community. Likewise, the occurrence of non-local material items, such as marine shell deriving from coastal regions and ground stone implements of granite from the Maya Mountains, attest to economic links to different areas within the lowland region. Participation in long distance exchange networks is suggested by the presence of material culture deriving from far-away raw material sources in Mexico and the highlands of Guatemala and Honduras, such as liquid mercury (Pendergast 1982) and items made of jade, obsidian, copper, gold and metal alloys (Pendergast 1990:173, 1992). Based on the artifact inventories and architectural styles typical of the Classic vs. Postclassic period it would appear that Lamanai developed stronger ties to the northern lowlands during the Postclassic period, most likely as a response to the loss of southern lowland trading partners as a result of the collapse.

Although the evidence is largely indirect, it would seem that a central component of Lamanai's continued resilience was the stability and effectiveness of community leadership. This is indicated by the fact that, unlike at most southern lowland centres, community leaders at Lamanai were still able to organize and see through to completion large-scale communal construction projects during the Terminal Classic and Early Postclassic periods, as the major reshaping of the Ottawa complex and the construction of new ceremonial structures in the N10-9 and N10-2 plazas groups demonstrate (Pendergast 1992:74, 1986:231-233). In addition, considering the huge effort dedicated to providing appropriate living quarters and settings for religious, ceremonial and administrative activities organized and directed by elites, it would seem that the ruling class continued to receive the full support of the community. A significant factor that might have encouraged the community's ongoing support of the local nobility was possibly that the ruling class at Lamanai demanded less labour tax in comparison to other centres. The tendency towards limited frontal modification of major ceremonial structures, as opposed to complete transformation, starting in Late Classic times may reflect a scaling down of public work projects that could be expected to accompany a reduction in labour tribute (Pendergast 1992:73). Pendergast (1992:66-67) observes that the comparatively small number of structures that can be considered to conform to the 'Palace' building type at Lamanai might imply the presence of fewer elites. Accordingly, the ongoing stability of the site from Late Classic times onwards might be due, at least in part, to the fact that the community was not plagued by the internal factional conflicts and associated facets of inter-elite competition that contributed to the instability and rapid decline of other southern lowland centres (Figure 2.1).

An important implication of the ongoing stability in community leadership and socio-political organization observed at Lamanai is that it suggests that local elites maintained essentially the same role in extra-local and community affairs as in the Classic period – i.e. to provide short-term and long term direction, organization and stability, as well as intercession with the gods and any other external forces in defence of the local community (Pendergast 1992:77). As the foregoing discussion demonstrates, however, continuity in these aspects of community life at Lamanai were accompanied by significant changes, innovations and shifts in direction in regard to behavioural patterns associated with religious and ceremonial activities, ritual practice, and the underlying beliefs, values and attitudes these activities reflect. Whether these patterns can be considered to reflect a strategy adopted by the elite class to achieve particular ends, a grass roots development, a more general shift in Maya religious beliefs and practices, or

Table 2.2: Aspects of continuity and change in community-level behavioural patterns associated with the Classic to Postclassic transition at the site of Lamanai.

Construction Activities

- Continuity in the maintenance and periodic renewal of public architecture through to Early Postclassic times.
- A shift in focus towards the frontal modification of existing structures starting in the Late Classic
- Construction of new ceremonial structures and facilities during the Terminal Classic and Early Postclassic
- New residential structures built on the periphery of the central precinct during the Terminal Classic and Early Postclassic.

Architecture

- Continuity in some architectural conventions, such as building techniques and materials
- Instances of the merging of established conventions with new design concepts
- Architectural affinities with the Peten during the Classic to Terminal Classic
- Emergence of the 'Lamanai Building Type' during the Late Classic
- Introduction of northern Yucatecan architectural features on elite residential/administrative structures starting in the Terminal Classic
- Introduction of new design concepts during the Terminal Classic and Postclassic

Use and Organization of Community Space

- Continuity in the function of some areas and structures within the central precinct through to Early Postclassic times (e.g. plaza groups N10-9 and N10-43
- Continuity in patterns of occupation associated with the northern part of the settlement through to Early Postclassic times.
- Changes in the pattern of use of particular areas of the central precinct during the Terminal Classic, as suggested by the intrusion of non-elite residential structures
- The focus of ceremonial and ritual activities shifts to the southern part of the central precinct starting in the Terminal Classic
- Ceremonial structures take on an additional function as burial repositories starting in the Terminal Classic
- Entry of visual expressions of religious values and notions of authority into elite residential contexts starting in the Terminal Classic
- Increased importance of religious and ceremonial activities within elite residential contexts starting in the Terminal Classic
- A general shift in the attitudes surrounding the maintenance and use of public areas during the Terminal Classic as suggested by the emergence of large midden deposits in association with public areas and ceremonial structures within the central precinct

Ritual and Ceremonial Practices

- Continuity in the physical settings of ceremonial and ritual activity and some basic aspects of ritual practice
- The introduction of the ball game as a new form of public ritual and ceremony during the Terminal Classic
- Reaffirmation of religious beliefs starting in the Terminal Classic as seen in clear symbolic and physical expressions of links to Classic period religious beliefs and ritual and ceremonial practices (e.g. the construction of the ballcourt)
- Changes in the specific content of conventional forms of offerings starting in the Late Classic
- A significant change in offertory patterns by Early Postclassic times, suggesting changes in associated ceremonies and beliefs (e.g. smash-and-scatter offerings)
- A significant change in some aspects of burial patterns starting in the Early Postclassic (e.g. positioning of the corpse and the nature and types of grave goods)
- A significant change in the depositional pattern of ceramics within burial contexts starting in the Terminal or Early Postclassic, suggesting important changes in burial practices, funerary rites and beliefs surrounding the function and appropriate treatment of pottery vessels.

some other phenomenon, is an issue that is explored in Chapter 11.

The role that external factors played in shaping local-level developments at Lamanai is less than clear based on the evidence at hand. Although some non-local influences, possibly deriving from northern Yucatecan conventions, can be seen in the architectural features of at least two Terminal Classic buildings at the site, the general construction of these structures, in one case, conforms to previously established patterns (N10-15 of the Ottawa group), and in the other, appears to represent an innovative local approach (N10-2). In addition, there is no direct evidence of military invasion, defensive structures are absent, and ceramic styles show a continuous development out of earlier precedents (Graham 1987, also see Chapter 7). Accordingly, although foreign influences are clearly present, there is little evidence to suggest that these influences relate to the direct control of the site by a foreign group, or even that the community was affected significantly by military pressures. Similarly, while there is some material evidence that disruptions in regional trade networks during the Late to Terminal Classic may have led to an alteration in Lamanai's trade patterns, it is also clear that these disruptions did not have a lasting negative impact on the community's ability to survive. In addition, there is little direct evidence that the community became more mercantile-oriented over the course of the Terminal Classic to Early Postclassic period. As Pendergast observes:

> Though it is not at all unlikely that waterborne transport played a large role in Lamanai's Postclassic commerce, as it had presumably done in the Classic and earlier, there is nothing at the site that points to specific traders, or even argues pervasively for a more dominant role for merchants after A.D. 1000 then they had done before. As in so many other respects, the community's passage from late Classic to Postclassic times seems to have not brought with it radical change (1990:173).

Another important issue surrounds the potential responsibilities and obligations that members of the community at Lamanai might have had to relations, friends, commercial contacts etc. residing at centres that were abandoned during the Terminal Classic period. For instance, did the community accept refugees? And if so, were these refugees from all levels of society and on what basis were immigrating groups accepted or rejected? The construction of new residential structures on the periphery of the central precinct starting in the Terminal Classic might be construed as evidence that they did, and certainly similar building activities at other sites in northern Belize are most often considered to reflect the influx of immigrant groups. However, at least at Lamanai, there is also an apparent lack of corroborating evidence that would justify identifying the inhabitants of these structures as immigrants from other communities, as opposed to locals, since architectural styles and the general approach to the construction of these structures, as well as associated pottery styles, do not appear to deviate, in any way, from the local pattern (David Pendergast 2004, personal communication). As a result, an equally plausible alternative explanation for this building activity might be that, for whatever reasons, members of the local community who previously resided in outlying areas either decided or were compelled to move closer to the central precinct.

The detailed reconstruction and examination of local patterns of ceramic production and consumption presented in the chapters that follow provide additional insights into the ways in which the local community was affected by regional-level developments and external pressures during the Terminal Classic to Early Postclassic period. The material-human interactions embodied in these patterns provide a means of examining the specific choices and responses to demand relating to economic, socio-political and ideological activities involving pottery items. The focus is on the patterns and interactions associated with areas and contexts of ritual and ceremonial activity, with particular attention paid to aspects of continuity and change in local approaches to the manufacture of ceramic vessels and the potential origins and relative occurrence of non-local pottery within these particular contexts over time.

CHAPTER 3
CERAMIC ECONOMIC PATTERNS AMONG THE LOWLAND MAYA DURING THE CLASSIC TO POSTCLASSIC TRANSITION

Human beings, the creators of that which archaeology seeks to investigate, are known for their creativity, reflexivity and intent – all non deterministic characteristics. The products and arrangements of their behaviors are conveyed as if part of a fragmentary document that is edited to a variable extent by numerous sources, both subtractive and rearranging. Our ability to make accurate inferences about past behaviours depends on understanding the relationship between the physical manifestations of things, material and nonmaterial aspects of culture and society (Bishop 2003:89).

As with other facets of the Maya economy, the characterization of ceramic economic patterns during the Late Classic to Postclassic period has been influenced heavily by conventional perspectives on the collapse period, and particularly Rathje's mercantile model (Rathje 1975; Sabloff and Rathje 1975), which draw a sharp contrast between Classic and Postclassic period economic patterns and infrastructures. Embedded within this perceived dichotomy, lowland ceramic economic systems are viewed as having undergone significant changes at the close of the Classic period in response to the decline and failure of elite culture and socio-political institutions and the concomitant rise in mercantile activity stimulated by the expansion of northern lowland hegemonies that emphasized trade and commerce. According to Rathje's model, the expansion of this new economic and social order and the participation of a growing segment of the population in commerce (particularly as consumers) promoted a trajectory towards economic efficiency and mass production as a logical outgrowth of the need to meet the demands of a growing consumer market. With regard to economic patterns relating to material culture, this meant the emergence of supra-local production distribution units (workshops under centralized control), which came to replace the local production units (households and villages) that had formed the base of regional economic infrastructures during the Classic period (Rathje 1975:440). These developments are considered to have had a significant affect on the production/consumption patterns of fine wares for domestic service and ritual activities such as funerary rites, in particular, since these items are viewed as having had an important integrating function as symbolic material expressions of active participation in trade networks, as well as the shared beliefs and values of network participants.

As discussed in Chapter 2, an important implication of Rathje's model is that the Classic to Postclassic transition, as a period of cultural development, encapsulates a major restructuring of economic institutions and infrastructure. With regard to the manufacture of ceramic items, this restructuring is purported to have involved significant changes in mode and organizational aspects of production and, in some cases, the location of production. Ceramic distributional patterns were also affected, with a greater volume of ceramics circulating outside of their area or place of manufacture and over wider geographical space. Since direct archaeological evidence of ceramic production in the form of workshops or production areas and their associated debris has remained elusive in the lowland region[1], current understanding of ceramic economic activities has been built necessarily on a range of indirect evidence (see reviews presented in Rice 1987a; Ball 1993; Lucero 1992; Becker 2003). This chapter presents current conceptualizations of ceramic economic patterns over the course of the Classic to Postclassic transition and discusses the evidence and assumptions upon which they are based. A main objective is to delineate the broader scholarly context within which the conceptual framework, methodology and findings of the present study are situated.

3.1. PARADIGMS OF CLASSIC PERIOD PRODUCTION AND EXCHANGE

Serving and Utilitarian Wares

The accepted characterization of Classic period ceramic production and exchange is largely based on the regional-level studies undertaken with respect to the ceremonial centres of Palenque (Rands and Bishop 1980), Tikal (Fry 1979, 1980) and Lubantun (Hammond et al. 1976; Hammond 1982). Although varied in scale and methodology, as well as in the specific kinds of evidence brought to bear on an understanding of production and exchange, these studies investigated the distributional patterns associated with ceramic products manufactured in particular geological or "resource procurement/manufacturing zones" (Rands and Bishop 1980:42) within these polities, and the types of behavioural phenomena that might be inferred. In brief, the results of these studies suggested that, with the

[1] The recent discoveries of a kiln structure dating to the Late Classic period at K'axob (Lopez Varela et al. 2001) and firing pits of Late Postclassic date (early facet) at Laguna de On (Masson 2000:81-86) have been interpreted as denoting local production of ceramics at these sites. The identification of these features as 'ceramic-related', however, remains highly speculative due to the lack of associated production debris in the form of ceramic wasters etc..

exception of painted polychromes, Classic period pottery production was a decentralized, part-time, seasonal activity, perhaps supplemental to agricultural production, that was in the hands of multiple sub-regional producers, working in dispersed, rural locations. It was argued that production was perhaps organized on the community level, as with present-day production in the Guatemala highlands (Reina and Hill 1978; Arnold 1978), with individual communities, or those situated within a particular raw material resource zone, specializing in the production of particular forms classes, or multiform assemblages similar to a ceramic *Group*[2] in the Type-Variety scheme (see Rice 1987a and Ball 1993 for opposing interpretations). The distributional patterns associated with the products of different sub-regional producers – products defined based on the co-variation of stylistic and technological attributes, and the latter characterized macroscopically or using scientific methods (NAA and petrography) and broadly linked to particular geological zones within a region – suggested highly localized consumption on a zone-specific basis. They also indicated that only a small amount of the pottery consumed within a particular region, as a whole, was manufactured outside of that region. The studies at Tikal and Palenque showed that these large ceremonial centres could be best characterized as 'consumers' rather than 'producers' of pottery items: consuming, often in large amounts, the products of various sub-regional producers, whilst manufacturing and distributing regionally themselves only small amounts of 'ritual ceramics', such as incense-burning paraphernalia. In contrast, Hammond et al.'s (1976) study of ceramics at the smaller ceremonial centre, Lubantun, showed that the trace element compositions of the numerically dominant ceramic types were similar to that of clays located within 6km of the site, which, they argued, suggests that most of the ceramics at Lubantun were manufactured locally. It has been argued by some (e.g. West 2002:161) that the different pattern indicated for Lubantun may imply that smaller ceremonial centres located outside of the Peten and adjacent areas were more self-sufficient in ceramic production than the major city centres situated within the heartland area (also see Fry's [1981:161] comments on the site of El Pozito in northern Belize).

Polychrome Serving Ware

The production of Classic period painted polychrome ceramics is most often viewed as presenting a sharp contrast to that of utilitarian pottery and less elaborately decorated serving wares (Rice 1987a; Ball 1993). Specialized production by full-time artisans is assumed, based on the high level of skill involved. Chemical analysis of the polychrome pottery typical of individual ceremonial centres has suggested local production, perhaps by several workshops operating within close geographical proximity (Reents-Budet 1994:195-203; Beaudry 1984). The glyphic texts that often appear on the more elaborate, figure-painted polychromes identify the artists who produced these vessels (or at least those who painted them) as being connected to and members of the elite class, "including sons not in direct line for the throne" (Reents-Budet 1994:218). Based on this evidence, it is often inferred that polychrome production was an 'attached' or elite-sponsored craft and perhaps organized as 'palace schools' (Reents-Budet 1994:218-219; Ball 1993). As West (2002:167) points out, however, the epigraphic evidence can also be considered as making more of a statement about the high status of the potters (or painters) involved in polychrome production than the direct control of production by elites.

The distributional patterns associated with painted polychrome ceramics are also different than that of utilitarian pottery. Polychrome ceramics generally have a more widespread, inter-regional distribution, although only comparatively small quantities of the polychromes associated with particular ceremonial centres appear to have circulated outside of their area or place of manufacture. Unlike utilitarian and serving wares, which are most often perceived as circulating through a regional distribution system based in market exchange (e.g. Fry 1979, 1980, 1981; Rands and Bishop 1980), polychrome ceramics, and particularly the figure-painted styles, are thought to have circulated on a more limited basis through redistribution or gift-exchange among elites, acting as social currency and items of prestige (Rice 1987a; Ball 1993; Reents-Budet 1994:234-236; Reents-Budet et al. 2000).

3.2. Terminal Classic Trajectories in Ceramic Economic Patterns

In pan-regional syntheses of lowland ceramic economic patterns, the Late to Terminal Classic period is characterized by three significant trends or shifts (e.g. Rands 1973; Rice 1987a; Ball 1993; Fry 1981; Sharer 1994). These include: 1) the regionalization of pottery styles and compositions (characterized either macroscopically or using scientific techniques), 2) a substantial decrease in the frequency and skill manifested in polychrome pottery and 3) the emergence of new, fine-pasted pottery types with a widespread geographic distribution that exhibit a high degree of standardization in form and surface treatment. Since these trends appear to represent clear departures from earlier Classic period patterns, they are considered to reflect changes in local and regional ceramic production and distribution patterns that occurred as a result of the broader political and socioeconomic developments of the Terminal Classic. The trend towards the regionalization of pottery styles and compositions has been inferred variously to imply a lack of centralized control over production, an increase in the number of pottery producers and/or the fluidity of membership in producing groups (e.g. Rice 1987a), or alternatively, an increased need to express and symbolize group identities and ties of affiliation (e.g. Ball 1993). Each of these factors are considered to be related directly

[2] A ceramic 'group' is defined in the Type-Variety scheme as "a configuration of closely related pottery types that demonstrate a distinctive homogeneity in range of variation concerning form, base color, technological, and other allied attributes" Gifford (1963:23).

to the escalation in intra- and inter-regional conflicts and competition among elites during the last centuries of the Classic period (see also Foias and Bishop 1997; Reents-Budet et al. 2000; Rice and Forsyth 2004) Similarly, the decrease in the amount and quality of polychrome pottery is thought to reflect a decline in elite-sponsored craft production, as part of the collapse of elite socio-political institutions and the concomitant economic decline experienced at southern lowland centres leading up to their abandonment (Rice 1987a; Rands 1973)

Unlike the first two trends, which are considered to relate to internal developments related to the collapse, particular to the southern lowland region, the emergence of new types of pottery at the end of the Classic period is conventionally viewed as being rooted more in external factors and pressures. On a very general level, the inception of these new types, which are regarded as constituting a fine ware sub-complex of domestic service ware and funerary pottery that circulated widely through trade, is considered as linked, either directly or broadly, to the spread new ideas as part of the expansion of hegemonies centred in the northern lowlands, particularly in the Gulf coast area of Mexico (e.g. Adams 1971; Sabloff 1975; Harbottle and Sayre 1975). Their rapid proliferation, however, seemingly at the expense of Classic pottery styles, is perceived as evidence of more fundamental changes in the lowland ceramic economy that occurred as part of this expansion. These new pottery types, which include the Fine Orange and Fine Gray wares that commonly occur at sites across the lowland region, particularly those located in the Peten and adjacent areas, and the Slate wares more typical at northern lowland sites, tend to have fine-textured, hard pastes and exhibit a high level of consistency in shape and surface treatment. Accordingly, they are commonly regarded as technologically superior to other Classic period ceramics. Specialized, full-time production by carefully organized groups of craftsmen working in workshops, perhaps under centralized control, is inferred, primarily based on the high level of technical skill these wares exhibit. In addition, the internal stylistic homogeneity within each ware, their widespread geographical occurrence, and the use of mould-made decorations has been taken as evidence of mass production primarily for export, and thus expanding trading economies.

Fine Orange and Fine Gray Ware

The introduction of Fine Orange and Fine Gray wares into the southern lowlands during the Late Classic period, and their widespread occurrence throughout the heartland area, was originally thought to signify the intrusion of northern groups into the southern lowland region (Rands 1973; Adams 1973). The superficial homogeneity of these wares, a characterization that was based primarily on observed similarities in shape and conspicuous surface features, was taken as a direct indication that these ceramics were produced at one or a few centres, perhaps located in the northern lowlands (Gulf Coast area), and traded widely (Sabloff and Willey 1967). It was the *perceived standardized production* of these wares, in addition to their widespread geographic distribution, that led Rathje (1975:433) to propose a general shift in the lowland Maya ceramic economy, starting in the Terminal Classic, towards the industrialization and centralization of ceramic production and the widespread distribution of ceramic products from fewer production centres, as part of the growth and expansion of regional networks of exchange and political affiliation driven by powers based in the northern lowlands. It was presumed that the widespread geographic distribution of ceramic trade wares, such as Fine Orange and Fine Gray ware, symbolizing participation in and ties to these networks, represented a dendritic circulation of these material items from a point or points of origin in the northern lowlands where the groups controlling and administering trade works were based. The results of subsequent chemical compositional studies of Late Classic to Early Postclassic Fine Orange and Fine Gray ware from different sites in the southern and northern lowlands, however, have revealed a more complex picture than Rathje's model suggests. For instance, a study conducted by Rands, Bishop and colleagues (Rands et al. 1982; Bishop and Rands 1982; Bishop et al. 1982) of Late Classic to Early Postclassic Fine Orange and Fine Gray ceramics from different southern lowland centres suggested that these wares were in fact produced in the southern lowlands in several 'resource procurement/manufacturing zones', and possibly workshops, situated in the heartland area and in the western periphery of the southern lowlands. In addition, it was argued that the distribution of the chemical compositional groups corresponding to different production loci suggests a regionalized pattern of consumption, rather than widespread distribution from one particular area or place of manufacture (Rands et al. 1982). Nonetheless, the results of the study also showed a decrease over time in the number of chemical compositional groups relating to Fine Orange ceramics, with Terminal Classic/ Early Postclassic Silo Fine Orange ceramics represented by only two dominant groups. It was suggested that this decrease in the number of chemical groups associated with Fine Orange ware might point to a reduction in the number of production loci producing Fine Orange ceramics by Early Postclassic times. This observation has been viewed by some Maya archaeologists (e.g. West 2002) as direct evidence supporting Rathje's model, implying a shift towards the centralized production of Fine Orange ceramics, at fewer production centres, for widespread distribution across the lowland region. A major problem with this interpretation is that it ignores other aspects of the results of the study, which point to an increase in certain aspects of compositional diversity during the Early Postclassic, as indicated by the substantial increase in the number of ungrouped samples during this time period (Rands et al. 1982). In addition, much of the compositional variability among the Fine Orange and Fine Gray ceramics that was revealed through chemical compositional analysis remained unexplained by the study since over half of the samples analysed yielded compositional signatures that fell outside the chemical groups that were established (Bishop 2003). More recent analyses conducted by

Bishop and Rands (Bishop 1992, 2003) on a greatly expanded corpus of material, including Fine Orange and Fine Gray ware ceramics from previously unsampled sites in both the southern and northern lowlands and earlier pottery types with similar decorative treatments, have helped to clarify the patterns that emerged in their initial study. These investigations have shown that Fine Orange and Fine Gray ceramics can be attributed to two geographically separated source regions in the southern lowlands – on the lower Usumacinta River and on the Pasion River – based on petrological similarities between the ceramic pastes associated with the two corresponding chemical groups and fired samples of clays from these particular regions. In addition, the main chemical group attributed to sources associated with the lower Usumacinta River was found to comprise three chemically distinguishable sub-groups, further distinguished by ceramic typological and chronological differences, suggesting multiple contemporaneous production loci as well as geographical shifts in production loci over time (Bishop 2003). Perhaps most significantly, these investigations have found that Fine Orange and Fine Gray ceramics with site provenances in northern Yucatan derive from production areas in the southern lowlands, and that both wares appear to represent an indigenous development out of earlier southern lowland manufacturing traditions, rather than the introduction of foreign products or production concepts (Bishop 2003:87-89). The important implication of these findings is that they appear to be at odds with Rathje's model. That is, since the compositional evidence does not support the contention that these wares originated from production areas located in the northern lowlands and that their superficial stylistic homogeneity does not simply reflect a common origin (i.e. the products of centrally controlled workshops situated in the northern lowlands), a straightforward relationship between the introduction and proliferation of these wares and the rise to power and expansion of hegemonies centred in the northern lowlands, which controlled and administered trade networks, as well as the production of ceramic trade wares, is not demonstrated effectively.

Slate Ware

Unlike Fine Orange and Fine Gray ware, whose stylistic homogeneity has been conventionally perceived as reflecting production at only one or a few production locations, Slate ware is characterized as representing a shared 'technological style' that persisted from Late Classic to Early Postclassic times. Centralized workshop production of Slate wares is assumed, and areal and chronological variation in macroscopic paste and surface treatment attributes (e.g. slip colour) are considered to reflect the expansion and contraction of spheres of political and economic interaction, affiliation and dominance as part of the rise and decline in power of competing Puuc and Toltec Maya groups (Rice and Forsyth 2004; Ringle et al. 2004; Cobos Palma 2004). For example, early Slate ware of the Cehpech complex (i.e. dating to A.D. 700-800), representing a significantly new and relatively standardized set of forms and characterized by a distinctive surface treatment (a 'slatey' grey-brown slip), is considered to correspond to the rising power of Pucc centres situated in the western area of northern Yucatan (Rice and Forsyth 2004; Ringle et al. 2004) Similarly, the stylistic and compositional homogeneity, or 'standardization', of Terminal Classic/Early Postclassic Slate ware is considered to reflect the emergence of a highly centralized government at Chichen Itza, headed by Toltec Maya, that controlled trade networks, as well as ceramic production (Kepecs 1998). In this case, standardization in macroscopic paste and stylistic characteristics is perceived as directly linked to high-intensity manufacture of the ware at Chichen Itza for distribution to a growing consumer market, although local production of Slate wares at Chichen Itza has yet to be demonstrated archaeologically by the presence of workshops or production areas or through definitive linkages between ceramic paste composition and local raw material resources.

It is generally assumed that Slate ware was produced, perhaps exclusively, in the northern Yucatan, based on its abundance at sites in this area, although no studies aimed at identifying potential manufacturing areas or centres through comparison of archaeological specimens with geological samples have been conducted to date. In addition, since detailed regional and local level compositional studies of Slate ware using scientific techniques are largely absent, local, regional and chronological variability among the recognized variants or *types* within this ware remain poorly understood in terms of potential technological and provenance relationships. This is especially with regard to the location and nature of production of variants. Nevertheless, differences in slip colour, paste colour and temper among and within recognized *types* suggest the use of different raw material resources, and possibly, multiple production loci, at the very least with regard to variants or *types* that appear to have been produced contemporaneously (c.f. Rice and Forsyth 2004; Kepecs 1998).

The Cultural-Historical Significance of Fine Paste Wares

It would seem significant that although Fine Gray and Fine Orange ware and Slate ware are presumed to have had similar practical and symbolic functions, as fine wares used in domestic service and ritual activities that symbolized active participation in a new and expanding social and economic order, associated production and distributional patterns appear to be different in fundamental respects. It is clear that Fine Orange and Fine Gray ceramics and Slate ware ceramics were primarily produced in different, geographically separated areas of the lowland region and that production occurred within the context of different political and economic circumstances and conditions in each case. The production of Fine Orange and Fine Gray ware in the southern lowlands and the proliferation of this pottery across the lowland region as a whole coincides with a period of economic decline and instability in elite socio-political institutions in the southern lowlands, whereas the production and increasing prevalence of different Slate ware variants appears to correspond to the rising power and influence of different Puuc and Toltec Maya groups. In addition, Fine Orange and Fine Gray ceramics appear to have circulated over wider geographical space, occurring at sites throughout the northern and southern lowlands, whilst the distribution of Slate wares is restricted primarily to sites in the northern lowlands.

The different production and distributional patterns indicated for these contemporaneously produced service wares suggest that other factors, perhaps unrelated to the growth of regional exchange networks, had a hand in shaping manufacturing priorities, as well as the demand for and distribution of different wares over social and geographical space. For example, central to Rathje's model is the notion that the internal stylistic homogeneity of the fine-paste wares reflects an increasing level of standardization accompanying mass production at one or a few locations, which is perceived as a development integral to growth and expansion of centrally administered exchange networks. While this conceptualization might be helpful for understanding spatial and temporal patterns relating to Slate ware, the more detailed studies of Fine Orange and Fine Gray ware have demonstrated effectively that wares that exhibit a high level of stylistic similarity can derive from multiple, and often geographically distant, areas of manufacture. Consequently, it would appear that, at least in the case of Fine Orange and Fine Gray ware, factors and mechanisms other than those strictly related to provenance and organizational aspects of production played a significant role in determining the level of similarity or dissimilarity within stylistic and/or functional classes of ceramics. Furthermore, when considering that the production of Fine Orange and Fine Gray ware occurred in areas where economic infrastructures and elite socio-political institutions were becoming increasingly unstable and eventually collapsed altogether, it is hard to imagine a shift towards centrally administered mass production. In this case, the regional trajectory towards standardization in shape and decorative treatment of serving wares might also be considered to reflect a change in attitudes or values regarding the ceramics used in particular social and ritual contexts. For example, it may reflect an increased concern for consistency and order – an interpretation that is perhaps more in keeping with the broader historical context of Fine Orange and Fine Gray ware production. The point is that both local and regional developments and circumstances need to be factored into interpretations of the behavioural and cultural-historical significance of patterning in the material record. Unifying explanations of these patterns run the risk of subsuming important cultural information.

3.3. Paradigms of Postclassic Ceramic Production and Consumption

Pan-Lowland Patterns

Postclassic ceramic inventories are most often characterized as exhibiting a continuing trend towards standardization and simplification, with widespread similarities within different stylistic and functional classes of ceramics on the regional level (e.g. Rathje 1975; Rathje and Sabloff 1975; Rice and Forsyth 2004). Stylistic trends in serving wares that first emerged during the Terminal Classic period continue into the Postclassic period. This continuity in serving ware styles, as well the general pan-lowland trajectory towards widespread regional stylistic similarities, is seen as directly linked to the growth of networks of exchange and communication, and the increasingly central role of mercantile activity in the lowland economy. It is generally thought that widely distributed trade wares, such as Fine Orange and Fine Gray ware, being material symbols of network participation and associated beliefs and values, prompted numerous local imitations, especially in regard to the decorative treatment of pottery regarded as locally produced, which is commonly considered to duplicate or be reminiscent of trade ware conventions (Rice and Forsyth 2004; D. Chase 1982). A concomitant supposition with regard to ceramic distributional patterns during the Postclassic period is that regional exchange networks promoted and facilitated the circulation of a greater volume of ceramics outside of their area or place of manufacture. When considering local level patterns of ceramic consumption, a related expectation, although not explicitly stated, is that the proportion of intrusive ceramics within local assemblages, possibly from quite distant production loci, increases significantly during the Postclassic period.

It is generally assumed that local production, at the household or community level, of most functional classes of ceramics, including serving wares, continued alongside centralized workshop production of ceramic trade wares during the Terminal Classic and Postclassic periods (c.f. Masson and Mock 2004; Rice 1980). In general, however, since the majority of the ceramics comprising Postclassic assemblages present a sharp contrast to Classic period ceramic inventories in terms of the skill and labour investment, they are commonly characterized as reflecting an overall decline in quality and workmanship, suggesting a period of decadence in ceramic production. When considering ceramic economic patterns in areas severely affected by the collapse, such as the central Peten Lakes District, this decline in the general quality of ceramic items is put down to a loss of skilled artisans or specialized knowledge that perhaps occurred as a result of a cessation in elite sponsorship of crafting activities (e.g. Rice 1980). With regard to sites in the northern lowlands and areas bordering that region, such as northern Belize, that experience continued or heightened prosperity into the Postclassic period, this trend is most often attributed to increased production; whether envisaged as a general compromise in quality in favour of increased output to meet rising demands (Pendergast 1990) or as a natural bi-product of the implementation of cost-saving measures associated with mass production (Masson 2000; Masson and Rosenswig n.d.).

Northern Belize

Ceramic analysts working with assemblages from sites in northern Belize often observe a general trend, starting in the Terminal Classic/Early Postclassic, towards increasing uniformity in vessel forms, compositions (an evaluation based on macroscopic paste attributes), design and design techniques, and also towards the simplification of decorative treatment (e.g. Masson 2000:43-57; Masson and Mock 2004; Masson and Rosenswig n.d.). Perceived as indicating a trajectory towards *standardized production*, these developments are most often viewed as reflecting the mass production of ceramics at fewer production centres, or alternatively, the existence of a rigid set of principles to which potters or workshops abided as part of their participation in regional trade networks. Masson and Mock (2004:374-375) characterize Terminal Classic to Late Postclassic ceramic assemblages in northern Belize as heavily influenced by northern Yucatecan manufacturing traditions, based on stylistic and technological commonalities with temporally comparative complexes at Chichen Itza during the Terminal Classic/Early Postclassic and at Mayapan during the Late Postclassic (also see Masson 2000: 47-57; Mock 1994). They suggest that these 'local emulations' of northern Yucatan conventions are part and parcel of a growing participation of local communities in regional exchange networks administered by these northern city states, and might also reflect a direct northern interference in local manufacturing activities. In addition, it is generally assumed that local assemblages contain a significant quantity of ceramics of non-local origin, including those deriving from production centres in northern Yucatan, although detailed paste compositional studies of vessel provenance have yet to be undertaken. Viewed from this perspective, Postclassic ceramic assemblages are portrayed as reflecting a passive and acquiescent response to political and economic developments emanating from northern Yucatan.

At many northern Belize sites, the Early Postclassic period is marked ceramically by a distinctive fine ware sub-complex of serving vessels and censers (the Zakpah Orange-Red group), characterized by a distinctive set of vessel forms (chalices, vessels with a segmented basal flange, frying pan/ladle censers, grater bowls etc.) and decorative traits (an orange slip often accompanied by incised and gouged decoration), that appears to represent a regional development particular to northern Belize, including the off-shore island of Ambergris Caye (Masson and Mock 2004; Walker 1990; Graham and Pendergast 1989). Masson and Mock (2004:375) see this sub-complex as a distinctly regional trajectory in serving ware styles and suggest that it may reflect a brief cessation of use of pottery stylistically similar to that of northern Yucatan circa. A.D. 900 - 1050, due to a disruption of trade networks as part of Chichen Itza's decline and prior to the rise of Mayapan. Alternatively, Valdez (1987, 1994), considers the sub-complex to correlate with an ethnic break in the occupation at the site of Colha, and sees new forms such as the grater bowl, or 'chilli grinder', as evidence of a significant change in local culinary traditions, perhaps deriving from a northern Yucatan precedent. The abundance of Zakpah Orange-Red-related vessels at Lamanai (referred to as Buk phase ceramics by Pendergast [1981] and Graham [1987]), has been interpreted by some (Walker 1990; Masson and Mock 2004) to suggest that Lamanai was the major producer of this pottery, with an important implication being the assumption that much of the Zakpah Orange-Red pottery found at other sites in northern Belize was produced at Lamanai.

Current Understanding of Ceramic Variation from the Terminal Classic Onwards

A significant aspect of current characterizations of ceramic economic patterns during the Postclassic period, on both the local and regional levels, is that they are based on evidence deriving from ceramic studies that have focussed almost exclusively on visual characteristics of pottery and macroscopic indicators of compositional and technological attributes. Detailed local and regional studies of compositional and technological variation using scientific techniques are entirely absent and in the few studies that do feature some sort of scientific examination of materials (e.g. Kepecs 1998; Rice 1987b; Smyth et al. 1995), analyses were performed on only a small sample of specimens presumed representative of typological groupings based on macroscopic, and most often, stylistic criteria, with the main purpose of augmenting the detail of descriptions, characterizations and interpretations built from the macroscopic evidence alone. This situation derives, to a large extent, from the general belief among Maya archaeologists (e.g. Bishop 1991; Masson and Rosenswig n.d.) that the absence of geological diversity in the Yucatan peninsula and adjacent areas to the south, coupled with close similarities in approaches to paste making, preclude the differentiation of ceramics produced in different locales based on compositional characteristics, despite the fact that several studies have shown that this is not the case (Iceland and Goldberg 1999; Arnold et al. 1999; Jones 1986; Angelini 1998; Bartlett et al. 2000).

To achieve behaviourally and technologically meaningful interpretations of compositional variation, including chemical composition, requires a detailed knowledge of the nature and mineralogy of the raw materials used in production and their geographic distribution, as well as the strategies and techniques employed in paste manufacture, which also have a significant influence on chemical compositional variation. Interpretations should be based ideally on range of complimentary information about vessel composition, including petrographic evidence of the raw materials used and the ways in which they were manipulated and prepared, and be grounded in a framework of comparative geological information

Figure 3.1: Conventional model of changes in the lowland ceramic economy as part of the Classic to Postclassic transition.

	Classic Period Patterns		⇨	Postclassic Period Patterns	Inferred Developmental Trajectory
PRODUCTION	Regionalization of pottery styles and compositions Dichotomy between the organizational form and location of production and distributional patterns of polychrome serving vessels versus other serving and utilitarian pottery			Pan-lowland standardization in vessel styles and compositions Increasing prevalence of mass-production in centrally located workshops Community craft specialization	General trajectory towards centralization and industrialization
	Polychromes	centralized workshop production involving highly skilled artisans			
	General serving and utilitarian pottery	household or community production; part time producers working in geographically dispersed, rural locations			
DISTRIBUTION	Polychromes	inter-regional distribution through gift-giving and/or redistribution among and by elites		Widespread inter-regional distribution of at least some pottery via extensive trade networks Increasing volume of ceramics circulating outside of their area or place of manufacture	Trajectory towards more widespread distribution involving appreciable amounts of pottery through expanding trade networks
	General serving and utilitarian pottery	regional or local distribution, coinciding with polities at most			
CONSUMPTION	Polychromes	generally localized to particular polities or centres, but with small quantities consumed on the inter-regional level		Regional consumption of at least some pottery A larger volume of non-local ceramics consumed on the local-level	Trajectory towards a regional pattern
	General serving and utilitarian pottery	local			

concerning the study area. The studies conducted by Rands and Bishop of Palenque ceramics and Fine Orange and Fine Gray ware approach this ideal in that their interpretations are based on a range of compositional and geological evidence. It is most often the case, however, that the geological context of the ceramics focussed upon is rarely examined beyond a very general level and detailed petrographic investigations of ceramic compositional variation that considers both the raw materials and human behaviours involved in production are largely absent. If a microscopic study of the ceramics analysed is conducted at all, the tendency has been to focus on the basic mineralogical composition of a small number of samples that are presumed representative of often geographically widespread stylistic categories, rather than investigating compositional variation within stylistic and typological categories as it relates to paste technology and provenance (e.g. Kepecs 1998; Rice 1987b). As a result, the technological and provenance relationships among and within typological, stylistic and functional classes of vessels, across time and geographical space remain poorly understood, and little is actually known about how levels of macroscopic similarity or dissimilarity among and within vessel categories relate to infrastructural changes in the organisational aspects of ceramic production and patterns of circulation and distribution of ceramic items. That this deficit exists in current understanding of the nature and meaning of ceramic variation in cultural and behavioural terms, from the Terminal Classic period onwards, as it relates to ceramic economic patterns and other human behaviours, has been widely recognized since Rathje first elaborated his explanatory model in 1975. As of yet, however, few attempts have been made to remedy the situation.

3.4. Current Characterizations of Terminal Classic to Postclassic Ceramic Economic Patterns

What We Think We Know

Current characterizations of ceramic economic patterns over the course of the Classic to Postclassic transition emphasize a fundamental transformation in ceramic economic activities and practices. As summarized in Figure 3.1, this transformation is considered to have affected various aspects of the lowland ceramic economy, involving significant changes not only in organizational aspects of production and the specific location of production units, but also in the patterns and mechanisms of distribution across geographical and social space. Also apparent is a perceived dichotomy between the production and distribution of polychrome ceramics and widely distributed trade wares and the production and distribution of the more common service and utilitarian wares that dominate local and regional ceramic inventories. The production and circulation of polychromes and trade wares is considered as physically and socially separate from that of most other kinds of pottery, being more closely attached to elite socio-political and religious activities and institutions, and involving individuals of a different and possibly higher, socio-economic status who possessed special knowledge. The observed pan-lowland trend towards increasing uniformity in vessel styles, starting in the Terminal Classic period, especially in regard to serving wares, is considered to relate, in part, to the imitation of products more closely attached to elite socio-political and economic institutions, but primarily to the increasing prevalence of more intensive organizational forms of pot making – i.e. mass production in workshops – which is seen as a likely response to the changes in demand that would have accompanied a general expansion and escalation in commercial activities. Centralization of production is suggested to have played a perhaps integral role in these developments.

At the same time, however, it is widely recognized that standardized production of different wares, as well as most other lower-level typological categories (certain Late Classic figure-painted polychromes *varieties* are perhaps an exception), has yet to be confirmed through detailed studies of paste technology and provenance. Hence a direct relationship between stylistic similarity, technological similarity and provenance cannot be taken for granted. Similarly, recent considerations of Maya political economy during the Terminal Classic to Postclassic period emphasize that there is little evidence at present to support the claim that ceramic production activities in fact became more centralized by Postclassic times (e.g. see recent discussions in Graham 2000 and Rice and Forsyth 2004). Accordingly, it would seem that current understanding of Terminal Classic to Postclassic ceramic economic patterns is limited significantly by missing information; direct archaeological evidence of the nature and location of production facilities, and equally as important, a working knowledge of the technological and provenance relationships and associations within and among the stylistic, functional and typological categories that form the basis of ceramic economic investigations.

While the discovery of ceramic workshops and production areas would undoubtedly lead to more confident reconstructions of the nature and organization of ceramic production, there is still much insight into ceramic economic patterns that can be gained through detailed studies of existing collections, especially when investigations of physical ceramic variation are set within an analytical framework that also considers depositional patterns and associated cultural and historical contexts of ceramic use and deposition. As Bishop (2003:89) observes, "our ability to make accurate inferences about past behaviours depends on understanding the relationship between the physical manifestations of things, material and nonmaterial aspects of culture and society". On a very basic level, achieving a more holistic understanding of ceramic economic activities during the Classic to Postclassic transition has been hampered by the fact that, more often than not, the analysis and interpretation of Terminal Classic and Postclassic ceramic materials proceeds from an *a priori* assumption that current characterizations of Terminal Classic to

Postclassic ceramic economic patterns represent a reasonably accurate conceptualization of the past behaviours and motivations of producers and consumers. It is taken for granted that economic restructuring is primarily responsible the changes observed in the ceramic record, and as a result, the potential contributing role of non-economic factors in shaping the manufacturing priorities, consumption patterns and changes in demand that are reflected in ceramic inventories have received less attention. Many of the studies included in the most recent consideration of the Terminal Classic period edited by Demarest et al. (2004a) demonstrate the centrality of conventional economic-based conceptualizations of ceramic change in a broad range of current research dealing with ceramic assemblages. By comparison, independent analyses designed to investigate the nature and potential factors (cultural, environmental and material) influencing areal and chronological ceramic variation, and thereby test the validity of accepted models, are few and far between. Such studies go a long away towards achieving more reliable inferences about the behaviours of pottery producers and consumers during the Terminal Classic to Postclassic period by situating the physical manifestations of Maya pottery within a broader and more meaningful framework of human-material interaction.

What We Don't Know

If one considers the body of evidence that forms the basis of current characterizations of Terminal Classic to Postclassic ceramic economic activities, it is clear that this knowledge base, as is currently exists, can be perhaps best described as more patchy than complete. For example, the level of information regarding Classic period vs. Postclassic period ceramic economic patterns is quite uneven, especially with regard to the southern lowlands. This situation has arisen primarily because ceramic economic activities prior to the collapse period, and particularly ceramic production, have traditionally received more attention, and consequently, have been more intensively investigated than Terminal Classic and Postclassic period ceramic economic activities. In addition, the most intensive studies of ceramic production and distribution patterns (e.g. Fry's studies of the Tikal region and Rand's and Bishop's work in the Palenque area and on Fine Orange and Fine Gray ware), which involve fairly detailed analyses of stylistic and compositional variation and the attribution of compositional groups to particular raw material sources, have focussed on delineating regional-level patterns. Comparable approaches have not been used extensively to investigate ceramic variation on the local level, and thus community-based patterns of ceramic production and consumption, as well as areal and chronological variation these patterns, remain poorly understood. This particular gap in current knowledge of lowland ceramic economic activities is particularly significant since reliable characterizations of regional-level patterns should be based ideally on a comprehensive understanding of local-level patterns.

Another issue concerns the nature of the evidence upon which generalizing statements regarding lowland ceramic economic patterns and trajectories are based. Firstly, there has been a clear focus on fine wares – serving wares in particular. Secondly, in depth investigations of serving ware production and distributional patterns have concentrated only on specific shape classes and/or typological and compositional categories of serving vessels. For example, Fry's (1979, 1980) study of patterns of ceramic production and exchange in the Tikal area focussed exclusively on monochrome slipped tripod dish/plates and vases, whilst a similar study undertaken by Rands and Bishop (1980) of the Palenque region, although considering a greater range of shape classes and surface treatment categories, only looked at one compositional category – 'sand tempered' pottery. Similarly, studies that have attempted to delineate patterns of ceramic production and exchange during the Terminal Classic to Early Postclassic period through detailed scientific analysis of compositional variability and provenance have focussed exclusively on either fine-paste wares (e.g. Fine Orange and Fine Gray ware) or specific serving ware *types* (in the type:variety sense) (e.g. Foias and Bishop 1997). In the case of Fine Orange and Fine Gray ware, it is also significant that these ceramics are comparatively rare at many sites, often being represented by only a handful of sherds.

That general statements and behavioural inferences concerning patterns and trends in ceramic economic activity are based almost exclusively on the findings of the few detailed studies that have been conducted on serving wares, is directly related to the focus on reconstructing regional-level patterns of ceramic production and exchange, deriving, at least in part, from an underlying assumption that this particular kind of pottery has the greatest potential for movement. Nonetheless, given the nature of the evidence upon which accepted characterizations of the lowland ceramic economy are based, some key questions that emerge are: 1) whether it can been presumed that the patterns that have been suggested for the limited corpus of serving ware pottery that has been studied in detail are representative of serving wares in general, 2) whether the patterns indicated for particular sites or regions can be applied to other sites or regions (the different patterns indicated for Tikal and Palenque vs. Lubantun suggest that they cannot), and 3) whether the patterns suggested for serving wares can be extended to all the pottery in Maya society during a given time period. For example, is it reasonable to assume that other functional classes of ceramics, such as utilitarian pottery, follow the same pattern and trajectory over time as serving wares in terms of their production (organizational form and scale)? Furthermore, is it reasonable to assume that utilitarian pottery, particularly jars, have a more restricted and localized pattern of circulation in comparison to serving wares, including trade wares, especially given the potential function of jars as containers for other commodities? A greater range of Maya pottery and other sites and regions need to be studied in detail before these questions can be answered.

3.5. Potential Limitations of Conventional Approaches to Study of Ceramic Economic Activities

The Absence of Detailed Analytical Studies of Whole Assemblages

The reconstruction of patterns of ceramic production, distribution and consumption across time and space depends upon a basic understanding of the technological and provenance relationships within and among different functional and stylistic categories of pottery. As the studies cited throughout this chapter demonstrate, analysis of ceramic materials using scientific techniques, such as thin-section petrography and NAA, have proved an effective means of studying these relationships when analytical results are combined with other typological and functional information. Nonetheless, to date, such studies have focussed on regional-level patterns associated with a limited range of selected kinds of pottery. Comparable published studies of whole ceramic assemblages, whether deriving from a particular site, architectural structure or complex, or particular kind of depositional context such as burials, are virtually non-existent for the Maya area (see Whitelaw et al. 1997; Day 1995; Day and Wilson 1998 for the application of a 'whole assemblage' approach to the examination of ceramic variability for Minoan pottery). Shepard's work at San Jose (Shepard 1939), Uaxactun (Smith 1955) and Mayapan (Smith 1971) is perhaps an exception. Even in these cases, however, the information on paste and decorative technology is not well integrated into the stylistic and frequency data, and vessel provenance is not fully investigated. Consequently, the information is of limited use in reconstructing site-level patterns of production and consumption. The insight that can be gained from a whole assemblage approach to detailed scientific investigation of ceramic variation is of fundamental importance to the reconstruction of patterns of ceramic production and consumption, since it provides a comparative base within which the potential significance of emergent patterns with regard to the stylistic, technological and provenance characteristics of different kinds of pottery vessels can be evaluated and interpreted. It also provides a means of quantifying the relative amount of ceramic products deriving from different production loci and/or manufacturing traditions that occur within particular contexts of use and deposition. In the absence of this wider ceramic context is difficult to judge the significance of patterns in the physical data as they relate to the behaviours of pottery producers and consumers.

The Central Focus of Ceramic Studies on the Socio-economic arrangements of Production

A central focus of research on lowland ceramic economic activities has been the delineation of the nature and structure of production/distribution and exchange systems as a means of gaining insight into social, political and economic relationships, as well as gauging relative levels of cultural complexity. In this context, 'pottery production' takes on a rather particular meaning, which Rice (1987c:168) describes as "the socioeconomic arrangements involved in practising the craft *as opposed to* the mechanics of building a pot" (*italics* are my emphasis). Accordingly, of principal interest is the definition of scale and mode in pottery production, as well as mechanisms of distribution (reciprocal exchange, redistribution, the existence of market-type practices), and the identification of craft specialization. In the absence of direct archaeological evidence of organization, in the form of production facilities, this is inferred indirectly through studies of ceramic variability. Particular attention has been paid to the level of standardization in morphological and surface treatment attributes, and the frequency and geographic distribution of different ceramic products.

The conventional focus on investigating the socio-economic basis of production and distribution has meant that documenting the actual techniques of manufacture has not been a primary concern. This general lack of interest in technological aspects of ceramics derives, in part, from an underlying assumption that production technologies and manufacturing procedures are directly determined by the nature and structure of production/distribution and exchange systems, with different organizational forms of production corresponding to specific technological correlates. There has been a tendency to view ceramic technology as a characteristic of pottery more worthy of basic description than explicit study in its own right.

Another contributing factor is the centrality of type:variety methodology in the analysis of ceramic variation. This conceptual framework is credited not only for providing a standardized system for the classification and description of ceramic assemblages, thereby facilitating inter-site comparative studies and promoting communication between analysts, but also for its explicit framework for the interpretation of ceramic data in cultural behavioural terms. A main underlying assumption of this scheme is that the classificatory units that form the basis of the typology – i.e. *types* and *varieties* – correspond to specific kinds of cultural behavioural phenomena. *Types*, consisting of one or several *varieties*, are recognized as reflecting material manifestations of the mental templates that relate to different categories of pottery within Maya society, whereas *varieties* are thought to reflect differences in the specific location of production and/or approach to the manufacture of pottery within these categories, or alternatively when the pottery was produced (Gifford 1960, 1976:8-11; see also Wheat et. al. 1958 and Smith et al. 1960). Since the distinction between different *types* and *varieties* most often rests on differences in visual appearance (surface treatment and morphology) (Smith et al. 1960:330-331), implicit in this conceptual framework is the expectation that ceramics that share a distinct set of visual and tactile characteristics also share a common technology and origin. Viewed from this perspective, *types* and *varieties* are seen as reflecting traditions of manufacture, or even production units, even in the complete absence of corroborating evidence of shared technological and provenance characteristics. Various investigations of the compositional characteristics of established types and varieties have shown repeatedly, however, that compositional groups often cross cut typological categories and *vice versa*. These findings suggest not only that a direct relationship between visual similarity and technological similarity, including provenance, cannot merely be assumed, but also that the presumed behavioural correlates of the typological units are, at least in some cases, invalid. (e.g. Shepard 1939; Smith 1955, 1971; Jones 1986; Angelini 1998; Howie-Langs 1999; Howie et al. 2005). It would seem, therefore, that the type:variety method, in some ways, limits understanding of ceramic technologies, and thus ceramic production, by subsuming variation that is of direct relevance to reconstructing associated patterns of cultural behaviour.

The Ambiguous Social Context of Pottery Production

A final issue concerns conventional conceptualizations of the social context of pottery production, which pose certain problems for reconstructing community-based patterns of ceramic economic activity. With the possible exception of certain fine ware pottery deemed to have been of central importance to the maintenance of elite socio-political institutions (polychrome ceramics and trade wares), current perspectives postulate that pottery production occurred primarily outside of ceremonial centres in essentially rural and geographically dispersed locations. This interpretation, however, is based primarily on negative evidence of production facilities in proximity to monumental architecture. The emergent image, therefore, is one of potters working largely in isolation, divorced from important nodes of social interaction and ceramic consumption in the ceremonial precinct, whether these nodes are envisaged as fostered by economic, ideological or socio-political factors. This conceptualization would seem to alienate potters and production activities from important contexts of activity that are fundamental to shaping demand, thereby down-playing the active role that both potters and pottery play in the creation, maintenance and reproduction of culture. A large part of potters' 'know how' derives from active participation in the day-to-day and communal activities and interactions that underlie patterns of consumption reflected in the ceramic record. Accordingly, it would seem that current conceptualizations serve to limit the potential of pottery to inform about the mores, values, beliefs and attitudes that lie at the heart of material patterning in the ceramic record.

A main reason why pervading perspectives of the location of pottery production pose certain problems for the reconstruction of local-level patterns in ceramic economic activities is that the concept of a 'community' as a meaningful context for social-interaction in which these activities are necessarily embedded is ambiguously defined. Communities are conventionally perceived as essentially spatially-delimited entities, demarcated by archaeologically visible boundaries and nucleated settlement. When considering sites such as Lamanai, where no obvious, archaeologically visible, community boundaries are identifiable, this restricted definition is not so easily applied. The key to resolving this apparent problem would seem to lie in adopting a perspective of community as a socially constituted entity. Following Yaeger and Canuto (2000:6), a community can be described as an "ever-emergent social institution that generates and is generated by supra-household interactions that are structured and synchronized by a set of places within a particular span of time". The central place functions of the ceremonial precincts of Maya city centres provide an important focal point for regular community interaction and thus the creation, affirmation and modification of a shared community identity. Since potters actively participate in this interaction through the creation of material culture that has a shared meaning and value, the specific location of production facilities is immaterial. What is important is that it is within the contexts of activities and interactions associated with the ceremonial precinct that the material-human interactions fundamental to ceramic production and consumption converge in the creation and maintenance of a community

3.6. Towards a More Holistic Understanding of Ceramic Economic Activities during the Classic to Postclassic Transition

As the foregoing discussion demonstrates, there is great potential for advancement in current understanding of the lowland ceramic economy during the Classic and Postclassic transition. Conventionally held assumptions regarding the nature and cultural and behavioural significance of ceramic change during this period are based on an incomplete knowledge of ceramic variation within Maya society, on both the local and regional levels. Detailed scientific studies of local-level ceramic variation provide a logical starting point for any attempt to begin to fill this information gap, since it is community-based studies that must necessarily form the building blocks of broader cultural-historical interpretation. To achieve holistic and reliable reconstructions of community-based patterns of ceramic production and consumption requires an approach to the study of local-level ceramic variation that situates the physical ceramic evidence within an explicit environmental and cultural framework of material-human interactions. Such an approach should ideally involve:

- A whole assemblage approach to the study of ceramic variation that integrates macroscopic methods of analysis with scientific techniques in order to elucidate the stylistic, technological and provenance characteristics and relationships within and among different kinds of pottery.
- An independent assessment of macroscopic ceramic variation that is mindful of technological aspects of pottery in order to shed the inherent biases of conventional classification methods and interpretive frameworks.

- A comparative geological study of local raw material resources to assist in a behaviourally meaningful interpretation of the results of compositional analyses of the archaeological specimens using scientific techniques.

- A consideration of complementary information regarding the nature of variation in the local ceramic record, such as the depositional patterns of ceramics within different contexts of cultural activity and observed patterns and trajectories in other aspects of the material record (e.g. building activities and the use and maintenance of communal and private living space).

The following chapter presents a methodological approach to the study of local ceramic variation for the reconstruction of community-based patterns of ceramic production and consumption that attempts to overcome the limitations of conventional approaches to the analysis and interpretation of Maya ceramics. This methodology seeks to provide a means of testing conventional characterizations of ceramic economic patterns during the Terminal Classic to Early Postclassic period, whilst generating new insight into the potential behavioural and cultural significance of ceramic change at Lamanai.

CHAPTER 4
AN INTEGRATED APPROACH FOR THE RECONSTRUCTION OF COMMUNITY-LEVEL PATTERNS OF CERAMIC PRODUCTION AND CONSUMPTION

Technology is by definition a part of culture; it cannot content itself with a vision of reality that creates an arbitrary separation between material, social, religious and economic categories. Only by studying concrete operational sequences in a way that preserves these different aspects of technical facts will it be possible to uncover their underlying logics and to grasp in all its subtlety just what makes them what they are (Mahias 1993:177).

Ceramic products are the end result of a complex interplay among various environmental, material, social, economic and ideological processes, considerations and constraints. They are as much expressions of ideas, conventions and priorities, in a material and social sense, as they are material facts reflective of human interaction with the natural environment. It should be our task to contextualize pottery manufacture, use and deposition, and ceramic analysis, in an interpretive framework that considers and examines these contexts, as well as the physical properties of the material objects produced. This chapter presents a conceptual and analytical approach for the analysis and interpretation of community-level patterns of ceramic production and consumption. Grounded within an anthropological interpretive framework that aims to explore and define the wider setting of ceramic economic activities, the analysis focuses on physical and technological variation, using an integrated approach that combines traditional macroscopic methods of analysis with scientific techniques.

4.1. CERAMICS AS WITNESSES TO COMPLEX MATERIAL-HUMAN INTERACTIONS

Pottery has rich potential as a witness to the complex material-human interactions that lie at the heart of the past life ways that we seek to illuminate from an often scattered and fragmented archaeological record. Ceramics are expressions of human creativity, innovation and tradition and, as such, they represent a material transformation of substances found in nature into cultural items that are permanent and useful. Within the routine of day-to-day life, ceramics are not only tools that facilitate a variety of activities and tasks that are often fundamental to human survival, but also cultural items that have value and meaning in their own right. They are items that are used in particular contexts of human activity and interaction, and that are displayed and exchanged. Accordingly, pottery plays an intrinsic role, both functionally and symbolically, in countless scenarios, everyday activities and events. Further, when viewed against the backdrop of these broader contexts of human activity and interaction, ceramics convey messages about the identity, status, affiliation and interrelationships – of both individuals and groups, and of both their producers and consumers. By recognizing that the relationship between humans and ceramics is necessarily highly complex, we acknowledge that such objects can provide a rich commentary on the nature of the activities, behaviours, transactions and interactions that form the fabric of day-to-day life.

A ceramic assemblage or deposit can be viewed as a materially-manifested convergence of knowledge, people, place and premise. Its depositional context, location and its relationship to a particular kind of architecture or other natural or man-made features can inform us not only about the nature and kinds of human activity the ceramics are associated with, but also about the wider social setting in which this activity took place. The depositional pattern (placement and fragmentation) of ceramics within a deposit provides further information regarding associated human behaviours, particularly concerning the condition and way in which the pottery was interred, as well as attitudes and ideas regarding the intended function and/or appropriate treatment of ceramics within particular contexts of use and deposition. The richest source of cultural information, however, is the ceramics themselves, providing unique insight into the kinds of activities that took place, as well as the shared cultural knowledge and values of both the users (consumers) and producers of pottery. For example, the morphology and decorative/surface characteristics of pottery offer clues as to their role or intended function within particular cultural contexts – for example, as implements of cooking, storage or serving food and drink – as well as their potential intrinsic and symbolic value as common tools, items used only on special occasions, prestige items, or social currency. Similarly, the composition of ceramic bodies can provide evidence of pottery provenance which, when compared to the location of deposition may provide an indication of its perceived value or significance. Perhaps most significant, however, is ability of the physical and functional attributes of ceramics to inform about the shared values and needs of human groups; for they not only provide evidence of the tastes and requirements of the people that used pottery, but also the behaviours, approaches and choices of potters in attempting to meet consumer demands. The plasticity of clay offers a medium of expression for ideas and attitudes and potters are at once problem solvers, artisans and individuals affected by and acting upon their economic, natural and social environment. To make pottery successfully requires knowledge and training and yet finished products must also fulfil the needs and wants of those who use them. Myriad choices are available to potters for solving any perceived ceramic need or problem, and there are usually a number of alternative pathways through the production process to achieve an acceptable or desired end result in terms of the final ceramic product. But potters do make specific choices, carrying out learned and repeated everyday practices with regard to the raw materials they prepare and use to make the ceramic body, in terms of vessel forms and surface and decorative treatments and in terms of their method of firing. These 'ways of doing' (conscious or subconscious) correspond to a technical logic that is as much determined by social factors – for example, received technical knowledge, tradition and the material and social necessities of society – as material and environmental ones[1] (Mahias 1993; Lemonier 1993; van der Leeuw 1993). In this way, therefore, the way in which pots are made, or the set of choices that define the manufacturing process, is a material expression of group identity, corresponding mutually to the values, ideas and attitudes of producers and consumers.

When viewed from this perspective, it is apparent that a ceramic assemblage represents the culmination of an array of material-human interactions that together form a 'cultural biography' of pottery items united by way of their shared depositional context. This cultural biography tells a story about the physical nature of particular kinds of ceramics – ceramic items deemed appropriate by those who made and used them – that were deposited in a specific place, and in a specific way, by particular community members, as part of particular kinds of human activity. This story, therefore, is essentially one about ceramic production and consumption. The different cultural and physical attributes of a ceramic assemblage – i.e. its depositional context, the way and condition in which the pottery was deposited and the physical and functional characteristics of the pottery comprising it – form multiple lines of evidence that provide different and complementary insight into the cultural and behavioural significance of both the deposit and the pottery itself. It is by considering these multiple lines of evidence that the kinds of factors influencing observed patterning in the material record can be investigated in a way that is mindful of the wider conditions within which ceramic production, use and deposition occurs.

4.2. The Reconstruction of Patterns of Ceramic Production and Consumption as a Means of Investigating the Factors Influencing Continuity and Change in Community Activities

A central purpose of the present study is to examine the kinds of factors and pressures that contributed to shaping community life at Lamanai during the Classic to Postclassic transition, especially in regard to economic, ritual and ceremonial activities involving ceramics. Fundamental to this objective is the reconstruction of patterns in the production and consumption of pottery items and charting continuity and change in these patterns over roughly a 450 year interval that spans the Terminal Classic and Early Postclassic periods. As the terms are used here, *production patterns* primarily relates to variability in local approaches to pot making or manufacturing practices, rather than the specific organizational form of production, and *consumption patterns* concerns the range and relative frequency of different categories of pottery, defined stylistically, functionally, or in terms of their origin of manufacture, that occur within the depositional contexts examined. As we have seen in Chapter 3, within the context of regional studies, the reconstruction of patterns of ceramic production and consumption involves discriminating among the ceramics manufactured at different locations and charting their distribution within a defined geographic area. When the focus is on reconstructing local-level patterns, however, more fine-grained distinctions regarding the nature and origin of ceramic products are often desired. The analyst must be able to make decisions about what pottery appears to be of local and of non-local origin, where non-local pottery might originate from, and to document the relative frequency of ceramic products deriving from different geographical regions. Also of central interest, is discriminating products deriving from different local manufacturing traditions, examining continuity and change in these traditions over time, and identifying local imitations of pottery styles typical of other sites or regions, instances of foreign influence on local manufacturing traditions, and so on, with an acceptable level of confidence. Ceramic researchers employ a number of different analytical approaches to makes these distinctions. Such judgments are most often arrived at, however, through examination of variation in the visual (style) and technological (paste composition and manufacturing techniques) characteristics of the vessels concerned, estimation of the relative quantities of different categories of vessels, defined according to specific functional, stylistic and/or technological criteria, and through the association of certain pastes or styles with particular raw material sources, specific sites or regions, and/or specific time intervals.

The body of evidence upon which these distinctions and connections are based can vary considerably, ranging from a focus on conspicuous attributes and features that can be evaluated with the naked eye (e.g. Fry 1979, 1980) to more detailed analysis of compositional and technological variability using scientific techniques, and drawing upon comparative geological and typological information (e.g. Rands and Bishop 1980; Bishop 1992, 2003; Neff 2003) By nature of the detailed information they generate, however, archaeometric studies of ceramic production technologies facilitate a more direct means of making connections and distinctions between manufacturing procedures, and of linking ceramics to

[1] By material factors, I mean considerations such as the physical and chemical properties of raw materials, the desired mechanical performance characteristics of vessels, manufacturing technology etc., whereas environmental factors include local geology, climate and geography.

particular locations or areas of manufacture through their raw materials. Another benefit of such studies is that they provide complementary information about ceramic variation at a high level of resolution, which, in turn, can be used to cross-check distinctions and connections made based on conventional forms of macroscopic evidence – e.g. the visual appearance of vessels – and consequently, to strengthen or refute claims made regarding the behavioural significance of such observations.

Conceptual Approach

The conceptual approach adopted in the present study to reconstruct patterns of ceramic production and consumption derives from *chaîne opératoire* methodology, which focuses on reconstructing "the technical chain of sequential material operations by which natural resources are acquired and physically transformed into cultural commodities" (Dobres 2000:154, also see Cresswell 1983; Sellett 1993; van der Leeuw1993). With regard to the manufacture of ceramics, this sequence of operations involves, at minimum, three necessary stages: 1) the selection and preparation of raw materials, 2) vessel forming and finishing and 3) firing[2]. A main premise of *chaîne opératoire* research is that the choices that potters make at each of these stages of the production process directly determine the nature of the final ceramic product and its effectiveness and suitability, in both a material and social sense. Ethnographic studies of the technical practices of contemporary potters have shown that these choices are anchored in a stock of technical, cultural and environmental knowledge and perceptions that limits perceived possibilities (e.g. Day 2004; Mahias 1993; van der Leeuw et al. 1991; van der Leeuw 1993). Accordingly, the choices that potters make (past and present) can be considered as being as much determined by social and ideological factors as by the natural conditions within which pot making occurs (for example, local geology, weather patterns and climate [Arnold 1985]). Ethnographic studies have also demonstrated that the sets of choices that different potters/groups of potters make over time (e.g. Arnold 1987; Arnold and Nieves 1992) and space (e.g. van der Leeuw et al 1991; Mahias 1993; Gosselain 1998) are observable and can be readily studied for their technical and ideational logic as they are manifested materially as *variants* of the same basic sequence of operations (in this case, the three basic stages of the manufacturing process). These *technical variants* (*sensu* Lemonnier 1980, 1986) not only constitute technical and material facts regarding the technical practice and ceramic repertoire of a given society or community that can be described and compared, but as 'ways of doing the same thing' that are embedded within a specified social, intellectual and environmental context, they can be considered to represent tangible expressions of custom/habit, and underlying perceptions, beliefs and values connected to ethnic, community and/or social identity. A great benefit of *chaîne opératoire* methodology as a conceptual and analytical framework, therefore, is that it enables the ceramic analyst to investigate and study both the material and the symbolic representational aspects of ceramic variation.

Within the context of ethnographic case studies, the focus of *chaîne opératoire* research is most often on delineating and comparing the specific sequence of technical operations followed by different potters/groups of potters in order to examine the functional linkages between various elements of technical processes, as well as the contextual, material, social or ideational dimensions of observed technical variation (Mahias 1993:162). When considering the technical practices and behaviours of prehistoric potters, reconstructing operational sequences to a comparable level of detail and linking these sequences to the hands of specific individuals/groups of individuals is virtually impossible. This situation not only derives from the fact that past technical processes and the people involved cannot be observed directly, but also has to do with the nature of archaeological ceramic assemblages, which in many cases are representative of activities and behaviours associated with ceramic consumption rather than production. Since consumption assemblages provide only an incomplete (and biased) record of the totality of ceramic products manufactured and in circulation during any given time period, the exact nature and total range of ceramic variation, and thus technical variation, that existed in reality is indeterminable from these assemblages. Furthermore, the reconstruction of the technical processes associated with archaeological ceramics must necessarily rely on the incomplete evidence that can be gleaned from final ceramic products, which are sometimes poorly preserved and often in a fragmentary state. This endeavour is further confounded by the fact that many individual technical actions leave no trace on the final ceramic product. As a result, it is most often only possible to reconstruct part or certain aspects of the sequence of operations that comprised the original manufacturing process. Moreover, explaining and interpreting the significance of observed variation in technical processes is often complicated by a lack of sufficient chronological control over the body of material under study as well as an incomplete understanding of production parameters (mode, scale and organization) and the wider social setting within which production occurred. Consequently, it is difficult to discern whether observed variations in specific physical or technological attributes of ceramic products (for example, textural differences among the pastes of morphologically and decoratively equivalent vessels or the colour of the slip and/or presence/absence of additional decoration on otherwise equivalent vessels) in fact relate to actions of different potters/groups of potters at any one given time, or if they simply define the range of products manufactured by a single potter/group

[2] In ethnographic studies technical processes are elaborated in greater detail, especially with regard to vessel forming and finishing, which often involves several independent steps that can be observed (c.f. Mahias 1993:159; van der Leeuw et al. 1991:153). The exact nature of this part of the manufacturing process is often indeterminable for archaeological ceramics, since evidence of earlier stages of manufacture (forming techniques) is often obliterated by subsequent procedures associated with surface finishing.

of potters, reflect chronological differences among the products manufactured by one or multiple potters/groups of potters, relate to fluctuations in workmanship, innovation and accidents of production or a combination of such factors (Shepard 1956:318). This is not to imply that *chaîne opératoire* methodology cannot be used profitably in the study of prehistoric technological practices, but rather to illustrate that by virtue of the nature of archaeological ceramic assemblages and the level of understanding of the human contexts within which technical processes are embedded, the focus and objectives of *chaîne opératoire* research when dealing with archaeological ceramic assemblages are necessarily different.

Analytical Framework

In the present study, the *chaîne opératoire* is invoked as a tool to facilitate comparison of the physical and technological attributes of pottery vessels deposited in association with ritual and ceremonial activities at Lamanai, that is, pottery vessels that were recovered from specific consumption contexts at the site. Hence, from the outset, the focus of the study was on examining the nature and behavioural significance of ceramic variation associated with particular kinds of contexts of use and deposition, rather than delineating and examining the total range of ceramic and technical variation that existed within the community as a whole, during the time period under investigation. The rationale for selecting only particular kinds of ceramic deposits for study was to attempt to control for variation due to significant differences in the social context within which the pottery was used and deposited, so that variation due to temporal factors alone could be studied more effectively. The main objective in using the *chaîne opératoire* as a comparative framework for the physical analysis of the pottery was to enable a characterization of ceramic products according to the basic sequence of technical operations they all share in common – i.e. the selection and preparation of raw materials, vessel forming and finishing, and firing – so that variation in the physical and technological characteristics of the pottery could be delineated and examined, on various scales, according a single set of criteria. Of central interest was the examination of the relationships between different aspects of physical variation in pottery, in order to identify patterns of technical similarity and difference within the assemblage and to investigate the technology and provenance of different ceramics. *Chaîne opératoire* was not used in order to identify, in an exhaustive fashion, specific sequences of technical operations followed by different potters/groups of potters. Rather, the aim was to illuminate physical and technical similarities and differences among the ceramics, and to study continuity and change in these patterns over time. These data were then contextualized archaeologically and geologically, aiding the reconstruction of community-level patterns of production and consumption, revealing new information on ceramic change for the Terminal Classic to Early Postclassic period.

The Analysis of Physical and Technological Variation within the Assemblage

The examination and characterization of physical and technological variability within the Lamanai assemblage was accomplished through an integrated approach to ceramic analysis that combines conventional macroscopic methods of analysis with natural-sciences-based analytical techniques, including thin section petrography, neutron activation analysis (NAA) and scanning electron microscopy (SEM). The underlying objective of this approach was to obtain a detailed picture of the manufacture of different stylistic and functional classes of vessels, thereby enabling the comparison of pottery groups meaningful in terms of provenance and technology, as well as visual appearance and function (see Day et al. 1999 and Wilson and Day 1994 for a comparable methodology).

The macroscopic phase of the analysis focussed on establishing groups of 'like' vessels. To this end, the material was characterized according macroscopic attributes reflecting choices made by potters at the three basic stages of the manufacturing process, based on decorative, morphological, paste/fabric and functional criteria. Then the minimum number of individual vessels was estimated for each group. A subset of material was then selected for detailed examination using scientific techniques, reflecting the range of stylistic, technological and compositional variability observed with the unaided eye. A total of 646 individual vessels were sampled for detailed scientific analysis, representing approximately 38% of the minimum number of vessels recorded. Thin section petrography was used to examine technological and compositional variation within the entire subset of material on the microscopic level, and chemical compositional variation was assessed for a smaller sample of vessels (N=213). A detailed examination of decorative treatments and firing with a scanning electron microscope was conducted on 31 vessels. The selection of samples to be included in each successive phase of the archaeometric study was based on a consideration of information revealed during prior phases of analysis, with the aim of sampling across the range of variation observed. All of the ceramics that were selected for analysis by the different scientific techniques are listed by sample number in Appendix I. A summary of the analytical methodology employed for the analysis of physical and technological variation within the Lamanai assemblage on the macroscopic, microscopic and chemical levels is presented in Figure 4.1.

4.3. Analytical Techniques and Procedures

Characterization of Stylistic and Technological Variability at the Macroscopic Level

The macroscopic analysis of the Lamanai ceramics forms the cornerstone of the analytical work, since it was during this phase of the analysis that the categories of ceramic products to be compared and contrasted through detailed scientific examination were initially defined. The

Levels of Analysis		Stage of Manufacture	Selection and Preparation of Raw Materials	Forming and Finishing (morphology and surface treatment)	Firing (approximate temperature reached/regime)
	Macroscopic	Visual inspection	- Paste colour and texture - Presence/absence and types of inclusions and their properties	- Surface and decorative treatment - Vessel shape, morphology and dimensions	Regime: Paste/Fabric colour; presence/absence and nature of firing horizons
	Microscopic	Thin section petrography	Grouped and characterized according to Whitbread's descriptive system based on: - Characteristics and properties of the microstructure -Characteristics and properties of the groundmass, including composition of inclusions (type, relative quantity, physical characteristics) -Textural concentration features and amorphous concentration features		Temperature: -optical activity of clay matrix Regime: - nature and colour of firing horizons - - Physical properties of the microstructure e.g. nature of voids
		SEM		Technology of decoration characterized by: - treatment of surface - microstructure and morphology of slip	Temperature and regime: - vitrification microstructure (nature and properties in comparison to standard)
	Chemical	NAA	Chemical composition based on major, minor and trace constituents		
		SEM		- semi-quantitative characterization of relative composition of slip surface vs. ceramic body - semi-quantitative characterization of chemical composition of slips and painted decoration	

Figure 4.1: Summary of the analytical methodology employed for the analysis of physical and technological variation using a combination of macroscopic, microscopic and chemical techniques.

methodology employed to sort and group the pottery represents a significant departure from standard practices associated with the type:variety system – the predominant methodology followed in Maya archaeology. It is well recognized that technological attributes are not well integrated into the ceramic units that form the basis of the type:variety scheme (e.g. Rice 1976; Rands 1991; Bishop 1991), with the implication that the system, as classificatory and interpretive framework, subsumes important information on technological variation that is of direct relevance to reconstructing patterns of ceramic economic activity. Since the examination of technological variation forms a central component of the present study, it was necessary to devise a means of grouping the pottery that considered technological as well as the standard stylistic attributes. As Irving Rouse (1960:313) stated: "classification, like statistics, is not an end in itself but a technique by means of which to obtain specified objectives, and so it must be varied with the objective." Accordingly, it is the nature of the issues being investigated in the present study that justifies this new classificatory approach.

The corpus of material examined in the study includes sherd assemblages deriving from temple middens as well as whole and reconstructed vessels from special contexts such as offerings and burials (Chapter 6). The macroscopic analysis of this material was conducted over the course of four field seasons, which varied in length from three weeks (1999) to 3 months (2000, 2001 and 2002). Only well preserved vessels from well defined, and most often, primary archaeological contexts were selected for study. The main objective of the analysis was to document variation in vessel surface treatment, morphological and paste/fabric[3] attributes, within both individual lots of material and the corpus as a whole. Each of three phases of analysis had specific objectives. The first, undertaken between 1999 and 2000, focussed on identifying all ceramics to be included in the study, based on existing catalogues, vessel descriptions, excavation records and publications. The material selected was located, compiled, washed and catalogued. In this way, groups of 'like vessels' were established based on attributes of vessel paste, surface treatment and morphology so that a representative sample of sherds and vessels could be selected for detailed examination using scientific techniques. During the second phase of the analysis, these groups were further examined in greater detail. This phase of the analysis focussed on variation in vessel morphology and surface treatment, with the major aim of further refining the established vessel groups and documenting the variation within them. The third phase of the analysis involved a detailed study of vessel fabric/paste at the macroscopic level. Variation was documented both within and between vessel groups considering its implications in terms of vessel composition and firing.

3 Within the context of New World archaeology the term 'paste' is synonymous with 'fabric' which, within the context of Old World ceramic studies, refers to the fired ceramic body. The more common usage of 'paste' in the context of Old World ceramic studies is to refer to the wet mixture prepared to make pottery as opposed to the end result after firing. This distinction is not made in New World ceramic studies.

ANALYSIS OF WHOLE AND NEARLY COMPLETE VESSELS

Many of the whole and reconstructed vessels included in this study have been presented elsewhere in publications by Pendergast (1981a, 1982) and Graham (1987). The descriptive system used by Pendergast and Graham summarizes and describes vessels according to a number of categorical, descriptive and metric attributes that relate to vessel context, surface treatment and morphology, and in terms of their condition upon recovery. This includes alterations or additions to the vessel surface caused by use wear or postdepositional processes. For the purposes of the present study, the original descriptive system was expanded to include paste attributes as well as a wider range of metric attributes. Detailed information about surface treatment and morphology was re-recorded, and these new descriptions were checked against those of Pendergast and Graham to identify discontinuities in terminology and instances of inter-observer error in the measurements taken.

Pendergast's and Graham's descriptions of vessel morphology follows closely the scheme detailed by Sabloff (1975:19-27) in his presentation of ceramic materials at Siebal, in terms of distinctions made between shape classes and their subdivisions based on vessel profile. The terminology they use to describe vessel forms, in general, as well as more specific aspects of a vessel's morphology, such as rim or lip form, is also very similar to Sabloff's[4]. Pendergast and Graham perhaps pay more attention, however, to the subtle variation in vessel form that usually exists even within the subdivisions of shape classes, and when warranted, they choose to describe in detail the exact profile of the vessel, from its base to its rim, rather than assigning it a general 'form

4 Significant departures from Sabloff's basic terminology include the following:
For bowls and dishes: 'composite silhouette' has been added as a profile or form type to describe vessels on which the curvature of the wall profile changes *vis a vis* an angular junction (following Shepard 1948: 4-5,12-13). The exact curvature of the profile and the positioning of the angular junction vary, and thus different profiles or forms are described in detail. Smith (1955:25) describes vessels with this basic morphology as a 'flaring or out-curving side bowl or dish angling to base', whereas Gifford (1976:241) terms them dishes or bowls 'with a basal break'. Similarly, Sabloff (1975:26-27) suggests the subdivision term 'basal angle' to describe similar profiles. At Lamanai, the positioning of the angular junction is not always low on the vessel wall and directly associated with the base, as is implied by Smith's, Gifford's and Sabloff's terminology. Hence Shepard's terminology was followed, as it most adequately accommodates the variations in curvature that occur within this profile type.
For jars: Sabloff's category 'outflared neck' is termed 'flaring neck'.
For rim forms: Sabloff's category 'outflared everted' is termed 'flaring everted'. Sabloff's categories 'interior folded' and 'exterior folded' are reserved for rims that actually show evidence of folding, i.e. the 'folding down' of the neck wall to form the rim has not been obliterated. The term 'bolstered' has been added to describe rims of a similar shape that have smooth margins and show no clear evidence of folding.

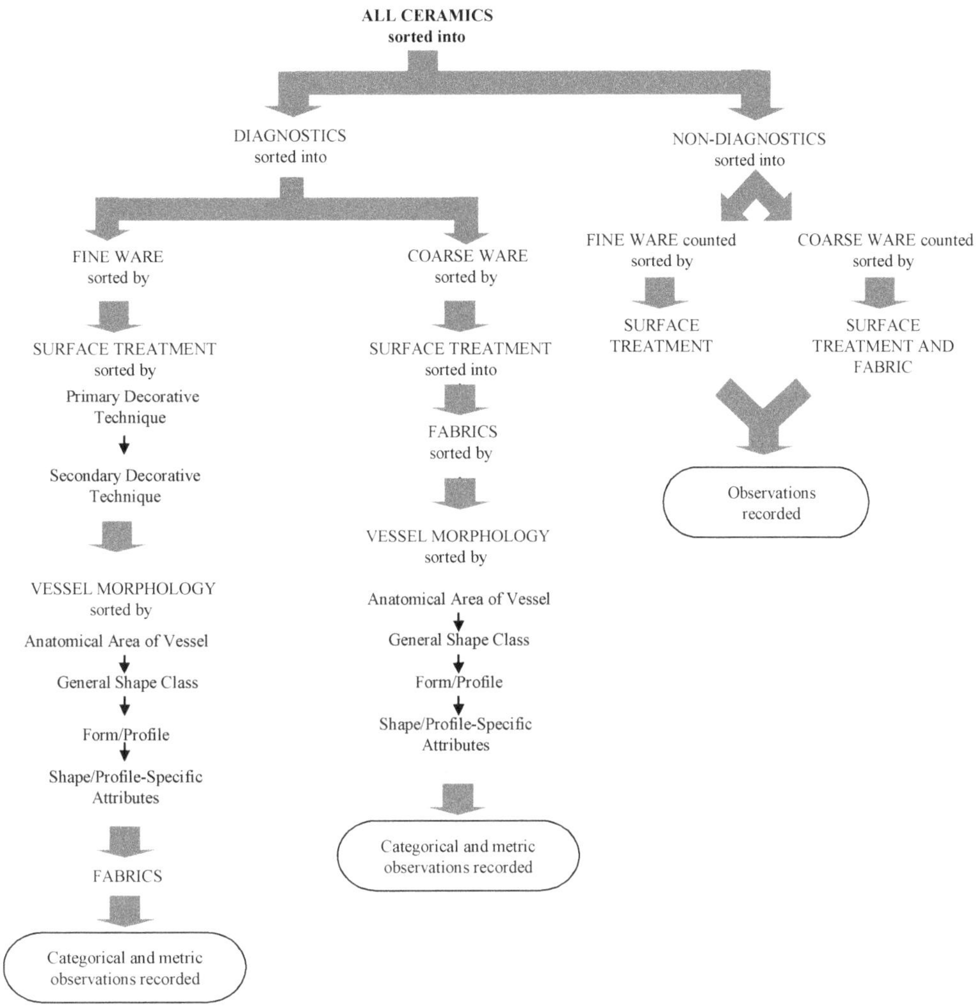

Figure 4.2: Procedure for sorting and grouping.

division' within a particular 'primary form class' that it resembles (see Sabloff's definitions (1975:22-27). In dealing with ceramic assemblages that undergo considerable change over time such as that at Lamanai, a more detailed consideration of vessel morphology enables a closer examination of the ways in which forms change and/or stay the same over time, thereby providing a more comprehensive picture of the development of ceramic styles from a morphological point of view.

The descriptive terminology of Pendergast and Graham to characterize aspects of a vessel's surface treatment and decoration follows that used by Thompson and Shepard (1939), Smith (1971) and others. Pendergast has tended to use the Ridgeway System (1912) for characterizing colour, whereas Graham has used both this system and Munsell method, which is now the standard in Maya ceramic studies. In the present study, Munsell Soil Color Charts (2000) have been used.

The detailed study and description of whole and reconstructed vessels facilitated the identification of many of the vessels forms represented in the sherd assemblages.

ANALYSIS OF SHERD ASSEMBLAGES

The midden assemblages examined as part of the present study had not been processed previously beyond the separation of sherd material. In addition to establishing types and quantities of vessels, the material was characterized in terms of stages of production, by attributes reflecting raw material selection and their preparation (or paste recipes), forming and finishing techniques and methods of firing. The sorting strategy employed is presented in Figure 4.2 and begins with a basic division of the material into *diagnostic* and *non-diagnostic* sherds. Diagnostic sherds are those which display clear evidence of the kind of vessel from which they derive or provide specific information about details of vessel morphology of surface treatment. These sherds are decorated or undecorated and include rims, bases and specific body parts such as necks, shoulders, feet, and handles. Non-diagnostic sherds predominantly include slipped and unslipped body sherds, but also heavily damaged and eroded sherds.

Subsequently, both diagnostics and non-diagnostics were sorted into *fine ware* and *coarse ware*, a distinction that

Table 4.1: Surface treatment attributes.

	Primary Decorative Mode ➔	**Colour** ➔	Secondary Decorative Mode
Fine Ware	Monochrome slipped	Black	Fluted
	Monochrome, interior slipped	Red	Gouge-incised
	Monochrome slipped and painted	Orange	Groove-incised
	Polychrome, slipped and painted	Red to Orange	Impressed
	Polychrome, interior slipped and painted	Red-Orange, too eroded	Incised
	Slipped-resist	Black on red	Incised and impressed
	Monochrome slipped or resist – too eroded	Red and black	Incised with punctuates
		Black and orange on natural buff	Absent
		Black and red on buff	
		Black and red on dark orange	
		Black and red on natural red-orange	
		Black and red on orange	
		Black, red and orange on buff	
		Black, red, orange and white on buff	
Coarse Ware	Slipped-unslipped	Brown/black/purple-grey/red-brown	Plain
	Unslipped/smoothed	Brown/grey/black	Striated
	Washed	Buff	Incised
		Buff/brown, white flecks	
		Buff-brown/brown/grey/black	
		eroded	

was based on the presence/absence of comparatively large rock fragments in the paste, but which also included such considerations as the presence or absence of surface decoration, the overall quality of the vessel and their intended function. In this sense, the term *fine ware* denotes vessels with surface decoration, that are well made, that tend to have comparatively fine-textured pastes and that probably functioned as tablewares or serving vessels in ritual and commensal contexts. In contrast, *coarse ware* refers to vessels that are generally undecorated, that are of relatively poorer quality, that have comparatively coarse-textured pastes and that were primarily used for food preparation and storage (i.e. utilitarian functions).

With regard to the non-diagnostics, the fine ware and coarse ware were sorted according to differences in surface treatment, and in the case of the coarse wares, also by paste. For example, the fine ware was sorted into sherds that were slipped on both the interior and exterior surfaces, only one surface (interior or exterior) and those that were too eroded to make this distinction. Along the same lines, the coarse ware was sorted into sherds with a washed, well smoothed, poorly smoothed or striated exterior surface and also by paste, according to the colour of a freshly fractured area of the sherd and the nature and type of aplastic inclusions. For both the fine ware and coarse ware material, general observations were recorded regarding the groups and relative quantities of fine ware and coarse ware sherds.

The sorting procedure for diagnostic sherds was more extensive, beginning with a separation of the material according to differences in surface treatment. For the coarse ware this involved a basic division into slipped, washed and unslipped sherds. For the fine ware this involved sorting the material according to general decorative technique, the presence/absence of a second technique and, if present, the type of technique, and also by colour. The division of the fine ware according to these attributes of surface treatment are summarized in Table 4.1.

Once sorted by surface treatment, the fine ware was then subdivided according to vessel morphology. The sherds comprising each group were first sorted according to the anatomical area of the vessel they represented, and then by general shape class. Each shape class was subdivided further according to profile type and then by shape class or profile-specific attributes such as rim or lip form. Figure 4.3 illustrates the process of sorting groups of diagnostic sherds into successive morphological subgroupings for monochrome orange slipped bowls.

Finally, each of the morphological subgroups were organized according to paste characteristics, based on the colour (including the nature of the firing horizon when present) of a freshly fractured area of the sherd (the rim for rim sherds) as well as the presence/absence, nature and type(s) of aplastic inclusions present. Other properties such as the overall look and feel of the fabric were also considered. Comparisons were made between the fabrics of each shape class and of each surface treatment group.

With regard to the coarse ware diagnostics, once sorted by surface treatment, they were then sorted according to paste characteristics and vessel morphology, following the same procedure as was applied to the fine ware. The division of the basic surface treatment groups according to paste characteristics was especially important for identifying subgroups within the large quantity of unslipped sherds with smoothed surfaces. The initial

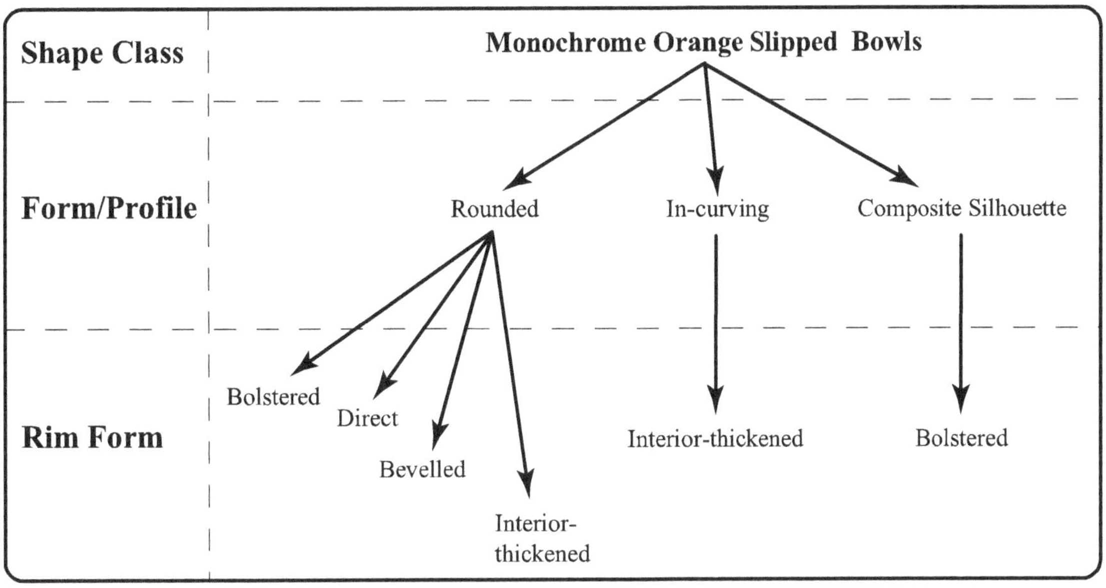

Figure 4.3: Bowls sorted by morphological attributes.

distinctions between pastes were based largely on the colour of fresh fractures along with the kinds of aplastic inclusions present, although specific textural differences such as the density of inclusions were also considered. The primary distinguishing attributes of colour included the nature of the firing horizon, if present, and the colour of the exterior margin of the fracture.

THE RECORDING OF ATTRIBUTES AND MEASUREMENTS

The diagnostic sherds of each lot of material were the focus of the detailed analytical work. Both descriptive and metric attributes were recorded for the sherds comprising each shape class within a particular surface treatment group, and in each case, vessel counts were recorded for each shape class as well as for each profile type within that class. In the vessel group descriptions that appear in Chapter 7 and in Appendix II, the data is organized according to the criteria of: *Frequency, Principal Identifying Modes, Paste and Firing, Surface Treatment and Decoration and Forms and Dimensions*. This format closely follows the standard used for the presentation of ceramic assemblages according to the type:variety scheme (Willey et al. 1967; Ford et al. 1992). The specific procedures followed in the collection of these different categories of data are summarized in turn.

Frequency - Vessel frequencies were calculated based on the minimum number of vessels represented by diagnostic sherds rather than from a count of the total number of individual sherds with particular surface treatments or morphologies. The minimum number of vessels represented within each lot of material was determined for each surface treatment group, based on a count of the diagnostic sherds occurring within each profile type and shape class. These counts were undertaken only after a concerted effort was made to identify all 'joins' (conjoining parts of the same vessel) within and between lots. For most shape classes and profile types, these counts were based on rim sherds. However, in cases where two diagnostic elements of a potential vessel were represented, for example dish rims and pedestal rims from pedestal-based dishes, only the element represented by the greatest number was included in the total count. The minimum number of vessels represented within each surface treatment group was calculated based on the counts for the shape classes comprising the group, and the minimum number of vessels represented in each lot of material was calculated based on the totals determined for each surface treatment group.

Paste and Firing - Variation in paste attributes at the macroscopic level was documented for each group according to the following criteria:
Paste colour including the presence/absence and nature of firing horizons
Paste texture – presence/absence , physical properties (for example, colour and relative size [fine, medium, coarse etc.]) and relative frequency (few, abundant etc.) of aplastic inclusions.
Paste colour was characterized descriptively since the limited range of colour standards included in the Munsell Soil Chart did not permit distinctions among the often subtle differences in colour that were observed. Although this approach is admittedly subjective, it was felt that it would enable a more accurate description of the observed variation in colour.

Surface Treatment and Decoration - The terminology for vessel surface treatment is consistent with the standard terminology used in Maya archaeology, and follows Smith (1955, 1971) and Gifford (1976). The specific attributes relating to surface treatment that were recorded include: primary decorative technique, secondary decorative technique and colour. Within the primary decorative techniques, the term 'resist' refers to bichrome vessels with relatively large, bold decorations. As it is used here, the term describes the surface treatment that Smith (1955:59-60) refers to as 'Classic resist'. Slip colour is difficult to assess in a systematic and standardized way, both on a group level and on a case by case basis, partly due to differences in the preservation of ceramics deriving from different postdepositional contexts. For example, in lots where vessel surfaces were severely eroded, an accurate assessment of slip colour was often impossible. Within lots of better preserved material, it was also observed that differences in the preservation and quality (relative glossiness and thickness of the slip layer) of the overall surface finish produced subtle variations in slip colour. Another problem was that the colour of the slip was often inconsistent or varied across the surfaces of individual vessels or sherds. In addition, it was often not possible to match the exact colour of the slip to any of the colours in the Munsell Soil Color Chart. This being the case, slip colours were characterized for each surface treatment group based on the best preserved and most homogenous examples.

In many instances, the full suite of surface treatment attributes of interest was not observable, most often because the sherd was too heavily eroded or too small to permit an accurate assessment. In the case of larger sherds, from which morphological information was obtainable, but which were either too heavily eroded or burnt to permit an accurate assessment of surface treatment attributes, only general information was recorded. For example, it was observed that particular vessel forms tended to have either had either a red or an orange slip and not any of the other surface treatments. In these cases 'colour' was recorded as *red-orange or too eroded* to account for the fact that these sherds were most likely originally red or orange slipped *and not* something else. A similar problem arose in trying to assess the colour of conformal red and orange slipped sherds that were represented in large numbers within the better preserved lots of material. Within such lots, it was observed that colour tended to grade between red and orange with only a few examples of what could be considered 'true red' or 'true orange'. Colour also varied on individual sherds. In these cases, sherds were only recorded as being 'red' or 'orange' if they were true red or true orange (the corresponding Munsell designations for all basic colours referred to in the text are listed in Appendix III). All other sherds were characterized as *red to orange* in order to account for their 'transitional' colour, while still distinguishing them from their 'true red' and 'true orange' counterparts, as well as the morphologically similar sherds that were either severely eroded or burnt. Surface treatment assessments were also problematic for some of the composite silhouette dishes as morphologically identical forms had a 'red slipped', an 'orange slipped ' or a 'resist' surface treatment. Severely eroded or more ambiguous examples of these dishes were treated in a similar fashion as the problematic red-orange sherds and were characterized on a more general level as *red/orange/resist.*

Forms and Dimensions - The terminology that is used to describe vessel morphology is consistent with the standard terminology used in Maya archaeology, and specifically follows Sabloff (1975) and Shepard (1948). The morphological attributes that were recorded include categorical attributes such as the anatomical area of the vessel represented, general shape class (Sabloff's 'primary form class'), profile category or form (Sabloff's 'divisions' within the primary form classes) and form/profile specific attributes such as rim form and lip form (Sabloff's 'subdivisions'). Metric attributes recorded include rim diameter (taken from the exterior margin of the rim), neck diameter (taken from the interior margin of the neck), rim to neck height (measured as the vertical distance between the lip and the inflection point of the neck) and wall thickness (taken from the upper shoulder of jars and 2 cm below the rim for all other shape classes). Diameters were determined using a diameter board and callipers were used for all other measurements. For the comparatively few base sherds, base diameter and base height (when determinable) were recorded for ring bases and pedestal bases, whereas wall thickness was measured on pedestal bases only.

Metric attributes were recorded for all of the whole vessels, but only for approximately 60% of the diagnostic sherds within the assemblage. The sherds that were not measured include appendages such as handles and feet, sherds that did not represent the anatomical areas of interest, and sherds that were too small to give accurate measurements. In many cases, not all of the desired measurements were obtainable, resulting in incomplete sets of metric data for many sherds. This unavoidable problem is reflected in the differences that are readily observable among the counts upon which the descriptive statistics and correlations are based. None the less, all possible measurements were recorded and included in the statistical analysis.

Information Obtained from Non-Diagnostic Sherds - In order to gain a more comprehensive understanding of the surface treatment and morphology of the original vessels represented by the diagnostic sherds, an attempt was made to associate groups of non-diagnostic body sherds with particular surface treatment groups and morphological subgroups identified through the sorting and analysis of the diagnostic sherds. This comparative exercise, which focussed on overall similarities in terms of surface treatment, colour, wall thickness and fabric, was particularly important for establishing body surface treatment accorded to particular jar forms, since complete examples are nonexistent. Such details were elucidated for other shape classes through comparison of the diagnostic sherds with more complete vessels displaying the same attributes.

THE ANALYTICAL PROCEDURES USED AND TYPE:VARIETY METHODOLOGY

The methodology employed in sorting and analysing the sherd assemblages differs from the standard practices of the type:variety method in a number of fundamental ways. At the most basic level, the working units of analysis are different. The groups of 'like vessels' that resulted from the sorting procedure employed here were produced through the application of a set of criteria (general attributes) for gauging similarities and differences between groups of vessels in these respects. All of the resulting groups, therefore, can be described according to the same set of criteria (or attributes) and likewise, the assemblages can be characterized in terms of the attributes themselves, allowing both the comparison of groups of ceramics in relation to any of the attributes as well as the study of attribute variability. The basic working units of analysis can be the attributes themselves, or any grouping within the hierarchy of attributes upon which the sorting of the material was based. The unit of analysis that is invoked will depend upon the question being asked or the issue being investigated. Within the type:variety scheme, however, the basic working unit of analysis is the *type*, which is defined according to the constellation of attributes that makes it unique. The content of these constellations, in terms of the individual attributes they comprise, is not always the same, meaning that on a very basic level, *types* are not always described according to the same set of criteria. This makes comparisons between *types* difficult, according to a standard set of attributes.

A second point of difference relates to the frequencies that are reported as part of the vessel group descriptions (Chapter 7 and Appendix II). According to the type:variety scheme, frequencies are determined at the level of the *type* and the *variety*, based on an overall count of all of the sherds identified as that particular *type* or *variety*. While these overall sherd counts give an idea of the quantity of a particular *type* in comparison to other *types*, they provide no indication of the minimum number of individual vessels representing a particular *type* or *variety*. If the underlying goal of the analysis and presentation of ceramic data is to document and describe what was recovered, then general frequencies of this sort may be a sufficient gauge of the quantity of one kind of ceramic against another. However, when the enquiry moves beyond the basic description of the material and into investigating and describing constructs related to human activity such as production and consumption, knowing the number of vessels becomes important, and thus more specific information is required. It is necessary to know how many individual vessels occur and what kind of vessel they are, whether perceived an example of particular *type or stylistic group* that is defined according to a particular set of attributes, or as a specific kind of vessel within that *type*; for example a particular vessel *form* that occurs within a particular *type*. A count of the minimum number of different kinds of vessels comprising both stylistic groups and the assemblage as a whole provides this basic information, enabling a conceptualization of an assemblage (or stylistic

The separation of the ceramic lots into fine ware and coarse ware as a first procedural step in the analysis also represents a departure from the established practice among ceramic researchers following the type:variety method (see Willey et al. 1967 for a discussion and Ford et al. 1992 for the most recent description of type:variety procedures). The decision to first sort the material by paste attributes and surface finish (the definition of *ware* as defined by Willey et al. 1967, Smith 1971, Rice 1976 and others) and only afterward by decorative treatment and form, closely follows the classificatory procedures outlined by Smith (1971:6-8, 13-14). However, the more common practice within type:variety methodology is to treat *ware* as a higher order analytical concept that is defined only after the entities of *group*, *type* and *variety* have been established based on attributes pertaining to decorative treatment and vessel form. Some have argued (e.g. Rice 1976) that attributes related to paste composition should comprise an independent analytical unit within the type:variety scheme that cross-cuts the existing units of analysis (*variety*, *type*, *group*). Certainly, compositional attributes are only rarely used as standard criteria for differentiating different *types* and *varieties*. The selection and preparation of raw materials, however, is a fundamental aspect of ceramic production, and thus it is difficult to make any statements about the production of ceramics without a consideration of vessel composition. Whether vessels that look alike share the same mineral and chemical composition and provenance, are issues to be investigated and not assumed. To understand the meaning of the variation in surface treatment and morphology in terms of cultural behaviour – i.e. the choices made by potters during various stages of the manufacturing process and the location of production – it is necessary to investigate the relationship between stylistic attributes and vessel composition, since vessel composition often provides more direct evidence of whether vessels derive from the same place/system. In the present study, basic compositional information has been considered in the initial division of the material into fine ware and coarse ware, and specific macroscopic indicators of potential compositional similarities/differences were examined again at a later stage in the sorting and grouping process. As the term is used here, *ware* denotes a higher order classificatory concept that permits a consideration of functional and compositional information from the outset of the analysis. The concept of fine ware and coarse ware is useful for bringing to light, at an early stage in the analysis, a variety of fundamental similarities and differences between vessels – basic technological and functional characteristics – that differentiate them as finished products that are produced and consumed.

A final difference between the methods employed here and the type:variety scheme resides in the basic presentation of the data. The descriptions of the vessel groups are presented in a format similar to type:variety descriptions and include much of the same types of descriptive information, such as *surface treatment*, *frequency* and *forms and dimensions* etc. The objective in using a familiar descriptive format was to ensure that the information was presented in a 'user-friendly' manner. Nonetheless, type:variety designations such as *group*, *ware*, *complex* etc. were deliberately omitted in order to avoid implying any unfounded production-related statements concerning the interrelationships between the

Lamanai vessel groups and established type:variety entities.

Petrographic Analysis

Thin section petrography is well-established as a powerful and effective means of analysing compositional variation among ceramic vessels on the microscopic level and of differentiating and characterizing them according to the geological characteristics of the raw materials used in their production (e.g. Shepard 1956; Peacock 1970; Shotton and Hendry 1979; Bishop et al. 1982; Freestone 1991; Whitbread 1995). This data, once integrated with comparative geological information (published maps and descriptions and the results of analyses of rock and clay samples prospected from the area of interest) enables the analyst to predict or, in some cases, to identify the area or region in which different fabrics (or pastes) were produced. The potential to differentiate among ceramic products based on their raw materials and to link different fabrics to the geological landscape has led to the technique's widespread application in studies aimed at elucidating patterns of trade, exchange and group interaction. As discussed below, however, petrography has had a wider range of applications in recent studies.

An important advantage that thin section petrography has over other types of compositional analysis is that it permits investigation of various aspects of technology, such as the treatment of raw materials, paste recipes, forming techniques and firing methods (Freestone 1991; Whitbread 1989, 1995, 1996; Tomkins et al. 2004). This information not only informs about production methods, in a general sense, but also variation in technical practices, enabling an additional basis for differentiating ceramic fabrics on technological grounds. Another advantage is that it permits examination of the association of minerals and, as a visual technique, textural criteria, enabling fabrics to be subdivided or discriminated even when they are mineralogically similar. Numerous studies have illustrated how an expanded focus of petrography to include an assessment of technological characteristics and a wider range of petrographic attributes greatly enhances the potential of ceramics to inform about a range of past human behaviours related to the production, circulation and use of pottery, on various scales of analysis, especially when the technique is used in concert with other macroscopic, microscopic and/or chemical means of characterization (for example, the extensive work done on Minoan Crete: c.f. Wilson and Day 1994; Whitelaw et al 1997; Day et al. 1999; Faber et al. 2001; Tomkins et al. 2004). Fundamental to this approach is the principal that a ceramic fabric does not only reflect the geological reality of a particular area, but also human habit/choice in manufacturing procedures (Day et al. 1999; Day 2004). Viewed from this perspective, ceramic fabrics not only inform about the ways in which pottery items were made, but also can be considered as material expressions of group identity and interaction.

A comprehensive understanding of both the mineralogy of a fabric and the technical procedures involved in producing it is also of fundamental importance to the interpretation of chemical compositional data. It is well recognized that a range of natural and human factors contribute to chemical variability among archaeological specimens. These include: behavioural variables associated with paste preparation techniques, such as levigation of clays, clay mixing and the addition of temper (e.g. Kilikoglou et al. 1988; Neff et al 1988, 1989; Carpenter and Feinman 1999); natural variation within clay beds (e.g. Neff and Bove 1999; Hein et al 2004a; Hein et al. 2004b); and postdepositional alteration (e.g. Shaw et al. 2001). To assess the potential influence of these factors requires knowledge of the microscopic compositional properties and characteristics of the ceramics analysed, as well as associated raw materials. Accordingly, petrographic analysis provides an important reference base for evaluating and interpreting the potential significance and meaning of emergent patterns in chemical data sets.

The level of insight that can be gained into both the geological and behavioural significance of microscopic compositional variability is greatly enhanced when the analysis of archaeological specimens is accompanied by extensive geological prospection and the experimental firing and analysis of clay samples from the area of interest. This approach is particularly useful in microregional- and local-level studies of production, enabling a deeper understanding of observed variation among fabrics, in both human and geological terms. It also allows investigation into aspects of production, otherwise inaccessible through other means of analysis, such as the kinds of raw materials selected for use amidst other alternative possibilities, the location of sources in relation to settlements and resource procurement strategies (e.g. Day 1989).

That thin section petrography, constituting an inexpensive and yet highly effective and informative method of compositional analysis, is underutilized in Maya archaeology is perhaps an understatement. Since Anna Shepard's study of Plumbate ware (1948), in which petrography figured prominently in the analysis and interpretation of Maya pottery, only a handful of published ceramic studies have incorporated the technique as an independent means of examining ceramic variation (e.g. Jones 1986; Iceland and Goldberg 1999; Bartlett et al. 2000; Bartlett and McAnany 2000). Of particular note is the work of Lea Jones (1986), which examined continuity and change in tempering materials at different sites in the southern lowland region, from Preclassic to Terminal Classic times. Jones' study represents the first diachronic investigation of regional level patterns in the compositional characteristics of lowland pottery, considering a wide range of stylistic categories. Nonetheless, her analysis concentrated mainly on the presence/absence of specific inclusion types and the nature and potential significance of petrographic variation beyond these general distinctions was not fully investigated.

The more common usage of petrography in Maya ceramic studies has been to provide supplementary information regarding the general mineralogy or temper found in particular pottery types or groups established through macroscopic classification or chemical compositional analysis (particularly NAA) (e.g. Rands and Bishop 1980; Rice 1987b; Kepecs 1998; Bishop 1994, 2003) . The current paucity of detailed petrographic studies of Maya ceramics at least in part derives from an ongoing misconception of the nature and level of information it can potentially generate regarding the production and provenance of pottery from the lowland region. It is generally assumed that the geological homogeneity of much of the area, being underlain by limestone, precludes the differentiation of fabrics based on their mineralogical composition (e.g. Bishop 1991). However, as Bartlett et al. (2000) have recently shown for the northern Belize region, and as the present study aims to demonstrate for the Lamanai area, considerable observable compositional variation exists among the soils and rock formations comprising the lowland region, even between clays separated vertically by as little as one metre. Once this geological variation is delineated and understood, it can provide a basis for differentiating fabrics that share the same general mineralogy. These findings underscore that the potential for petrography to inform about the manufacture and provenance of lowland pottery is enhanced significantly when the analysis of archaeological specimens is accompanied by a detailed and systematic study of raw material resources.

SAMPLE PREPARATION

Petrographic analysis was conducted on 646 ceramic specimens and 43 geological specimens (35 fired clay samples and 8 rock samples) prospected from the area surrounding the site (Chapter 5). Cut samples of the rock specimens were sent to a commercial laboratory, Spectrum Petrographics, for thin sectioning, whereas the ceramic and fired clay thin sections were prepared by the author at the University of Sheffield[5]. Prior to thin sectioning, samples 0.5 – 1.0 cm in thickness were cut from each of the ceramic specimens and the clay briquettes (preparation method is described in Appendix IV) using a circular saw and each sample was impregnated with Epothin epoxy resin. The ceramic samples were impregnated on a hotplate by immersing heated samples in a warmed solution of the epoxy resin and drizzling additional epoxy resin onto the cross-sectional surface to be mounted on the slide. The clay samples were impregnated in a vacuum oven in epoxy resin disks to ensure the integrity of the samples throughout the preparation process. Prior to being mounted on glass slides with Norland 38 UV optical adhesive, the impregnated samples were ground on a diamond grit lap wheel and then polished on a glass plate using a 600μm grit carborundum slurry, thoroughly washed and dried. The mounted samples were then lapped to a 20-30μm thickness on a Hillquist thin section making machine, polished with a carborundum slurry, washed, dried and cover-slipped.

ANALYSIS AND DESCRIPTION

The petrographic analysis of the archaeological and geological specimens was conducted on a Leica Laborlux polarizing microscope. Each sample was examined at various magnifications (x25 – x400) in plane polarized (PPL) and under cross polars (XP). Whitbread's (1989, 1995:365-396) system for the characterization and description of ceramic fabrics (pastes) was used to investigate compositional variability among the archaeological specimens. Unlike approaches to petrographic analysis that estimate proportions of inclusion compositions (e.g. point counting), thereby focusing on the mineralogy of ceramic fabrics, Whitbread's system provides a standardized framework for the classification and characterization of fabrics according to a range of physical properties and features, with the underlying aim of identifying technological processes as well as provenances. The system combines elements of standard descriptive methods used in sedimentary petrography and soil micromorphology to achieve a comprehensive account of the nature and variability of ceramic fabrics in regard to the clay component, added constituents and manufacturing procedures contributing to the properties and appearance of fabrics under the microscope. A major benefit of this approach is that it enables the analyst to investigate the similarities and differences among fabrics with regard to both the nature of the raw materials used and human behaviour.

Following Whitbread's method, the ceramic fabrics were first grouped and then characterized according to a standardized descriptive framework (see Whitbread 1986, 1989, 1995: 379-387). The fabric groups were established based on observed similarities and differences in their textural and physical properties, as well as the specific nature, frequency and proportion of the aplastic inclusions they contained. Since the grouping was done without reference to the stylistic and NAA data, the analysis constitutes an independent investigation of variation within the Lamanai assemblage. The descriptive system used to characterize the fabric groups focuses on their distinguishing characteristics in terms of the nature of the raw materials used in their production, properties and features reflective of manufacturing procedures and variability in these respects. This information provides a means of differentiating fabric groups according to the provenance of their raw materials, paste recipes and firing. The results of the petrographic study were quantified by counting group members and calculating fabric group frequencies for the total number of specimens examined.

The geological specimens collected from the Lamanai area provided an important comparative base for the interpretation of the ceramic petrographic data, and especially for identifying fabric groups consistent with raw material resources from the local area. The rock

[5] Due to the breakdown of the Hillquist thin section making machine at the University of Sheffield, 150 cut and impregnated samples were sent to the Fitch Laboratory in Athens, Greece for preparation.

samples were characterized according to Folk's (1974) classification scheme for limestones and served as a means of identifying and differentiating carbonate tempers used in the production of the ceramic fabrics. The fired clay samples were characterized according to similar criteria as the ceramic fabrics, with a focus on identifying distinguishing physical and compositional properties and features. This comparative baseline was used to identify and differentiate fabric groups connected to clay/soil types occurring in close proximity to the site.

Neutron Activation Analysis

Instrumental neutron activation analysis (INAA) is the dominant and standard technique used in Maya archaeology to investigate ceramic compositional variability. Focusing on chemical variation among ceramics, this method is used to detect similarities and differences among pottery related to source (clays used or paste recipes), along multiple dimensions of major, minor and trace elemental variation (Bishop et al. 1990). The principals and procedures of neutron activation and gamma ray measurement have been described in detail in numerous publications (e.g. Bishop 1980; Bishop et al. 1982; Glascock 1992; Neff 2000; Glascock et al. 2004). Once the elemental concentrations have been determined for the specimens under study, statistical methods are used to search for structure in the data, form groups of samples with like compositions, and to define and evaluate these groups.

The main advantages of INAA in pottery studies are: (a) high accuracy, precision and sensitivity for many elements; (b) the capability to measure more than 30 elements simultaneously; (c) the small sample size required (less than 200mg); and (d) the demonstrated inter-comparability of results obtained at different laboratories (Bishop et al. 1982; Neff 2000; Glascock et al. 2004,). The technique is also credited for its potential to discriminate among pottery produced at multiple centres within areas characterized by widespread geological homogeneity, for which mineralogical distinctions on the microscopic level might not be discernable (Bishop et al. 1982; Bishop 1991). It has also proven effective in investigations of compositional variability among fine-pasted fabrics that contain comparatively few mineralogical inclusions and course-textured pottery lacking diagnostic mineralogical inclusions, which are, consequently, difficult to differentiate through petrographic analysis (Tite 1999:196). Within the context of Maya ceramic studies, the use of often sophisticated methods of statistical analysis to group pottery chemically and large sample sizes have contributed to the successful application of INAA in provenance studies, through the establishment of chemical reference groups well defined in terms of the correlations of element concentrations to one another that make them unique (Bishop 2003). Likewise, the potential for chemical compositional studies to inform about a range of cultural-historical issues related to the production, distribution and consumption of Maya pottery has advanced significantly, through approaches to ceramic analysis and interpretation that integrate a range of typological, stylistic, art historical, technological, contextual, geological and compositional (including general microscopic) information (e.g. Reents-Budet et al. 2000; Bishop 2003; Reents-Budet and Bishop 2003; Neff 2003).

The traditional application of INAA in Maya pottery studies has been in provenance determination (Beaudry 1991; Beaudry-Corbett 2003), through comparison of the compositional signatures of pottery with unknown provenances with those of reference groups for which provenances are known or presumed, based on archaeological criteria (particularly criteria of abundance) or through chemical similarity to contemporary clay deposits and experimentally replicated pastes (e.g. Neff 1989a, 1989b, 2001; Neff and Bishop 1988). The underlying premise of this approach, commonly referred to as the 'Provenance Postulate'[6], is that if the parent rock contribution and weathering histories of two clay sources are sufficiently different, the chemical difference between the two sources will exceed the variation within each source, enabling the sourcing of pottery to one source or the other through chemical compositional analysis (Glascock et al. 2004). Although it is well recognized that the chemical profile of a pottery clay (which in and of itself may vary to some degree) is modified to a certain extent through paste preparation procedures and postdepositional processes, it has been argued, regarding ceramics produced in Mesoamerica, that is still possible to discriminate pottery deriving from different clay sources based on their trace element constituents (elements present at concentrations below 1000 ppm) (e.g. Bishop et al. 1982; Glascock et al. 2004).

Observations made by researchers working among contemporary Maya potters (Arnold 1971; Howry 1978) have provided some rationale for this assumption. For example, Reents-Budet et al. (2000) argue that since the paste recipes (the use specific clays and tempers as well as proportionately consistent mixtures thereof) used by contemporary potters are often particular to individual workshops or potters, the products of different workshops or potters should be chemically distinguishable. Applying this reasoning to archaeological situations, they suggest that "pottery that shares a distinct paste-chemistry compositional profile should represent the output of a group of potters working closely together within one community and, possible, even within one workshop" (Reents-Budet et al 2000:101). Although the reasonableness of this assumption often has been taken for granted in Maya ceramic studies and treated as a 'given' in the interpretation of chemical compositional groups, its validity remains to be demonstrated through

[6] Although this proposition was originally termed 'the Provenience Postulate' by Weigand et al. (1977:24), Neff (2000:107) has argued recently that the term 'provenance' be used instead of 'provenience' "to avoid confusion arising from the fact that the term 'provenience' is also used as a synonym for archaeological context". Glascock, Neff and Vaughn (2004) also use 'Provenance Postulate' in their recent review of INAA of archaeological ceramics.

the analysis and comparison of pottery known to be produced in different, archaeologically documented workshops. To date, the absence in the lowland region of definitive production localities in the archaeological record precludes this type of study.

INAA was used in the current study for two main reasons. Firstly, it is the standard technique used in Maya archaeology to characterize and investigate ceramic compositional variability. Accordingly, the application of the technique in the current study integrates both the methodology and the results, which are based on the combined information derived from the analysis of ceramics by multiple techniques, into the wider framework of literature regarding the production and scientific analysis of Maya pottery. The second aim in using INAA was to enable an investigation of compositional variation within the assemblage on the macroscopic, microscopic and chemical levels. These different levels of analysis provide complementary information regarding the nature and significance of variability in the compositional and physical properties of the pottery assemblage, enabling a comprehensive understanding of the relationship between variation in the visual attributes of ceramic products and their technology and provenance. In addition, since the geology of northern Belize has been conventionally regarded as lacking the level of diversity required to discriminate fabrics produced by different potters on the local and regional levels, on petrographic grounds (e.g. Bishop 1991), it was felt that INAA might provide a means of making such distinctions on the chemical level.

SAMPLE PREPARATION AND ANALYSIS

A total of 215 ceramic samples and 32 fired clay samples were analysed by INAA. The selection of archaeological samples was based on a consideration of stylistic and petrographic data, with the general aim of capturing the range of stylistic and petrological variability within the assemblage. Multiple vessels were selected for analysis from among each of the stylistic and petrographic groups, whenever possible. Nonetheless, the sample set also includes many vessels that are stylistically and/or petrographically unique within the Lamanai assemblage. All but three of the clay sample were included in the analysis. The standard that was used to determine elemental concentrations for the archaeological and clay specimens was SOIL-7, supplied by the International Atomic Energy Agency.

To prepare samples of the pottery and clays for irradiation, a small fragment (circa. $1cm^2$) was removed from each sherd and briquette. For each sample, all surfaces of the fragment were first cleaned with a tungsten-carbide burring tool to remove all traces of slip, soil or other contaminants and then the fragment was crushed into a fine powder using an agate mortar and pestle. The powders were put into clean glass vials, dried at 100°C for 24 hours, and then sealed and stored in a desiccator. Approximately 130g of each powdered sample was transferred into clean polyethylene vials, which were capped and heat-sealed. All weights were recorded to the nearest 0.01 mg. The SOIL-7 standards were prepared similarly.

The samples were irradiated in batches of 10, including two SOIL-7 standards, in the reactor at Demokritos, for 45 minutes at a thermal neutron flux of approximately $3x10^{13}$n cm $^{-2}$ s^{-1}. Each sample was measured eight days after irradiation for determination of the short-lived radionuclides (Sm, Lu, U, Yb, As, Sb, Ca, Na, La), and then again after 20 days for determination of the long-lived radionuclides (Ce, Th, Cr, Hf, Cs, Tb, Sc, Rb, Fe, Ta, Co and Eu), using a Ge □-detector covering an energy range of 80-1600 keV. Of these elements, As Sb, U, Tb and Ta were not included in the cluster and principal components analyses of the data set (see below) due to their naturally high heterogeneity and poor counting (c.f. Hein et al. 2002). For information purposes, the values and summary statistics (mean, standard deviation and % standard deviation) for all 21 elements are given in the tables and Appendices (V and VI) accompanying the discussion of the results of the statistical analysis of the clay and ceramic data sets (Chapters 5 and 9).

STATISTICAL ANALYSIS OF THE DATA

The elemental data set was evaluated using standard statistical procedures, including cluster analysis and principal components analysis (PCA), on log-transformed data (log base 10). The application of these techniques to pottery studies is discussed in detail in numerous publications (e.g. Bishop 1980; Glascock 1992; Neff 2000; Glascock et al. 2004). A standard computer program obtained from Brookhaven National Laboratory (described in Glascock 1992) was employed for the cluster analysis of the ceramic specimens, whilst the cluster analysis of the clays, and all analyses by PCA, were performed using the statistical software package SPSS. Cluster analysis was performed separately on the ceramic and clay specimens, in order to partition the respective data sets, remove outliers and to characterize the emergent chemical groups. Both data sets were evaluated further using PCA to examine the relationships between the specimen groups established through cluster analysis and to identify the elements most responsible for the differences between the groups. PCA was subsequently performed on the complete data set (i.e. the clay and ceramic groups) to examine the relationships between the ceramic and clay groups from the same perspective.

SEM Analysis

Although seldom used in Maya pottery studies, scanning electron microscopy has proved an effective means of investigating manufacturing procedures associated with the surface/decorative treatments and firing methods, and has been widely employed by analysts working with Old World ceramics over the past three decades (e.g. Noll 1978; Noll et al. 1975; Maniatis and Tite 1981; Aloupi and Maniatis 1990; Kilikoglou 1994; Faber et al. 2001; Shaw et al. 2001; Buxeda et al. 2003b; Tomkins et al.

2004). The temperature at which ceramics were fired in antiquity can be estimated through examination of the morphology and internal microstructure of the ceramic body. This evaluation is based on the degree and kind of vitrification observed and comparison with microstructures for which firing temperatures are known. Such a standard has been developed by Maniatis and Tite (1981). The basic principles behind this methodology are summarized by Tite et al. (1982:109):

> The continuity or extent of the glassy phase [in a fired ceramic body] increases with increased firing temperature and the firing temperatures employed in the manufacture of ceramics can therefore be estimated by comparing the vitrification microstructure in their as received state and after refiring in the laboratory at known temperatures for a standard period.

Maniatis' and Tite's standard, which is based on extensive examination of archaeological specimens and fired clays, provides a means of estimating maximum firing temperature, within 50-100 °C, for calcareous (> 5% CaO content) and non-calcareous bodies subjected to reducing and oxidizing atmospheres (Maniatis and Tite 1981; Tite and Maniatis 1975; Tite 1995). In combination with macroscopic observations regarding paste colour and firing horizons, SEM analysis of body microstructures also provides information on firing regime.

Manufacturing procedures associated with decorative and surface treatments can be investigated through examination of the micromorphology of the surface finish and semi-quantitative (bulk and point) chemical analysis of the body, slip and paint layers (e.g. Noll et al. 1975, Noll 1978; Aloupi and Maniatis 1990; Faber et al. 2001). The micromorphology of the surface finish provides information regarding the presence/absence of slips and their relative thickness; the degree of vitrification of slips and paints, their particle nature (course and fine) and their degree of adherence to the ceramic body. Semi-quantitative chemical analysis of slips, paints and the ceramic body to which they were applied enables a general investigation of the nature of the materials used for different paints and slips and the type of clay used to make slips vs. the ceramic body. Taken together, this information enables a characterization of the way in which different surface finishes were achieved and also, a comprehensive understanding of the visual and tactile similarities and differences observed within and among pottery groups at the macroscopic level.

SAMPLE PREPARATION AND ANALYSIS

A total of 32 ceramic specimens were included in the SEM analysis. Multiple specimens were selected from each of the main stylistic groups, taking into consideration the different fabrics represented within each group, as well as variation in physical attributes of surface treatment such as colour. The main objective was to examine similarities and differences in decorative technology and firing within and among the main stylistic and petrographic categories. For each specimen, a freshly fractured cross section encompassing the body and the interior and exterior surfaces of the original vessel was mounted, at approximately a 45° angle, on a stub, using a carbon-based adhesive. The mounted samples were illustrated, noting the location of painted decoration, when present, and then coated with carbon in an automated machine.

Each sample was examined with a Philips 515 SEM, outfitted with the energy dispersive system EDAX 9900. Photomicrographs were taken of the body and finished surface of each specimen, as well as any distinctive feature observed in individual specimens. All photomicrographs characterizing the micromorphology and/or microstructure of the body and surface were taken at the same magnification (x2000 for the surface and x500 for the body). Multiple semi-quantitative chemical compositional readings were taken for the finished surface and body of each specimen. Whenever possible, both bulk and point readings were obtained. In some cases, however, it was impossible to obtain accurate point readings for the clay component of the body due to the high inclusion content of the fabric and the nature of the fracture. Representative compositional spectra based on point readings were obtained for all paints, slips and, whenever possible, bodies. The photomicrographs and spectra obtained for each specimen were compared in detail to identify similarities and differences among the bodies and finished surfaces. Of central interest, was examining the variation in decorative technology and firing methods within and among the different stylistic and fabric groups represented in the sample set.

4.4. SUMMARY

This chapter has presented a conceptual and analytical approach for the analysis and interpretation of community-level patterns of ceramic production and consumption that is grounded within an anthropological interpretive framework which aims to explore and define the wider setting of ceramic economic activities. It has argued that technical practices are expressions of group identity that can be identified and studied for their cultural and ideological content. The identification of technical practices begins with examination and comparison of the choices made at different stages of the manufacturing process in the creation of different pottery vessels. This task is accomplished through an integrated approach to ceramic analysis that combines macroscopic methods of analysis with scientific techniques that facilitate a detailed examination of different aspects of vessel technology. By situating emergent variation and patterns in the physical evidence within a wider framework of information regarding the use and deposition of pottery , as well as the cultural and environmental setting in which it was made, the factors contributing to shaping patterns in the physical evidence can be investigated from a more informed perspective. The next chapter focuses on documenting and describing the environmental setting of pottery production as it exists today. This information will form an important basis for the interpretation of the results of the petrographic analysis, especially for the determination of provenance.

CHAPTER 5

GEOLOGICAL AND ENVIRONMENTAL SETTING OF THE TERMINAL CLASSIC TO EARLY POSTCLASSIC CERAMIC ASSEMBLAGE

...ceramics are not the simple products of 'culture' in an analytically isolated sense but have significant interrelationships with the environment (Arnold 1985: 231)

Central to the reconstruction of ceramic technologies is the characterization of ceramic fabrics, which not only reflect a geological reality, but also choices made in the way of raw material selection and preparation, forming, finishing and firing. To document the technological aspects of ceramic fabrics and to attribute them to geological sources based on their petrographic characteristics requires a detailed knowledge of the geology of the local area and surrounding region, as well as other aspects of the environment. This chapter presents the environmental context of the Terminal Classic to Early Postclassic assemblage that is the focus of this study. It looks at regional climate and weather patterns and the geology and topography of the lowland region. It presents the results of the geological survey conducted of raw material resources available for pottery manufacture in the surrounding local area and discusses their physical, chemical and geological characteristics, as relates to their potential identification in ceramic bodies and their suitability for pottery production.

5.1. ENVIRONMENTAL AND GEOLOGICAL SETTING

Belize is situated along the north east coast of Central America and together with Guatemala and the Mexican States of Campeche, Yucatan and Quintana Roo comprises the Yucatan Peninsula (Darch 1981) (Map 5.1). It is bounded on the east by the Caribbean Sea, bordered on the west and south by Guatemala and on the north by the Mexican State of Quintana Roo. In the north, the Rio Hondo, which lets into Chetumal Bay, forms a natural boundary between Belize and Mexico, while the Sarstoon River forms the natural border with Guatemala in the south.

Regional Climate

The climate of Belize is described as sub-tropical (Wright et al. 1959: 15). Mean relative humidity is 80-88% and the mean annual rainfall ranges between 1524mm in northern areas of the country and 4064mm the extreme south (Walker 1973, King et al. 1992). Drier conditions prevail in northern Belize and daily temperatures and humidity generally decrease in areas with higher altitudes such as the Maya Mountains (Robinson 1983:23-26).

Throughout the country, a marked climatic feature is the dry season, which normally occurs for a three month period from February to April. In northern Belize, which is the area most severely affected by the dry season, the deficiency in rainfall often produces drought-like conditions (King et al. 1992:20-24).

The dry season is followed by a wet season, from May to November, during which 70% of the annual precipitation is received. Instability in weather patterns in August to November give rise to tropical storms and violent hurricanes, which are extensions of the Caribbean hurricanes moving westward from their source area. These storms and hurricanes can affect Belize between September and November and have been known to cause severe damage in coastal areas and on the Cays, as well as extensive flooding in inland regions (Robinson 1983:23-26).

Seasonal and regional variability in temperature is much less marked than for rainfall. November to January are the coolest months, with a mean daily temperature of 23.9°C (at Belize City). During the hottest months (May to September), the mean daily temperature is slightly higher at 27.2°C (Robinson 1983:27). Comparatively higher temperatures are reached in the northeast of Belize close to the Mexican border and on the Cays. At Santa Cruz, for example, the mean maximum temperature during the hottest months is 32.9°C and the highest recorded temperature is 37.5°C.

Weather and climate can influence ceramic production patterns in a number of different ways. Rainy weather can affect the accessibility of raw material resources, cause malformation, cracking and breakage during the drying and firing stages of manufacture, and increase the time required to make ceramic products. For these reasons, dry climates interrupted by a substantial rainy season with heavy rains and high relative humidity have a regulatory effect on pottery making by promoting a seasonal pattern of production that is limited to drier times of the year (Arnold 1985:77-90). In addition, these conditions often prompt the emergence of technological innovations and strategies to mitigate the ill effects of weather (Arnold 1985:97). In his comparative study of contemporary pottery making communities, Arnold notes that in Central America production activities tend to be restricted to the dry season, most often to avoid the adverse effects of rain (Arnold 1985:71-76, Table 3). Assuming that regional climate and weather patterns (in northern Belize) have not changed significantly since the Terminal Classic period, it can be expected that these factors would have had a least some influence on production patterns in the past.

Map 5.1: Map of Central America.

The Regional Landscape and Geology

The landscape of Belize may be described simply as a low-lying shelf that is embayed on the eastern side and that has been uplifted on the southern side to form the mountain mass known as the Maya Mountains (Wright et al. 1959:22-23). Situated off the coast of the mainland in the embayed area are isolated carbonate platforms and shoals that are locally called 'Cays'. These Cays, which comprise Holocene to Recent sediments overlaying karstic Pleistocene limestones, are part of the large reef system which includes the Belize Barrier Reef, the fringing reefs off the eastern coast of Quintana Roo, the island of Cozumel and Arrowsmith Bank (a drowned platform) (Gischler and Lomando 1999; Purdy et al. 1975; Pusey 1975). The mainland area of Belize can be divided into two techno-morphological regions: the low relief northern region which is composed primarily of flat-lying Tertiary and Quaternary limestones, the principal outcrop of the Yucatan Peninsula; and the mountainous southern region where out crops of Paleozoic shales, gneisses, granites, porphyries and Cretaceous limestones form the Maya Mountains (Map 5.2).

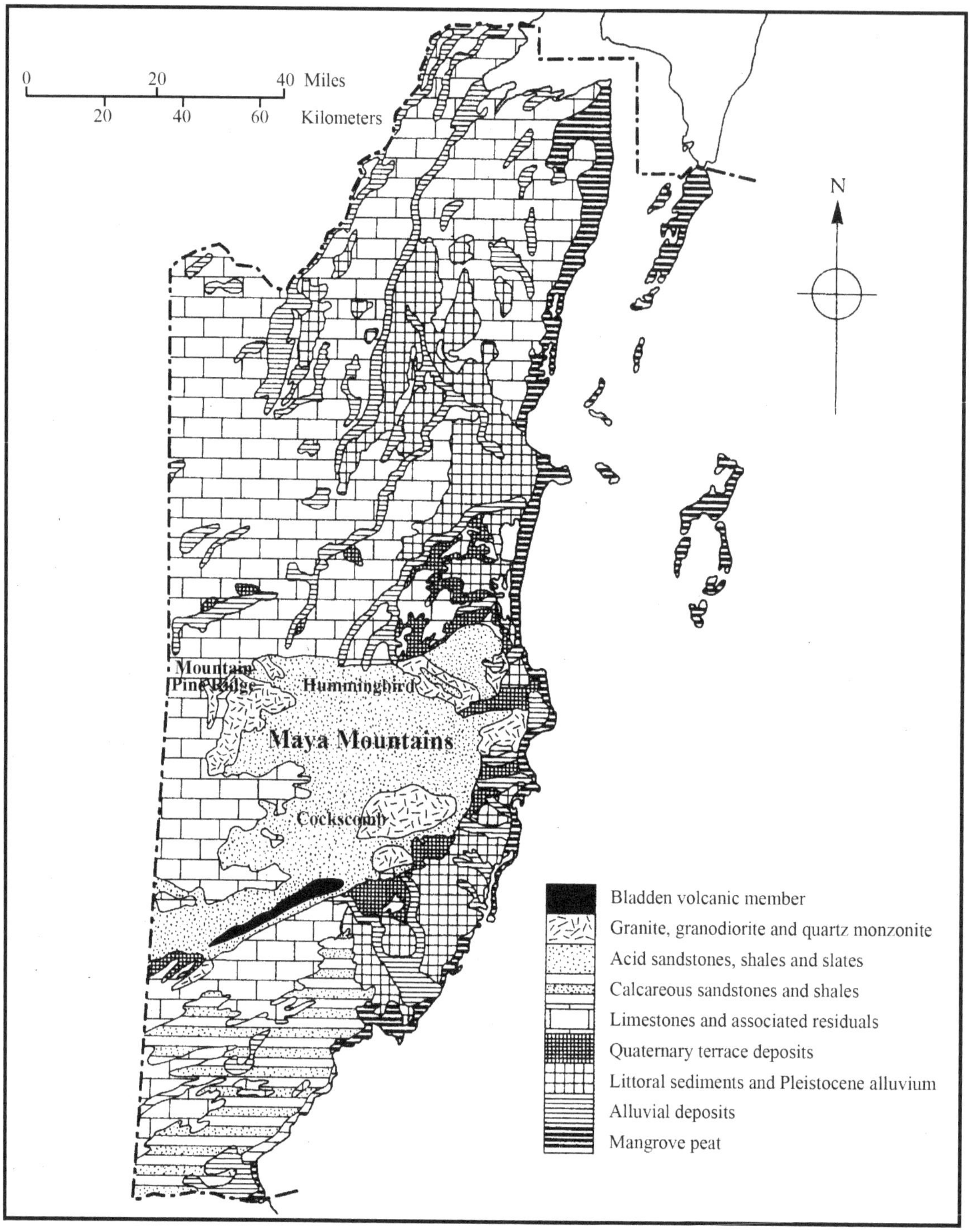

Map 5.2: Geological map of Belize (after Furley and Crosbie 1974).

THE MAYA MOUNTAINS

The Maya mountains are described as 'an upthrown fault block of Paleozoic metasediments and volcanic rocks intruded by granites' (Hall and Bateson 1972:950) The faulting that produced the uplifted block appears to have occurred at the close of the Paleozoic and produced an upland plateau with a steep approach on the northern and eastern sides and a gentle fall away on the south and west. During the Cretaceous, the plateau was partly submerged under sea water at its western end and thick deposits of limestones and dolomites were laid down (Wright et al. 1959:23; Bateson 1972:957). Since the Cretaceous, part of the overlying limestone has been eroded away and streams have cut deeply into the Paleozoic rocks exposing the underlying granite. The eroded sediments have been carried away by a network of streams and rivers and deposited as sandy stream terraces in the internal valleys or as gravelly fan deposits on the coastal plain (Wright et al.1959:24).

The main mass of the Maya Mountains is made up of sedimentary and metasedimentary rocks, primarily sandstones, mudstones, siltstones and shales. Dixon (1956) subdivided this group of rocks into an upper (dominantly arenaceous) and lower (dominantly argillaceous) series exhibiting different folding styles and separated by a regional unconformity. More recent work in the area (Kesler et al. 1971; Bateson and Hall 1971; Bateson 1972; Hall and Bateson 1972), however, has shown that there is little evidence to support Dixon's subdivision. Consequently, the stratigraphy and associated nomenclature has been revised and the two series have been combined to form the Santa Rosa Group, which is regarded as the time-equivalent to the sedimentary sequence of adjacent areas in Guatemala that bears the same name. An integral part of the Santa Rosa Group is the Bladen Volcanic Member, which is a belt of volcanic rocks that crop out on the southern margin of the Maya Mountains. These rocks comprise lavas, typically rhyolites and porphyries, which are surrounded by pyroclastic rocks and volcanic sediments that grade into and interbed with the normal sediments of the Santa Rosa Group (Hall and Bateson 1972:950-953).

In the Paleozoic, the older mass of sedimentary rock forming the Santa Rosa Group was intruded by igneous rocks in three major areas. These batholiths generally consist of granitic rocks, although detailed petrographic and chemical investigations (Kesler et al. 1974) have found their individual compositions to be distinct. The Mountain Pine Ridge batholith primarily consists of 'true granite' where as the Cockscomb and Hummingbird batholiths consist of granodiorite that is rich in muscovite. Also present in the Hummingbird batholith, is muscovite quartz monzonite and rocks that have been catacastically deformed, producing gneissic textures (Kesler et al. 1974:550). This later characteristic, in particular, is distinctive of the Hummingbird batholith.

The main mass of the Maya Mountains is flanked on the north and south by subsidiary masses of Cretaceous limestones of a far lesser height. The landscape in these areas is characterized by conical hills, and outcrops of tower karst occur in several places. South of the Maya Mountains the limestone hills have been faulted in such a way that their bedding is nearly vertical. In this position normal erosion has carved them into sharp angular ranges (Wright et al. 1959:27). In contrast, the limestone hills of the north comprise a more rolling topography. Well developed karst systems occur in both areas, often forming extensive series of subterranean caverns and waterways.

The Cretaceous limestones of Belize form part of an extensive mass that is widely exposed in adjacent Guatemala (Coban and Campur Formations). The Coban Formation, which underlies the land regions west and southwest of the Maya Mountains, comprises thick-bedded karstic limestones and dolomites of Lower to Middle Cretaceous age with evaporites in the lower part. At the base of this formation is clayey microcrystalline dolomite, which has been termed 'Hill Bank Formation' by Gulf Oil Geologists (King et al. 1992:27). The Campur Formation is Late Cretaceous in age and primarily consists of fine grained limestones, dolomitic limestones and dolomites, tending to have a karst-type physiographic expression, that locally interbed with the shale, siltstone, calcareous conglomerates and breccias of the Sepur Formation (Dickerson and Weisbord 1931:483; Flores 1952:408; Vinson 1962:433). The Campur Formation flanks the Maya Mountains on the north, and from there, gently dips northward.

THE NORTHERN REGION

Extending north from the Cretaceous limestone formations flanking the Maya Mountains to the Mexican border is the low relief northern region of Belize. The landscape in this area can be described as a lowland plain comprising coastal and inland swamps, stranded reef deposits, sandy plains, lagoon complexes, and a large area of gently sloping limestone hills and plains that rises to a slight plateau in the west (Robinson 1983:27-29). The geological history that gave rise to this landscape is characterized fairly simplistically as having involved a series of sea level fluctuations and the associated deposition of sediments. Wright et al. (1959:24-26) propose a possible sequence of deposition which begins in late Pliocene to early Pleistocene times, when the entire area was submerged beneath the sea and the ancient shoreline (equivalent to the 55m contour) formed a bay immediately north of the Maya Mountains. They suggest that erosion of the sheet of Cretaceous limestone that covered much of the upland region would have filled the bay with limestone rubble and sand. The main rivers draining the upland regions were north tending at this time, letting directly into the bay and bringing down from the upland region large quantities of quartz gravel and flint, which would have been deposited fan-wise over the coastal shelf and subsequently reworked by wave activity. During Quaternary glaciation, the shoreline gradually receded to the 30m contour (15m above the modern sea level), exposing sediments contiguous with the old shoreline and promoting the emergence in the bay of two shoal areas composed of quartz sand. A subsequent change in sea level brought about the emergence of areas of bare limestone to the north of and flanking the islands of quartz sand. Wright et al. (1959:25) describe these areas of exposed limestone as two parallel 'tongues' that were "produced by faulting of the coastal shelf and associated with a slight upward movement which gives each tongue a low scarp along the eastern edge and a gentle slope to the west". In the troughs created between the tongues of uplifted limestone, typical lagoon sediments including silt and clay subsequently accumulated. The Rio Hondo and the New River, which form the principal river systems draining the northern region in modern times, follow these fault-guided valleys.

Wright et al. (1959) suggest that, prior to the final drop in sea level that brought about the present day shoreline, conditions along the northern coast were comparable to those found in the offshore cays today. Hence the deposits and sediments found in coastal areas of the northern mainland were produced through reef building processes. In the Corozal area, the final drop in sea level

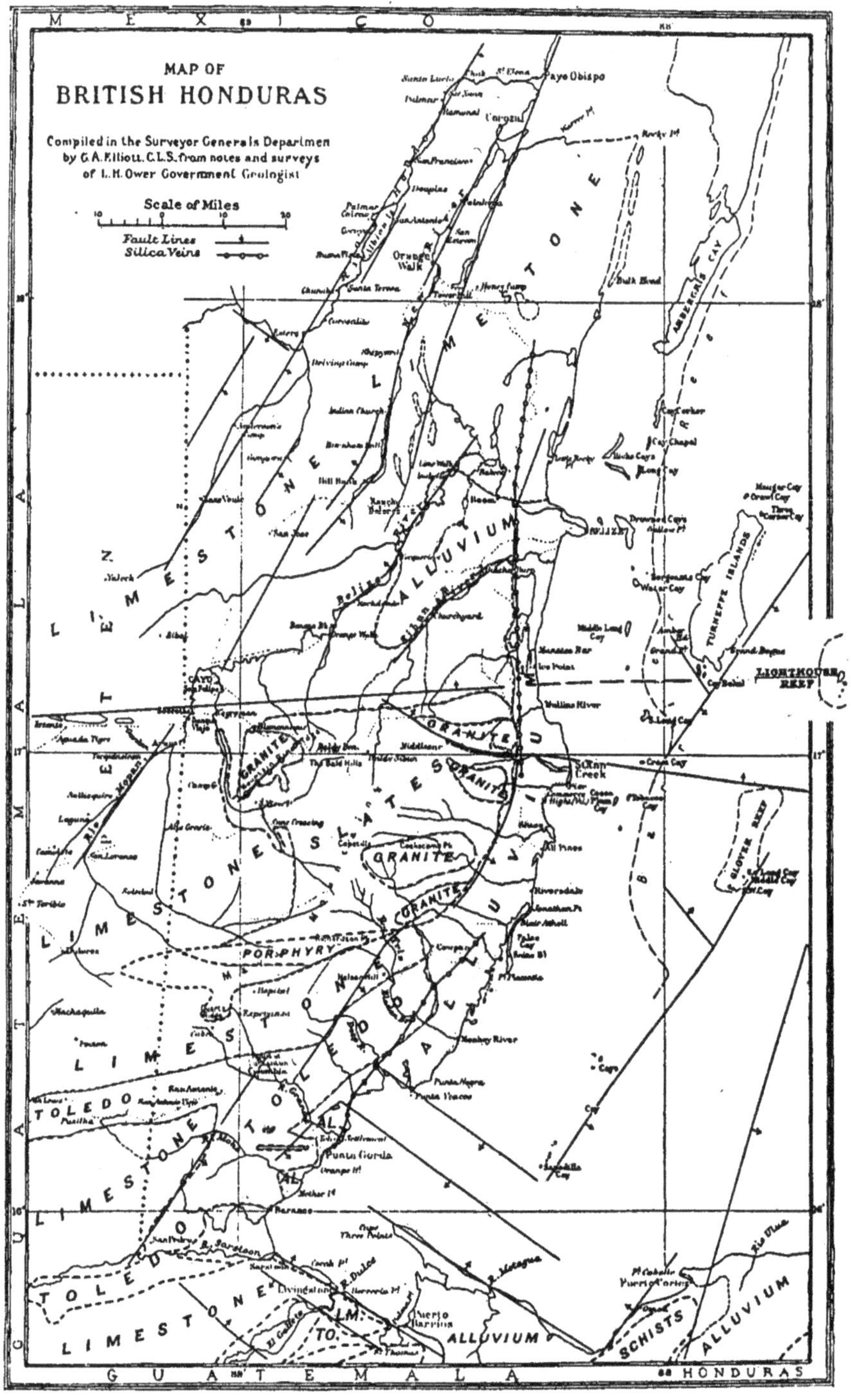

Map 5.3: Geological map of Belize by Ower (1928a: 497, FIG. I).

effectively exposed "a strip of stranded reef resting on limestone gravel and sands", which occurs today as deposits of unconsolidated chalky rubble (Wright et al. 1959:25-28).

The geological history proposed by Wright et al. for the northern mainland region represents the first attempt to provide a comprehensive and detailed description of the area in historical geological terms that accounts for the various geological formations and deposits comprising the northern landscape. Since the primary aim of their study was to assess the agricultural potential of land resources in Belize, the description of soil forming parent materials formed an integral part of their study. Prior to this, geological fieldwork in Belize had been primarily confined to surveys aimed at locating petroleum related and mineral recourses that were available in exploitable quantities (e.g. Ower 1927, 1928a; Flores 1952). As a result, the Maya Mountains, the area containing mineral bearing rocks and older limestones, was the primary focus of extensive field studies. In contrast, the northern region, having less economic potential, was deemed "of little geological interest" (Ower 1928a:496). Work done in this area was localized and confined to limited traverses, giving rise to generalized descriptions of particular sediments and formations and only a vague characterization of the geology of the region as a whole.

The more generalized treatment of the northern region in comparison to the south is vividly captured in the first geological map of Belize, presented by Ower (1927, 1928a, 1928b), which shows the entire northern region as comprised of limestone, apart from a strip of alluvial sediments situated between the Belize and Sibun Rivers (Map 5.3). The limestone of the north is also undifferentiated from the formations flanking the Maya Mountains. In his accompanying description, however, Ower (1928a:497-500) provides some further details about local geology in the north, noting the occurrence of 'chalky marls' on the New River around Orangewalk and Corozal (presumably the chalky rubble described by Wright et al.), the presence of flints in the alluvium around the Northern River and in the northeast corner of the region, and also that quartzose veins traverse the limestone in the area adjacent to the northeast coast. Nonetheless, these details are treated as incidental to his basic geological characterization as the extent of the various formations and deposits was not studied in detail.

Prior to the publication of Wright et al.'s (1959) geological assessment of Belize, Giovani Flores, a geologist working for the Bahamas Exploration Company Ltd., undertook a comparative study looking specifically at the age of the limestones comprising the northern region. Flores (1952:404-409) recognized that there was no evidence to support Ower's contention that the limestones of the Maya Mountains and the northern region were part of the same formation and presented the results of a field study which showed, based primarily on paleontological evidence, that there were at least four different formations and sediments comprising the northern region that were of geological importance. Although his sampling was confined to particular localities and his coverage of the area was by no means systematic or extensive, he was able to show that the limestone flanking the Maya Mountains on the north, that adjacent to the coast north of Belize City and that surrounding the New and Northern Rivers in the northeast of the country were of different ages and character. He found that the limestone flanking the Maya Mountains was Upper Cretaceous in age and actually consisted of limestones, dolomitic limestones and dolomites, whereas the formations adjacent to the north coast comprised limestones and dolomitic limestones of Paleocene to lower Eocene age that contain black chert. The formations in the northeast of the country, however, were found to consist of a dense, commonly fragmental, marine limestone of middle Eocene age that contains large inclusions of chert. Above the marine limestone he found Mio-Pleistocene deposits resembling a powdery weathered limestone, which consisted of "a sequence of mottled bentonitic clays with gypsum veins, and poorly consolidated sands, overlain by white sandy marls" (Flores 1952:409). He also observed that the Mio-Pleistocene deposits were widespread in the northeast, although according to his map he principally encountered them between the Rio Hondo and the New River (Flores 1952:406, Fig.I).

Flores' study was the first to move beyond the homogenized view of the geology of northern Belize by demonstrating effectively that there was regional variation in the age and mineralogical composition of the formations and deposits comprising the north. Unfortunately, the contribution of this study to achieving a more comprehensive understanding of the geology of the region as a whole is limited by the absence of detailed mapping of the different formations and deposits he recognized. It is presumably for this reason that Flores' study is not listed as a source that was considered as part of the geological assessment presented a few years later by Wright et al., despite the fact that the publication would have been available when they were conducting their field work between 1952 and 1954 (1959:6-9).

The geological assessment of the northern region presented in Wright et al. largely derives from a reconnaissance survey that was conducted by the authors, partly in conjunction with C.E. Dixon and his assistant, Mr Cunha, of the Colonial Geological Survey Department (1959:8). The culmination of this work is their map of SoilForming Parent Materials (Map 5.4) in which the authors divide the area of Belize north of the Maya Mountains into three principal limestone formations of different ages (Cretaceous, Eocene and Pliocene) and three distinct sediment types (coastal sediments, calcareous clays and stranded coral limestone and mangrove peat). These general limestone and sediment types are further subdivided on the local level based on differences in mineralogical composition, thereby providing a basic pictorial summary of geological variability across the region as a whole. A major contribution of the study is the detailed mapping of the different deposits and formations in relation to one another as well as the extent of the mineralogical variation that was observed therein. While previous authors such as Ower and Flores had reported on the distinctive mineralogical or lithological features of particular deposits and formations they had encountered at specific localities, Wright et al. were the first to attempt to map the geographic distribution of such variation in detail.

The extensive survey conducted by Wright et al. also yielded new information on mineralogical variability within the principal formation and sediments types. For example, in addition to observing the localized occurrence of flints in the limestones and chalky sediments east of the Northern River (as Ower did), they distinguish similar deposits situated between the New River and the Rio Hondo based on the prevalence of silica sand and the deposits situated southeast of the New River Lagoon based on the occurrence of tuff. They also identify local differences in the composition of the Pliocene limestone formations typical of the northeast corner of the region.

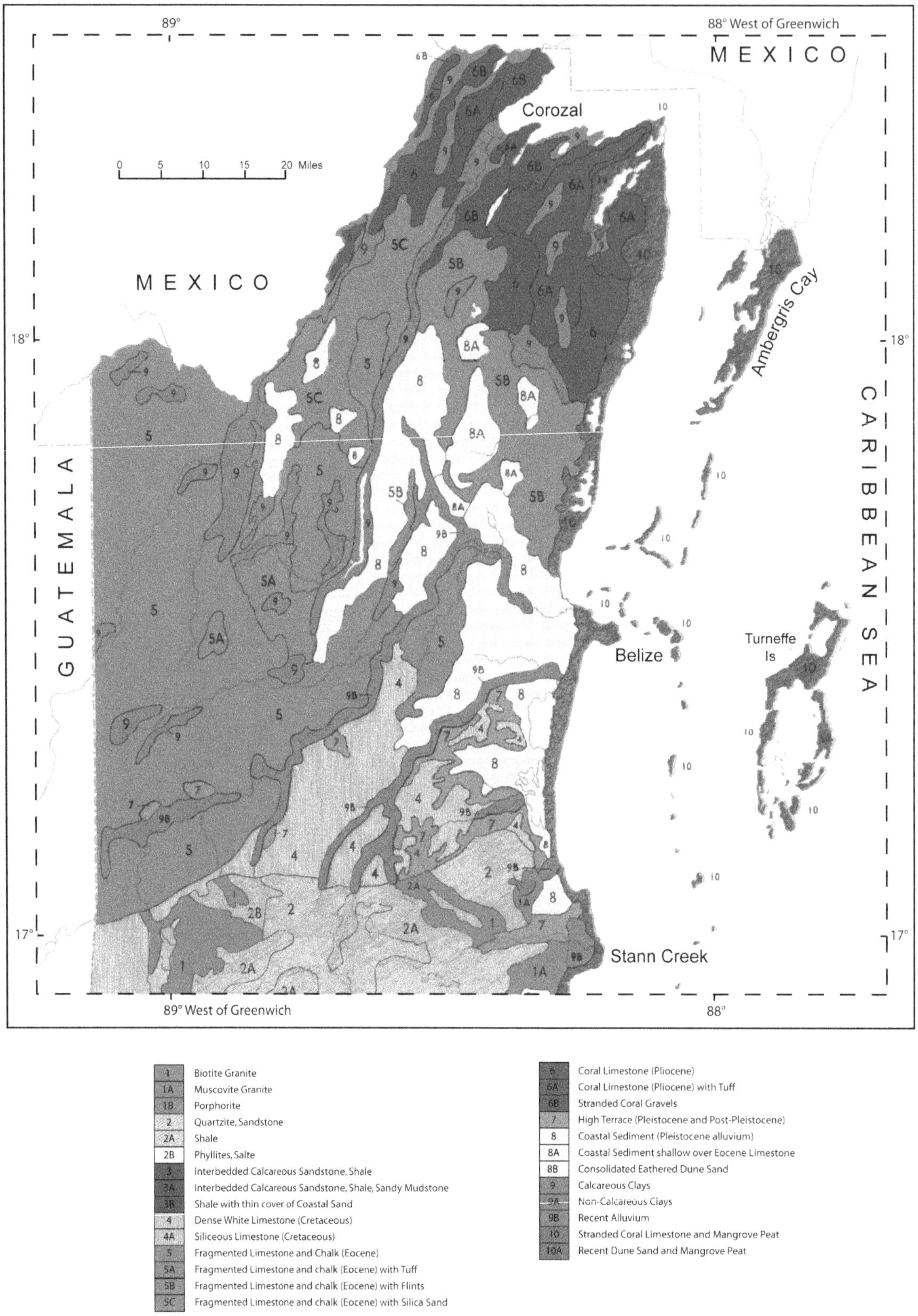

Map 5.4: Geological map of northern Belize (after Wright et al. 1959, FIG. VII) (see colour version on accompanying CD).

Since the publication of Wright et al.'s assessment of land resources in Belize, work done in the northern region has continued to focus on the characterization and description of soils and evaluating their eco-agricultural potential (e.g. Robinson and Furley 1983; King et al. 1992). The most recent study, conducted by King et al. (1992), identifies 15 different soil suites, defined according to differences in parent materials and characteristics such as colour and mineralogy, and these suites are sub-divided into a total of 56 different subsuites. Studies such as this have helped to move current understanding of the geology of northern Belize even further beyond the traditional view of the geological homogeneity of the region, by documenting the variation that exists in the character and mineralogy of soils as a direct result of differences in soil forming parent materials.

Despite this considerable advancement in knowledge in regard to the soils of the northern region, the characteristics and spatial interrelationships of the parent formations and deposits remain only ambiguously defined. Detailed consideration of the relationship between the compositional variation seen in the soils and local and regional variation in the underlying sedimentary sequence has played, perhaps necessarily, only a secondary role in soil surveys and assessments. The few detailed geological studies that have been undertaken have tended to concentrate more on delineating the characteristics and origins of specific types of deposits such as white marl or 'sascab' (Darch 1981; Darch and Furley 1983) and lime mud (Reid et al. 1992), rather than achieving a more comprehensive understanding of the sedimentary sequence, and variation therein, across the region as a whole. As a direct result, although a detailed and sophisticated characterization of variation in soils based on differences in parent materials and mineralogy has been achieved, it is one that is only vaguely understood in terms of the characteristics, distribution and spatial interrelationships of the parent materials themselves. The latter information remains piecemeal and often contradictory in the literature. This problem is well recognized and addressed by Darch (1981) in her study of the characteristics and origins of sascab, but has yet to be resolved on a more general level for the geology of the northern region as a whole.

Although the geological literature on the northern region is ambiguous and contradictory in some respects, when the information presented in various studies is taken together, what emerges is a fairly detailed account of the principal deposits and formations. The ages, characteristics and approximate geographic distribution of these deposits and formations are discussed in turn with particular attention to regional differences in their mineralogical composition and character. The geology of the region as a whole is summarized in Map 5.5.

Limestone and Overlying Deposits and Sediments = The entirety of the northern region is underlain by limestones, which decrease in age from the Cretaceous formations that occur immediately to the north of the Maya Mountains and at a few specific locations in the central part of the region, to the Pleistocene-Holocene formations of the northeast coast and off-shore Cays. The Cretaceous formations found in the central part of the region are considered to be the northern facies equivalents of the upper Coban and lower Campur Formations (the Yalbac Formation) and the upper Campur Formation (the Barton Creek Formation). These two formations were mapped together and found to occur at three principal locations north of the Maya Mountains: below an escarpment situated west of the southern end of the Rio Bravo, below the Booth's River escarpment, and along the western edge of the New River Lagoon (King et al. 1992:28). The Yalbac Formation underlies the Barton Creek Formation and comprises a sequence of anhydrite-carbonate alternatives, whereas the Barton Creek Formation consists of fine-grained limestones, dolomitic limestones and marls (King et al. 1992:27-28). Apart from the specific localities where these Cretaceous Formations occur, the central part of the northern region is underlain by Early Tertiary (Paleocene to Eocene) limestones of the Cayo and Dubloon Bank Groups (Wright et al. 1959; Flores 1952; King et al. 1992). The older Cayo group contains limestone and dolomite and this is the common rock type found in areas west of the New River drainage system. The bedrock in these areas, whether Cretaceous or Tertiary in age, is overlain by a horizon containing limestone at various stages of weathering and ferromanganiferous concretions, and often deep deposits of calcareous clays (King et al. 1992:222). The Dubloon Bank group, which contains limestone with chert, is more typical of the areas adjacent to the coast in the eastern part of the region. Subsurface, the Dubloon Bank limestones are soft and interbed with clays, whereas the outcropping limestones are hard, suggesting that the group may have had a thick carapace. Chert is ubiquitous in both the limestone and the overlying sandy clays (King et al. 1992:28, 244-247).

The younger Pleistocene-Recent limestones and dolomitic limestones that are found around the northern coast and on Ambergris Cay are dense coral limestones that are commonly fragmental, porous and cavernous (Flores 1952; King et al. 1992:29). In the vicinity of Chetumal Bay these limestones are often gypsiferous and weathered (King et al. 1992:213). The mainland limestones are overlain by shallow fine-textured calcareous deposits that are characterized by high levels of sodium and magnesium (King et al. 1992:188) or calcareous sands, which mainly consist of cryptocrystalline grains composed dominantly of micrite (Reid et al. 1992; Pusey 1975). The Pleistocene limestones of the off shore Cays are overlain by deep deposits of fossiliferous sand in most cases. In the low relief 'wet' flats of Ambergris Cay and the Cangrejo shoals, however, it is overlain by massive dolomite beds, surface and recent dolomite crusts, and unconsolidated sediments (sandy muds) of Holocene to Recent age (Mazzullo and Reid 1988; Gregg et al. 1992; Mazzullo et al. 1995). These dolomites appear to have precipitated under peritidal conditions despite the fact that they exhibit characteristics more typical of subtidal formation, and the dolomitization from sandy sediments into biowackestones involves neomorphic recrystallization. Detailed petrographic analysis has also revealed that dolomitic and non-dolomitic sediments from these areas are both texturally and mineralogically identical (Mazzullo et al. 1995:343).

Neogene-Pleistocene Deposits of Calcareous Material = The terms 'Chalky Marls' (Ower 1926, 1928a; Flores 1952), 'unconsolidated chalky rubble' (Wright et al.

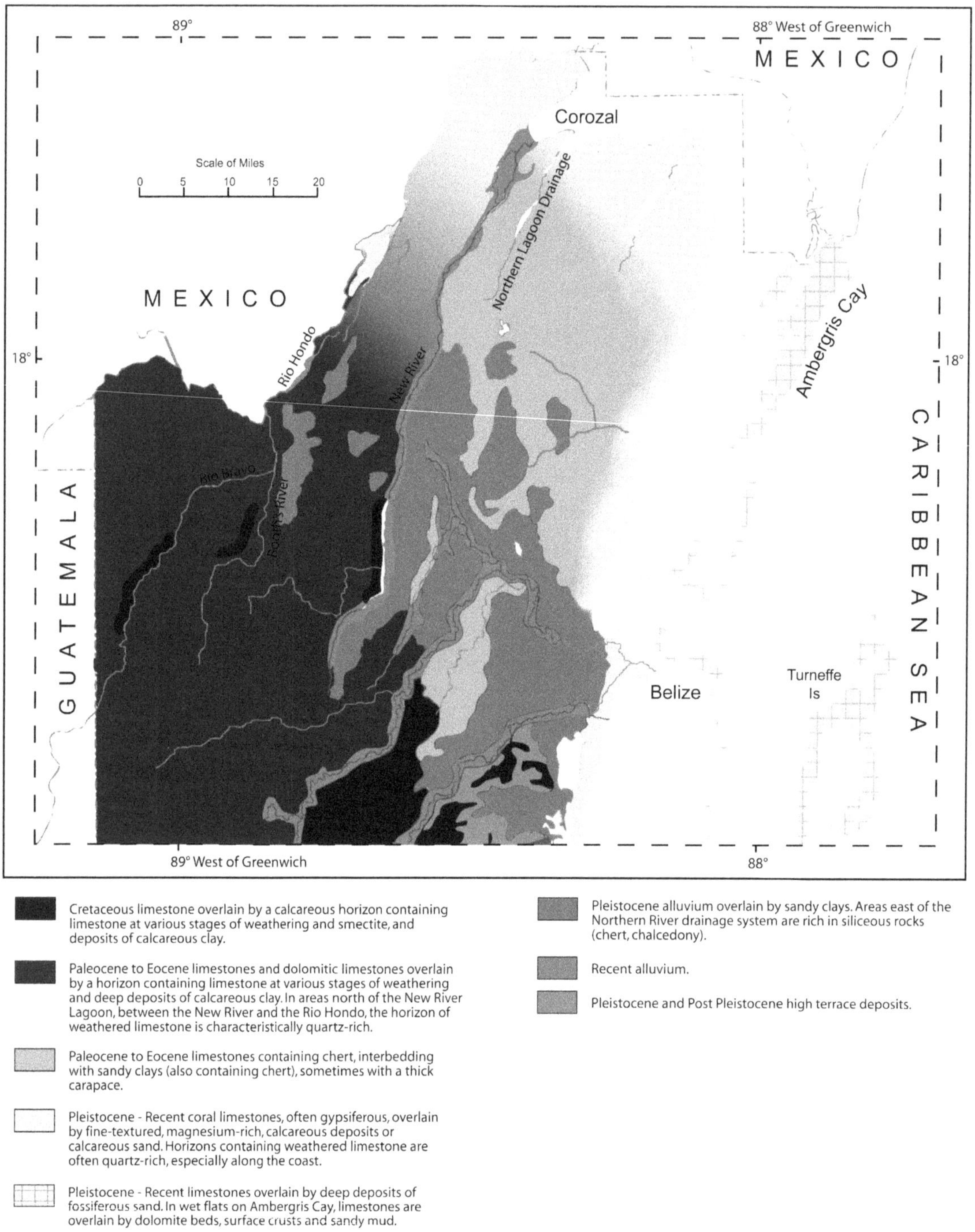

Map 5.5: Geological map of northern Belize showing principal deposits and formations (see colour version on accompanying CD).

1959) and 'sandy chalks' (Versey 1972) have been traditionally used to describe distinctive deposits of unconsolidated calcareous material of Neogene to Pleistocene age that occur in the northern region. More recently (e.g. King et al. 1992) geologists have recognized that these deposits comprise two different formations – the Tower Hill Formation and the Orange Walk Group – based on differences in lithology and composition. The Tower Hill Formation consists of 'sandy chalk' or 'sascab' (the traditional Maya term), which is described by Darch (1981:411) as "a white to cream coloured, unconsolidated deposit of silty fine sand texture, consisting primarily of calcite and quartz". In her study of the sascab found in the Orange Walk District of northern Belize, she observes that these deposits are locally heterogenous in texture and colour and contain layers comprised of fine sand, silt and coarse gypsum crystals. In areas where they outcrop the deposits are indurated, although unlike calcrete, they contain little clay. Bartlett et al. (2000) also note that samples of sascab taken from a modern quarrying site close to Water Bank, in the Orangewalk district, (north of Lamanai, on the New River) characteristically contain inclusions of mono- and polycrystalline quartz. A subsequent study by

Darch (Darch and Furley 1983) distinguishes the northern sascab, which contains calcite, gypsum and quartz, from similar deposits found in central Belize associated with Cretaceous limestone, in which smectite is the dominant mineral and quartz inclusions are comparatively rare. The Orange Walk Group differs from The Tower Hill Formation in that the former consists of sascab, interbedded with clay, and commonly has a lithified carapace (King et al. 1992:29).

Wright et al. (1959) hypothesized that these calcareous deposits comprise sediments deriving from an ancient fan deposit of limestone rubble, which was produced through erosion of the Cretaceous limestone cap that once covered the Maya Mountains. Accordingly, they also assumed that most of the northern region was underlain by these deposits. Subsequent investigations into the age, characteristics and origin of sascab, however, have suggested that there is little evidence to support this interpretation, and that it is more likely that the deposits were formed through in-situ deep weathering of the Tertiary limestones (Darch 1981; Darch and Furley 1983; Cornec 1985). Most recently, King et al. (1992:28-29) have suggested that these deposits may have a more restricted geographic distribution within the northern region than has been previously recognized. They observe that the Tower Hill Formation is typically encountered in areas north of the New River Lagoon, around the New River and between the New River and the Rio Hondo. The Orange Walk Formation also appears to be restricted to the northern part of the region, but they suggest that it occurs principally in the land areas surrounding the northern coast.

Alluvial Deposits - Scattered throughout the northern landscape are alluvial deposits comprised of siliceous sand that consists mainly of quartz. In most cases these deposits appear to constitute reworked and redeposited old alluvium that was transported into place during the late Pliocene to mid-Pleistocene (Wright et al. 1959; King et al. 1992). The parent material of the deposits is thought to originate from the quartz-rich crystalline or metasedimentary rocks of the Maya Mountains. Fragments of these rocks are sometimes found in alluvial deposits south of the Belize River, where the deposits are in closer proximity to their place of origin (King et al. 1992:200). Since these rocks are virtually absent in the deposits further north, it would appear that they were completely comminuted by the greater distances travelled from the source. The position and alignment of the deposits situated between the New River and Northern Lagoon would seem to support Wright et al.'s (1959:25) hypothesis that these deposits were brought down by north tending rivers that were draining the upland area. During Quarternary glaciation these transported sediments emerged as a large shoal, of which a perhaps substantial portion has been eroded away since. Similarly, King et al. (1992:29) suggest that the deposits bordering the Booth's River in the western part of the region are most likely remnants of an ancient strand plain associated with the Late Pliocene or very early Pleistocene shoreline. In a few areas east of the New River, however, the deposits appear to derive from the adjacent limestone, since "only about 10% of the quartz sand consists of fresh, angular grains typical of alluvium" (Wright et al. 1959:25, footnote 1). These deposits appear to have a localized occurrence along the drainage systems that dissect the larger deposits of alluvium, and hence they are thought to have formed through the *in situ* weathering of limestone within these particular environments. (King et al. 1992).

5.2. LOCAL ENVIRONMENT AND GEOLOGY

Lamanai is situated on the west bank of the New River Lagoon, the head waters of the New River (Plate VII.1, Appendix VII). The New River rises in the hilly south western part of the northern region and flows via creek (Irish and Ramgoat) and swamp into the New River Lagoon. North of the lagoon, the river takes a sinuous, and often braided, north-tending path until it lets into Chetumal Bay. The river system follows an ancient fault scarp and at Lamanai, the jagged outcrops of Cretaceous limestone that occur in some places along the lagoon shore are testaments to past tectonic activity in this area. The site and the surrounding area encompass three distinct geological and environmental zones that give rise to contrasting soil and vegetation regimes (Map 5.6). The site itself is located within an area underlain by Tertiary and Cretaceous limestone, whereas Pleistocene alluvial deposits characterize land areas situated to the north of the site and east of the lagoon. The intervening areas bordering the lagoon and river system are dominated by more recent fluvial deposits.

Land Areas West of the New River Lagoon

Soils west of the New River Lagoon underlying the site belong to the Yalbac Subsuite of the Yaxa Suite. In uncleared areas, the Yalbac soils support a high, semi-deciduous broadleaf forest, which includes species such as mahogany, sapote, and cahune (King et al. 1992). Much of the settlement at Lamanai lies within a conservation area and thus has been protected from destruction and clearance as part of lumbering and farming activities. As a result, the protected area of the site is covered by dense forest except for a few structures and plazas that have been cleared and maintained for tourist viewing (Plate VII.2). It is unclear at present how much of the settlement was forest covered during any phase of its occupation, or whether specific areas, such as the main the plaza groups within the ceremonial precinct, were ever entirely cleared of trees or other vegetation.

The Yalbac Subsuite is described as comprising 'black, dark grey and dark brown clays with blocky topsoils' (King et al. 1992:273). Below the top soil, the clay horizon often grades into lighter colours and sometimes contains limestone fragments and fine ferromanganiferous concretions. The clays overly limestone at various stages of weathering, and in the vicinity of Lamanai, the limestone is Cretaceous in age

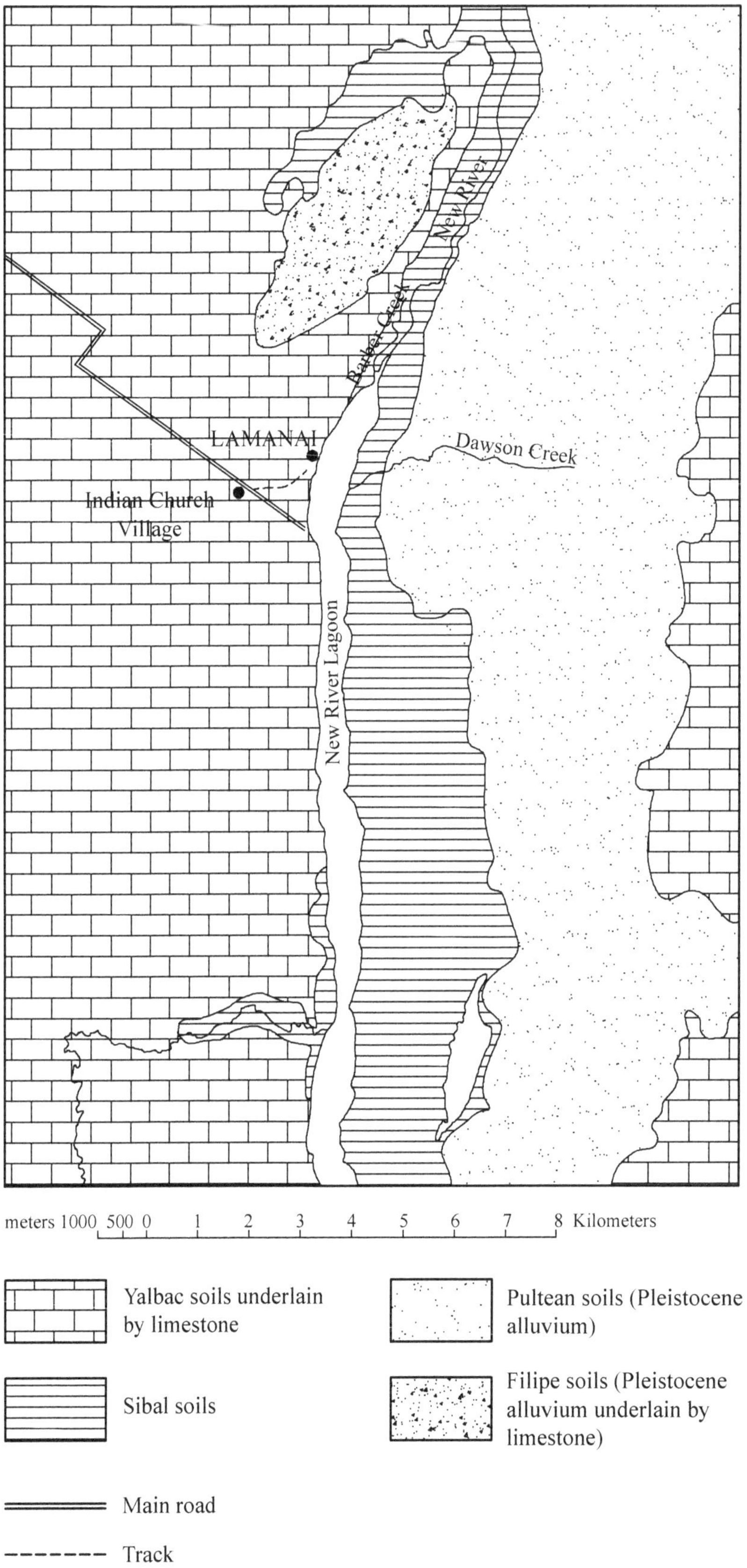

Map 5.6: Geological map of the local area.

(King et al. 1992:28, 222). Granulometric and XRD analysis of the Yalbac clays indicate that they tend to have a very high clay content, which increases with depth, and that they mainly consist of smectite and vermiculite (King et al. 1992:222-223). King et al observe that these clays show considerable shrink-swell capabilities and suggest that they form "*in situ* or in locally transported post-solution impurity residues from the limestone".

The soil profile observed in a north-south trench dug just inland from the lagoon shore, at the southern edge of the ceremonial precinct of Lamanai, generally agrees with King et al.'s description of the Yalbac soils (Figure 5.1, Plate VII.3). In the trench profile, black clayey topsoil containing occupation debris overlies olive grey clay, which grades into light brown to white clay. In a few specific areas within the trench, the white clays contain ferromanganiferous stains. Beneath the clay horizon, there is a horizon containing limestone at various stages of weathering, which consists primarily of sandy textured clays with a crumb consistency and common fragments and large nodules of dense limestone (Plate VII.4). The frequency and size of the limestone nodules appears to increase with depth.

Principal rock out crops in the area are either of dense marine limestone (Cretaceous in age) or indurated, sascab or soft limestone[1], the exposure of both being a direct result of faulting in ancient times. The outcrops of marine limestone are identical to sub surface nodules of the same rock and are cream-coloured, often with rust to purple coloured, ferromanganiferous viens, and extremely hard, exhibiting a slight concoidal fracture when broken (Plate VII.5). Outcrops occur mainly in areas adjacent to the lagoon and river system, such as the Harbour area of the site, in association with clayey soils containing limestone at various stages of weathering (Plate VII.6). The main outcrop of sascab (soft limestone) in the local area is situated further inland, approximately 1km directly southeast of the ceremonial precinct at Lamanai and 1km inland from the lagoon, on the south side of the main road leading to the modern village of Indian Church (Plate VII.7). This outcrop shares many features in common with the better documented deposits of sascab located further to the north (Tower Hill Formation), although it appears to lack the siliceous sand component that is characteristic of the northern deposits. In areas where the deposit is exposed, the limestone is white to cream coloured, and although it is fairly hard, it can be easily broken down into chunks with a pick or a machete. Subsurface, the deposits are still consolidated but comparatively soft and can be easily cut into blocks with a machete and removed intact (Plate VII.8). Once exposed to the sun and air these blocks become as hard as the surface limestone in as little as an afternoon.

It would appear that these local limestones were important resources in prehistoric times as they appear to have been the primary materials used in a range of activities related to the construction and maintenance of structures and public areas within the ceremonial precinct. For example, nodules of dense marine limestone were used as core fill in platforms and buildings, whereas the soft limestone was used to make cut facing stones (Plate VII.9). The soft limestone would have also provided a calcium-rich raw material that could be burnt to make lime for plaster and stucco.

Although not encountered *in situ* geologically, nodules of two forms of crystalline calcite also occur in core fill at the site. Presumably deriving from calcite veins in the limestone bedrock, these nodules are of comparable size to the more frequent nodules dense marine limestone and are visually distinctive both with the unaided eye and at the microscopic level. One type of crystalline calcite is colourless and comprised of very coarsely crystalline to extremely coarse crystalline spar (see Folk's [1974:163-165] crystal size distinctions based on the Wentworth divisions). In comparison, the other type is white in colour and has a sugary appearance, owing to the smaller crystal size, which is generally medium to coarsely crystalline.

Land Areas North and East of Lamanai

Land areas east and northeast of the site, could hardly provide a more vivid geological and environmental contrast. Directly across from the site, the eastern shore of the lagoon is dominated by fresh water swamp, as are areas to the north of the lagoon flanking the New River (Plate VII.10). Soils in these areas are of the Sibal Subsuite of the Tintal Suite, which are described as permanently waterlogged clays and silty clays with a 'muck surface' and a high mineral and organic content (King et al. 1992:212).

Lands immediately to the east and inland from the swampy areas are dominated by deep deposits of siliceous alluvium. The associated soils, which belong to the Boom Subsuite of the Pultean Suite give rise to a Pine Ridge savannah characterized by open sedge and grasslands with occasional clumps of palmetto and calabash trees (Plate VII.11). The characteristic profile of Boom soils comprises dark grey to grey sand-loamy sand topsoil overlying light grey, pale yellow, yellow or pale brown sand or loamy sand. The lower subsoil consists of red and white mottled sandy clay with brownish and yellowish patches (King et al. 1992:203). King et al. found that clay content within the profile often increases abruptly over short depth intervals (30cm). The soils tend

[1] Petrographic analysis of the marine limestone classifies it as a microcrystalline biogenic rock – a biomicrite, which conforms to type IIb of Folk's (1974:161-162) classificatory scheme – containing veins of medium crystalline to very coarsely crystalline calcite spar. The soft limestone classifies as a microcrystalline allochemical rock – an intramicrite conforming to Folk's type IIi –containing very coarsely crystalline calcite spar and mosaics of medium to coarsely crystalline spar. Staining with Alizarin Red indicated that both rocks are comprised entirely of calcite.

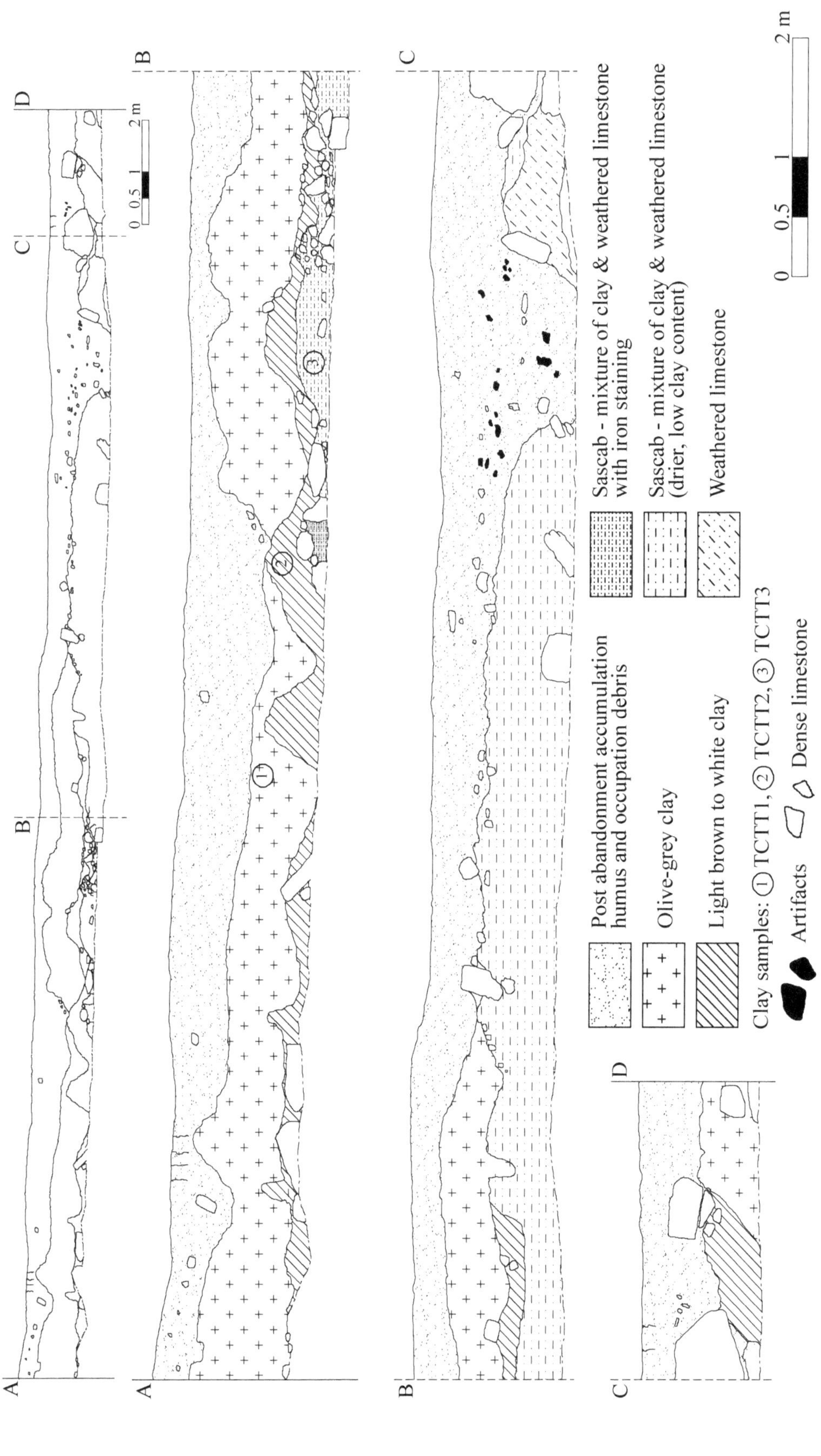

Figure 5.1: West section of North-South trench showing subsurface soil horizons and clay sampling locations.

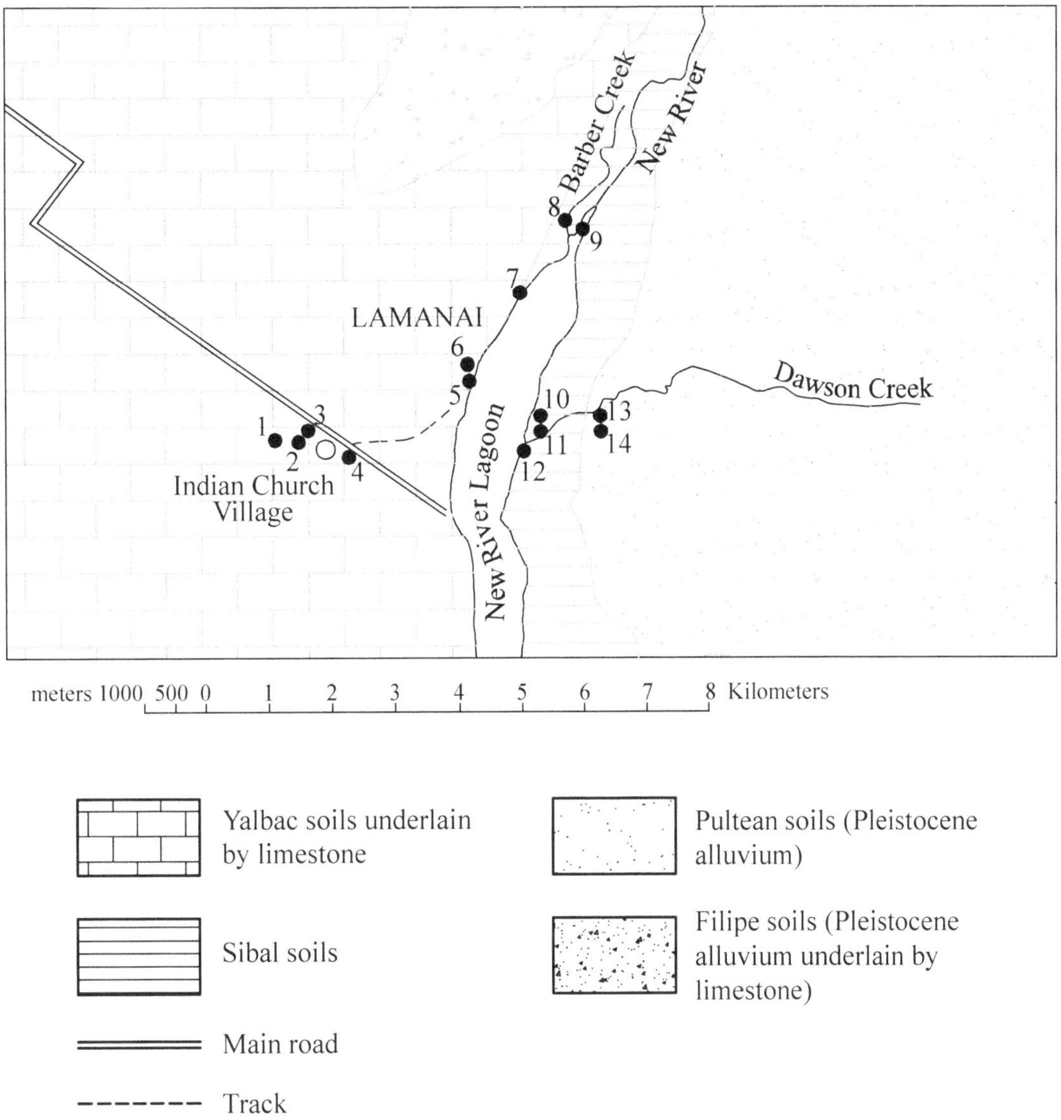

1 Arevalo Milpa (MIL5, MIL6)
2 Aguilar Milpa (MIL3, MIL4)
3 Ruano Milpa (MIL1, MIL2)
4 Ku property (NAS)
5 North-south trench (TCTT1, TCTT2, TCTT3)
6 N10-9 (JAG1, JAG2, JAG3, JAG4)
7 Harbour area (HAR1, HAR2, T HAR1, T HAR2, N HAR, E HAR)
8 Barber Creek (BC)
9 Bar Mouth (BM1, BM2, BM3)
10 Unnamed stream (UNC)
11 Dawson Creek north bank (DCB)
12 Mouth of Dawson Creek (DCM)
13 Dawson Creek south bank (DC1, DC2, DC3, DC4)
14 Inland location within area of Boom soils (DSV1, DSV2, DSV3, DSV4)

Map 5.7: Geological map of the local area showing clay sampling locations.

to be moderately acid and are formed through clay translocation with ferrolytic weathering producing further textural segregation. Test pits dug approximately 50m inland from the southern bank of Dawson Creek as part of the clay sampling programme show similar profiles. In these, light grey sandy topsoil overlies yellow to red brown clay, which grades into a lighter, pale yellow clay at depths of 50cm to 1m. The clays have a sandy texture throughout the subsurface profile.

An area of siliceous alluvium also occurs north of the site, to the east and northeast of Barber Creek. In this area, however, the sandy horizons overly calcareous clay subsoils and a horizon of weathering limestone. As a result, these soils (Filipe Subsuite of the Revenge Suite) differ from Boom soils in that they are neutral and base-saturated (King et al. 1992:243). King et al. (1992:241-242) suggest that the patchy occurrence of Filipe soils in areas east of the New River might equally be due to the vagaries of the original deposition of alluvium in the area or because these deposits accumulated only in specific areas through modern wash processes.

Clay Prospection and the Compositional Characteristics of Local Clay Resources

Since detailed geological studies of the Lamanai area have yet to be undertaken, conducting a survey of raw material resources available in the local area for pottery manufacture was of fundamental importance to the present study. Since clays and clayey soils are widespread throughout the local area, the clay sampling programme focussed on documenting the compositional and textural variability between clays from the different geological and environmental areas described above, and also among those formed in different environmental or geological contexts within the same area. A total of 35 clay samples were prospected from 14 different locations, all of which were within a 3km distance of the ceremonial precinct at Lamanai (Map 5.7). Locations were chosen due to their accessibility, close proximity to the site and, in some cases, based on information obtained from the local villagers as to where particularly visible deposits occurred. Multiple samples were taken at most locations, from varying depths below the ground's surface and/or at varying horizontal intervals. Fired samples of each clay were anlaysed petrographically and chemically by neutron activation to examine compositional and textural variability within the sample set. The results of the firing experiment are presented in Appendix IV, whilst the results of the petrographic and chemical analyses are presented below.

CLAYS FROM THE SITE AND INLAND SOURCES WEST OF THE NEW RIVER LAGOON

Clay samples were taken from seven different locations west of the lagoon within the area underlain by Yalbac soils and limestone (Table 5.1). Three of these locations were within the main settlement along the lagoon and one was within an area of more dispersed structures approximately 1km southwest of the central precinct. The remaining three locations were further inland within the agricultural lands or 'milpas' of the residents of Indian Church Village, which lie 2km – 2.5km west southwest of the central precinct at Lamanai.

Locations within the main area of the settlement included the north-south trench (described above), which was dug as part of modern construction activities at the site; an area south of structure N10-9, from which clay had been quarried during a previous monument reconsolidation project (Plate VII.12); and the 'Harbour' area of the site, where there were outcrops of Cretaceous limestone and the associated soil horizons appeared to be comparatively shallow. Multiple samples were taken at all three locations, and in the 'Harbour' area, most came from excavation units dug to the bedrock (Plate VII.13). Samples taken from the north south trench included two different coloured clays within the subsurface clay horizon and a sample from the underlying horizon containing weathered limestone (see Figure 5.1). Only one clay sample was taken from the area of dispersed structures situated 1km south of the central precinct. The modern village of Indian Church is situated in this area, and the clay sample was taken from the property of a local resident (Nasario Ku) when he was putting in a new septic tank. Clay sampling further inland was confined to three different milpas, the properties of Enrique Ruano, Mauricio Augliar and Raul Arevalo, and in each case the samples came from waterholes previously dug using a backhoe. The depth of the waterholes ranged from 1.34m to 1.9m below the level of the surrounding ground surface and two samples were taken from the each waterhole (Plate VII.14).

Petrographic Characteristics of Yalbac Clays - Thin section analysis of the Yalbac clays showed significant textural and compositional variation among them, which appears to relate directly to differences in the environmental or geological contexts the clays were formed in. The clays that form in association with horizons of weathering limestone, as with the samples taken from the Harbour area and the bottom of the north-south trench, are clearly distinguished from those that form directly below the ground's surface, by differences in texture, mineralogy and other compositional characteristics. In addition, differences were also observed between the lagoon side 'Harbour clays' and the sample taken from the bottom of the trench, which is situated further inland.

The samples taken from the subsurface clay horizon in the trench (TCTT1 and TCTT2), the area south of structure N10-9 (JAG1,JAG2, JAG3 and JAG4), the village property (NAS) and the milpas (MIL1, MIL2, MIL3, MIL4, MIL5 and MIL6) are very similar in terms of their overall texture and basic composition. In all cases, aplastic inclusions comprise approximately 15% of the groundmass and range in size from very coarse sand to very fine sand, with a mode size of fine sand. Subrounded to rounded inclusions of quartz are frequent and are accompanied by very few to very rare inclusions of crystalline calcite, finely to coarsely crystalline calcite mosaics and micrite lumps. Ferromanganiferous nodules

Table 5.1: Clays from the site and inland sources west of the New River Lagoon.

Sample Number	Location	Depth from ground surface (m)	Dry Colour	Moist Colour	Inclusions / Size (Udden-Wentworth grain size scale)	Fired Colour (750°)	Mineralogy (in order of abundance) and other petrographic features
TCTT1	north-south trench east of structure N10-9 (SS2)	0.8	light brownish grey (2.5Y/6/2)	grey (2.5Y/6/1)	few white rock fragments / medium sand	yellowish red (5YR/5/6)	calcareous, 15% inclusions - quartz, crystalline calcite, sparite, micrite - concretions and staining
TCTT2	north-south trench east of structure N10-9 (SS3)	1	light grey (2.5Y/7/2)	light grey to light brownish grey (2.5Y/6/2;7/2)	few white rock fragments / medium sand	yellowish red (5YR/5/6)	calcareous, 15% inclusions - quartz, crystalline calcite, sparite, micrite - concretions and staining
TCTT3	north-south trench east of structure N10-9 (SS4)	1	pale yellow (2.5Y/7/4)	pale yellow to yellow (2.5Y/7/4;7/6)	few white rock fragments / medium sand with very few granules and pebbles	light reddish brown (5YR/6/4)	calcareous, 50% inclusions - crystalline calcite, quartz, sparite, micrite - concretions and staining
JAG1	area south of structure N10-9	0.5	dark grey (5Y/4/1)	dark grey (5Y/4/1)	common white rock fragments / coarse sand with few granules and pebbles	strong brown (7.5YR/5/6)	calcareous, 15% inclusions - quartz, crystalline calcite, sparite, micrite - concretions and staining
JAG2	area south of structure N10-9 - 6m south southeast of JAG1	0.5	light greyish brown to greyish brown (5Y/5/2;6/2)	olive grey (5Y/5/2)	common white rock fragments / coarse sand with few granules and pebbles	brown (7.5YR/5/4)	calcareous, 15% inclusions - quartz, crystalline calcite, sparite, micrite - concretions and staining
JAG3	area south of structure N10-9 - 8m south of JAG1	0.35	light olive grey to olive grey (5Y/5/2;6/2)	olive grey (5Y/5/2)	common white rock fragments / coarse sand with few granules and pebbles	brown (7.5YR/5/4)	calcareous, 15% inclusions - quartz, crystalline calcite, sparite, micrite - concretions and staining
JAG4	area south of structure N10-9 - 5m south southwest of JAG1	0.5	very dark grey to black (Gley 1/2.5/N;3/N)	black (Gley 1/2.5/N)	common white rock fragments / coarse sand	brown (7.5YR/5/4)	calcareous, 15% inclusions - quartz, crystalline calcite, sparite, micrite - concretions and staining

Table 5.1: Clays from the site and inland sources west of the New River Lagoon (continued).

Sample Number	Location	Depth from ground surface (m)	Dry Colour	Moist Colour	Inclusions / Size (Udden-Wentworth grain size scale)	Fired Colour (750°)	Mineralogy (in order of abundance) and other petrographic features
HAR1	Harbour area - 7m west of excavation unit 1	0.5	dark greyish brown (2.5Y/4/2)	light olive brown (2.5Y/5/4)	few white rock fragments / medium sand with few granules and pebbles	reddish yellow (5YR/7/6)	calcareous, 50% inclusions - crystalline calcite, quartz, sparite, micrite, shell
HAR2	Harbour area - excavation unit 1	0.8	pale yellow (2.5Y/8/4)	pale yellow (2.5Y/7/4)	Few white rock fragments / medium sand with few granules and pebbles	light red(2.5YR/7/6)	calcareous, 50% inclusions - crystalline calcite, quartz, sparite, micrite, shell
T HAR1	Harbour area - excavation unit 2	from pit bottom	light olive brown (2.5Y/5/4)	dark grey (2.5Y/4/1)	common white rock fragments / very coarse sand with few granules and pebbles	reddish brown to yellowish red (5YR/5/5;5/6)	calcareous, 50% inclusions - crystalline calcite, quartz, sparite, micrite, shell
T HAR2	Harbour area - excavation unit 3	from pit bottom	light grey (5Y/7/1)	grey (2.5Y/5/1)	very few white rock fragments / medium to very coarse sand	white (2.5Y/8/1)	calcareous, 5% inclusions - crystalline calcite, quartz, micrite, shell
E HAR	Harbour area - embankment south of limestone outcrop	0.35	light olive brown to olive brown (2.5Y/4/3;5/3)	olive brown (2.5Y/4/3)	common white rock fragments / medium - to very coarse sand with few granules and pebbles	reddish yellow (5YR/6/6)	calcareous, 50% inclusions - crystalline calcite, quartz, sparite, micrite, shell
N HAR	Harbour area - embankment north of limestone outcrop	0.35	dark greyish brown (2.5Y/4/2)	dark greyish brown (2.5Y/4/2)	common white rock fragments / medium - very coarse sand with very few granules	reddish yellow (5YR/7/6)	calcareous, 50% inclusions - crystalline calcite, quartz, sparite, micrite, shell
MIL1	milpa - Enrique Ruano's water hole - directly below MIL2	0.40 below 1.55 level	light grey to grey (5Y/6/1;7/1)	grey (5Y/7/1)	common white rock fragments / coarse - very coarse sand with few granules and pebbles	light yellowish brown (10YR/6/4)	calcareous, 15% inclusions - quartz, crystalline calcite, sparite, micrite - concretions and staining

Table 5.1: Clays from the site and inland sources west of the New River Lagoon (continued).

Sample Number	Location	Depth from ground surface (m)	Dry Colour	Moist Colour	Inclusions / Size (Udden-Wentworth grain size scale)	Fired Colour (750°)	Mineralogy (in order of abundance) and other petrographic features
MIL2	milpa - Enrique Ruano's water hole	0.10 below 1.55 level	dark grey to very dark grey (Gley 1/3/N;4/N)	black (Gley 1/2.5/N)	few white rock fragments / coarse - very coarse sand with few granules and pebbles	light brown to brown (7.5YR/5/4;6/4)	calcareous, 15% inclusions - quartz, crystalline calcite, sparite, micrite - concretions and staining
MIL3	milpa - Mauricio Augliar's water hole	0.10 below 1.34 level	light olive brown to olive brown (2.5Y/4/3;5/3)	light olive brown (2.5Y/5/3)	few white rock fragments / coarse - very coarse sand with few granules and pebbles	yellowish red (5YR/4/6;5/6)	calcareous, 15% inclusions - quartz, crystalline calcite, sparite, micrite - concretions and staining
MIL4	milpa - Mauricio Augliar's water hole - directly below MIL3	0.40 below 1.34 level	light olive brown (2.5Y/5/3)	greyish brown to light olive brown (2.5Y/5/2;5/3)	very few white rock fragments / coarse - very coarse sand with very few granules	red (2.5YR/4/6;5/6)	calcareous, 15% inclusions - quartz, crystalline calcite, sparite, micrite - concretions and staining
MIL5	milpa - Arevalo water hole	0.10 below 1.90 level	light brownish grey (2.5Y/6/2)	greyish brown (2.5Y/5/2)	very few white rock fragments and black concretions / coarse - very coarse sand with very few granules	brown (7.5YR/4/4;5/4)	calcareous, 15% inclusions - quartz, crystalline calcite, sparite, micrite - concretions and staining
MIL6	milpa - Arevalo water hole - directly below MIL5	0.40 below 1.90 level	grey (2.5Y/5/1;6/1)	grey to dark grey (2.5Y/4/1;5/1)	very few white rock fragments and black concretions / medium - coarse sand	light brown to brown (7.5YR/4/4;6/4)	calcareous, 15% inclusions - quartz, crystalline calcite, sparite, micrite concretions and staining
NAS	Nasario Ku's property	1.0-2.0	greyish brown to dark greyish brown (2.5Y/4/2;5/2)	dark greyish brown (2.5Y/4/2)	common white rock fragments/ medium - coarse sand	light red to red (2.5YR/5/6;5/8)	calcareous, 15% inclusions - quartz, crystalline calcite, sparite, micrite - concretions and staining

Table 5.2: Clays from sources along creeks and rivers within areas of fresh water swamp.

Sample Number	Location	Depth from ground surface (m)	Dry Colour	Moist Colour	Inclusions / Size (Udden-Wentworth grain size scale)	Fired Colour (750°)	Mineralogy (in order of abundance) and other petrographic features
DCM	mouth of Dawson Creek at New Rive Lagoon, north bank	0.10-0.20	very dark grey to black (Gley 1/2.5/N;3/N)	black (Gley 1/2.5/N)	very few white rock fragments/ medium sand	very pale brown to light yellowish brown (10YR/6/4;7/4)	micritic matrix, 10% inclusions - quartz, chert, shell
DCB	Dawson Creek (0.4m deep deposit) - 100m upstream, north bank	0.10-0.20	light grey (2.5Y/7/1)	grey (5Y/5/1)	few shell fragments and whole specimens	pinkish white to grey (10YR/8/2 TO 2.5Y/5/1;6/1)	micritic matrix, 10% inclusions - quartz, shell
UNC	un-named creek north of Dawson Creek (0.3m deep deposit) - 100m upstream, north bank	0.2	pale yellow (2.5Y/8/2)	grey (2.5Y/6/1)	few shell fragments and whole specimens	white (2.5YR/8/1)	micritic matrix, 5% inclusions - quartz, shell
BM1	Bar Mouth (2m deep deposit), westt bank	0.5	light grey (2.5Y/7/1)	grey to light brownish grey (2.5Y/6/1;6/2)	few shell fragments and whole specimens	white to reddish grey (2.5YR/8/1 TO 2,5YR/5/1;6/1)	micritic matrix, 10% inclusions - quartz, shell
BM2	Bar Mouth, west bank - 4m south of BM1	0.25	light grey (2.5Y/7/2)	(2.5Y/6/1;6/2)	common shell fragments and whole specimens	white (2.5YR/8/1)	micritic matrix, 20% inclusions - quartz, shell, chaladonic quartz, chert
BM3	Bar Mouth, west bank - 4m south of BM1 directly below BM2	0.4	light grey to light brownish grey (2.5Y/6/2;7/2)	greyish brown (2.5Y/5/2)	common shell fragments and whole specimens	white to grey (2.5Y/8/1 TO 2.5Y/5/1;6/1)	micritic matrix, 20% inclusions - quartz, shell, chaladonic quartz, chert
BC	Barber Creek (0.5m deposit), west bank	0.10-0.20	dark grey to very dark grey (2.5Y/3/1;4/1)	very dark grey (2.5Y/3/1)	common white rock fragments/ medium sand	very pale brown (10YR7/4;8/4)	micritic matrix, 30% inclusions - quartz, chaladonic quartz, chert, shell, amphibole

and segregations are common and most samples have a mottled appearance (Plate VII.15).

The sample taken from the horizon of weathered limestone within the trench (TCTT3) is similar in some respects to the above clays in that it also has a mottled appearance and contains common ferromanganiferous nodules and segregations and frequent inclusions of subrounded to rounded quartz. However, in this sample, aplastic inclusions comprise approximately 50% of the groundmass, with finely to coarsely crystalline sized grains of calcite as the dominant inclusion type (Plate VII.16). These grains appear to be the terminal grades of the larger calcite mosaics which are also present. In general, inclusions of calcite (mosaics, crystalline calcite and micrite) have a more frequent occurrence in this sample and they also tend to be larger in size. The largest inclusions are the size of small pebbles, whilst the mode size is of medium sand. In all but one case (T HAR2), the Harbour clays (HAR1, HAR2, T HAR1, E HAR and N HAR) share the same characteristics except that they also contain very few inclusions of shell (pomacia) and ferromanganiferous segregations are completely absent (Plate VII.17). The Harbour sample that is somewhat different (T HAR2) contains fewer (5% of the groundmass) and smaller (fine sand sized) inclusions of quartz (Plate VII.18). The higher frequency and greater range of sizes of the calcite inclusions, in addition to the presence of the terminal grades, are features which clearly distinguish the Harbour clays and the sample from the bottom of the trench from the other Yalbac clays. These features would appear to be a direct reflection of the geological context of these clays, which form through *in situ* weathering of the underlying and adjacent limestone. The presence of shell along with the absence of ferromanganiferous segregations in the Harbour clays would seem to reflect the fact that this area of the site becomes waterlogged and partially inundated during the rainy season. This swampy environment is the habitat of snails and the high water content of the soils would contribute to the removal of ferromanganiferous residues.

CLAYS FROM SOURCES ALONG CREEKS AND RIVERS WITHIN AREAS FRESH WATER SWAMP

Clay samples were taken from five different locations within the areas east and north of the site characterized by Sibal soils, and in all case the samples came from the banks of water ways (Table 5.2). During the dry season, and to a lesser extent during the rainy season, deep deposits of clay are exposed at various locations along the shore of the lagoon and the banks of the creeks and streams that drain into it, where the flow of the water has cut into the bank during periods of higher and faster-moving water. These deposits are easily accessed by canoe or boat, although some are not exposed all year round, depending on water levels at any given time. Since it was known that the geology of inland areas east of the lagoon was different from that on the west or 'site side', exposed clay deposits were sought out along two creeks flowing into the lagoon from the east (Dawson Creek and an unnamed stream to the north) (Plates VII.19-20), where the lagoon narrows to form the New River (Bar Mouth) (Plate VII.21) and along a creek to the west of the New River (Barber Creek) (Plate VII.22). The unnamed stream that runs parallel to Dawson Creek, approximately 150m to the north, only becomes filled with water during the rainy season and in the dry season after heavy rains, and so the deposit that was sampled is only accessible by water at certain times of the year. At the time the deposits were sampled, their depths ranged from 0.3m – 2m and at all locations, except Bar Mouth (3 samples), one clay sample was taken.

Petrographic Characteristics of Sibal Clays - Petrographic analysis of the Sibal clays showed that, with the exception of the Barber Creek sample (BC), the clays are broadly similar in terms of their texture and basic composition. All samples (DCM, DCB, UNC, BM1, BM2 and BM3) except BC have a micritic fine fraction with aplastic inclusions comprising only 5-20% of the groundmass. Medium to very fine sand sized inclusions of quartz are frequent and shell fragments (pomacia) are common to rare (Plate VII.23). Inclusions that are very rare to absent in the samples include chert, chalcedony (chalcedonic quartz) and crystalline calcite, and with the exception of crystalline calcite, all of the rock and mineral fragments are subrounded to well rounded. The Barber Creek sample is similar in that it also has a micritic matrix and the types and relative frequencies of aplastics are basically the same, except that it contains very rare amphibole inclusions. The distinguishing features of this clay are that it contains a greater quantity of siliceous inclusions (30% of the groundmass), which are larger (medium to coarse sand sized) and well sorted, and the presence of the amphibole (Plate VII.24). The greater amount of rock and mineral inclusions in this particular sample appears to reflect the close proximity of Barber Creek to deposits of siliceous alluvium situated directly northwest, as the creek would certainly receive runoff from this area. The roundness and well sortedness of the inclusions is in keeping with the fact that the clay derives from a high energy, riverine environment. It is this characteristic, along with the micritic fine fraction that clearly distinguishes this clay from the samples taken from the area of siliceous alluvium east of the lagoon.

CLAYS FROM SOURCES EAST OF THE NEW RIVER LAGOON WITHIN AREAS OF PLEISTOCENE ALLUVIUM

Clay samples were taken from two different locations within the area east of the lagoon characterized by Boom soils: the south bank of Dawson Creek, 1km upstream from the lagoon, and an inland area 50m south of the creek deposit (Table 5.3) (Plate VII.25 and Plate VII.26). These locations were selected in order to examine similarities and differences between clays that form in different environments within this geological zone. To examine internal variation within these deposits, samples were also taken from varying depths.

Petrographic Characteristics of Boom Clays = The Boom clays are clearly differentiated from the other local clays by their mineralogy as well as other textural and compositional attributes. These clays are non-calcareous,

Table 5.3 Clays from sources east of the New River Lagoon within areas of Pleistocene alluvium.

Sample Number	Location	Depth from ground surface (m)	Dry Colour	Moist Colour	Inclusions / Size (Udden-Wentworth grain size scale)	Fired Colour (750°)	Mineralogy (in order of abundance) and other petrographic features
DSV1	savannah, 50m inland (south) of creek deposit	0.5	light yellowish brown (2.5Y/6/4)	light yellowish brown (2.5Y/6/4)	common white and grey rock fragments/ medium - coarse sand	yellowish red (5YR/5/6)	non calcareous, 30-40% inclusions - quartz, chaladonic quartz, chert, clinopyroxene, amphibole - concretions and staining
DSV2	savannah, 50m inland (south) of creek deposit - 2m west of DSV1	0.5	light yellowish brown (2.5Y/6/4)	light yellowish brown (2.5Y/6/4)	common white and grey rock fragments/ medium - coarse sand	yellowish red (5YR/5/6)	non calcareous, 30-40% inclusions - quartz, chaladonic quartz, chert, clinopyroxene, amphibole - concretions and staining
DSV3	savannah, 50m inland (south) of creek deposit - 2m west of DSV1	1	yellow to olive yellow (2.5Y/6/6;7/6)	yellow to olive yellow (2.5Y/6/6;7/6)	common white and grey rock fragments/ medium - very coarse sand	yellowish red (5YR/5/6;5/8)	non calcareous, 30-40% inclusions - quartz, gypsum, chaladonic quartz, chert, clinopyroxene, amphibole - concretions and staining
DSV4	savannah, 50m inland (south) of creek deposit - 2m north of DSV2	0.5	light yellowish brown (2.5Y/6/4)	light yellowish brown to olive yellow (2.5Y/6/4;6/6)	common white and grey rock fragments/ medium - coarse sand	yellowish red (5YR /4/6;5/6)	non calcareous, 30-40% inclusions - quartz, chaladonic quartz, chert, clinopyroxene, amphibole, sandstone - concretions and staining
DC 1	Dawson Creek (0.8m deep deposit), 1km upstream, south bank	water level	light grey to grey (2.5Y/6/1;7/1)	grey to light brownish grey (2.5Y/6/1;6/2)	few white and grey rock fragments/ medium - coarse sand with very few granules	light yellowish brown (10YR/6/4)	non calcareous, well sorted, 20-25% inclusions - quartz, chaladonic quartz, chert, micrite, mudstone - concretions and staining

Table 5.3 Clays from sources east of the New River Lagoon within areas of Pleistocene alluvium (continued).

Sample Number	Location	Depth from ground surface (m)	Dry Colour	Moist Colour	Inclusions / Size (Udden-Wentworth grain size scale)	Fired Colour (750°)	Mineralogy (in order of abundance) and other petrographic features
DC2	Dawson Creek, 1km upstream, south bank - 2m north east of DC1	0.5	light grey to grey (2.5Y/6/1;7/1)	light brownish grey (2.5Y/6/2)	common white and grey rock fragments/ medium - coarse sand	light yellowish brown (10YR/6/4)	non calcareous, 30-40% inclusions - quartz, gypsum, chaladonic quartz, chert, clinopyroxene, amphibole - concretions and staining
DC3	Dawson Creek, 1km upstream, south bank - 2m south of DC1	0.5	light grey to grey (2.5Y/6/1;7/1)	light brownish grey (2.5Y/6/2)	few white and grey rock fragments/ medium - coarse sand	light yellowish brown (10YR/6/4)	non calcareous, well sorted, 20-25% inclusions - quartz, chaladonic quartz, chert, micrite, mudstone - concretions and staining
DC4	Dawson Creek, 1km upstream, south bank - 2m west of DC3	0.5	grey (5Y/5/1;6/1)	olive grey (5Y/5/2)	common white and grey rock fragments/ medium - coarse sand	reddish yellow (7.5YR/6/6)	non calcareous, 30-40% inclusions - quartz, chaladonic quartz, chert, clinopyroxene, amphibole - concretions and staining

with very few to absent inclusions of micrite and no crystalline calcite, and generally contain a greater amount of aplastic inclusions (20% to 40% of the groundmass), which are predominantly angular to subrounded fragments of monocrystalline quartz (Plate VII.27). The quartz is accompanied by lesser quantities of other siliceous rocks – polycrystalline quartz (few), chert and chalcedony (very few) – and rare inclusions of clinopyroxene, amphibole, sandstone and mudstone (these are absent in some samples). One sample (DSV3) also contains gypsum (Plate VII.28). The inclusions generally range in size from very coarse sand to fine sand, with a mode size of fine sand, and the smaller inclusions of quartz commonly appear to be the terminal grades of the larger inclusions of polycrystalline quartz. Ferromanganiferous nodules (rust and black coloured) and segregations are other prominent features.

Among the Boom clays there is some compositional and textural variation, which appears to relate directly to differences in environment in which they were formed. Inclusions of micrite and mudstone are entirely absent in the samples taken from the inland location, whilst clinopyroxene and amphibole are not present in the creek samples. In addition, two of the creek samples (DC1 and DC3) generally contain fewer aplastic inclusions (20-25% of the groundmass), lacking the larger rock and mineral fragments (coarse sand sized) and with far fewer fine sand sized inclusions. Consequently, these samples are comparatively well sorted (Plate VII.29). Among the samples from the inland location, the sample taken at a depth of 1m (DSV3) is the only one that contains gypsum. Gypsum is also entirely absent in the samples from the creek deposit.

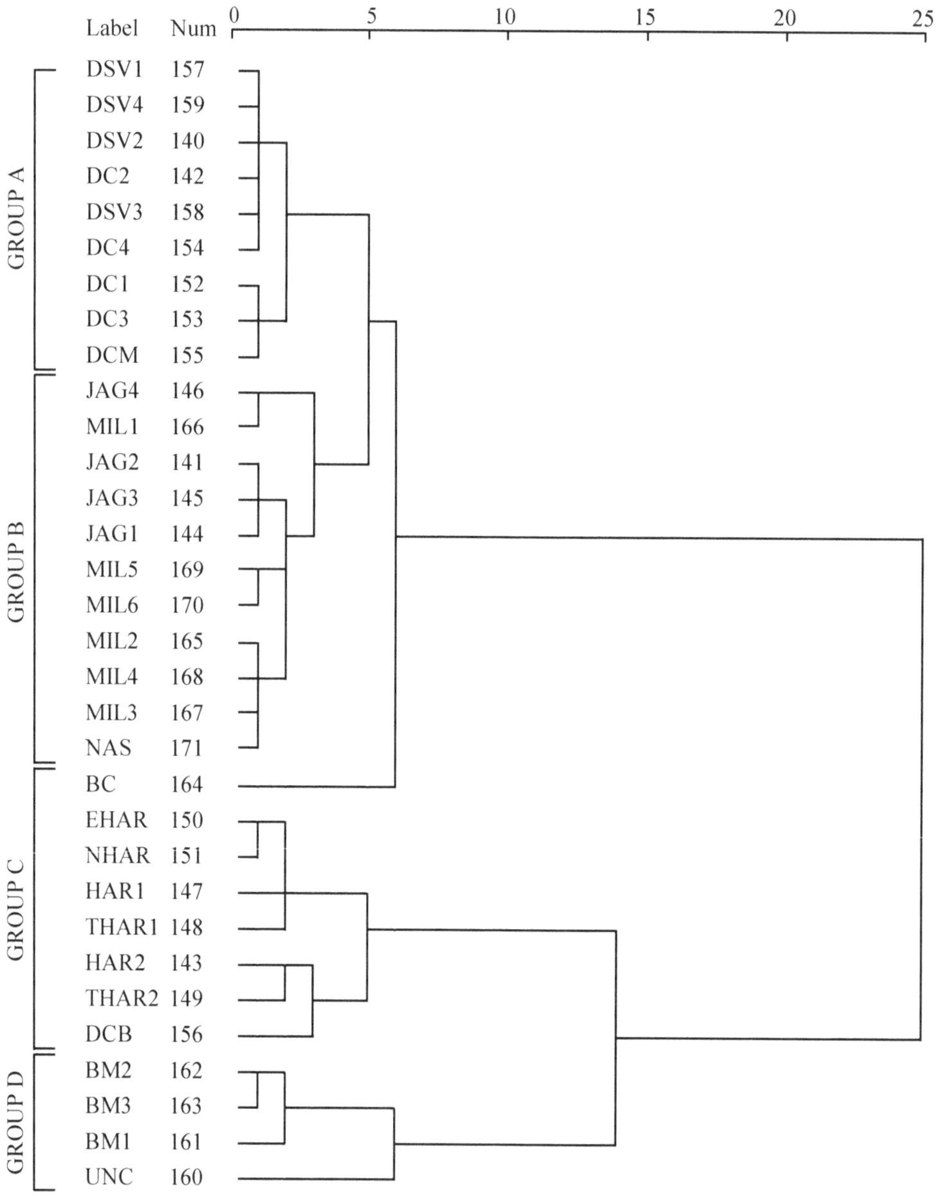

Figure 5.2: Dendrogram of clay samples prospected from the Lamanai area showing the four principal chemical groups. Elements included were Sm, Lu, Yb, Na, Ce, Th, Cr, Hf, Cs, Sc, Rb, Fe, Co, Eu.

CHEMICAL COMPOSITIONAL VARIATION AMONG LOCAL CLAYS

The results of the chemical analysis of local clay samples support the petrographic findings, discriminating four chemical groups that correspond to differences in the geological and environmental contexts in which the clays were formed. As shown in the dendrogram (Figure 5.2), cluster analysis produces four clusters of samples (Groups A, B, C and D), with the sample from Barber Creek (BC) occurring as a chemical outlier between and Groups B and C. Group A comprises samples taken from the east side of the lagoon in the area dominated by Pleistocene alluvial

Table 5.4: A summary of average compositions and standard deviations (actual and percent) for the four chemical groups identified and the Barber Creek clay sample. Concentrations are given in ppm apart from Ca, Na and Fe. Specific element values mentioned in the text are marked in bold.

	Group A			Group B			BC	Group C			Group D		
	Ave. Value	S.D.	% S.D.	Ave. Value	S.D.	% S.D.	Values	Ave. Value	S.D.	% S.D.	Ave. Value	S.D.	% S.D.
Sm	10.35	2.88	27.82	10.63	1.98	18.64	8.83	3.58	1.03	28.65	1.94	0.77	39.52
Lu	0.64	0.15	23.93	0.78	0.08	10.89	0.60	0.27	0.08	31.06	0.13	0.04	29.69
U	4.19	4.33	103.33	1.60	0.18	11.27	2.28	1.31	0.78	59.36	0.74	0.22	29.59
Yb	5.19	0.90	17.41	6.55	0.84	12.78	5.27	2.21	0.68	31.04	1.02	0.33	31.96
As	3.05	2.48	81.04	12.42	3.72	29.93	1.63	10.65	11.54	108.35	0.33	0.26	76.92
Sb	1.33	0.15	11.63	2.21	0.36	16.43	1.13	1.02	0.37	36.34	0.23	0.06	27.45
Ca	1.68	2.10	125.12	5.18	3.88	74.92	1.13	26.74	4.94	18.48	33.40	2.93	8.77
Na	0.08	0.01	10.98	0.06	0.01	25.56	**0.02**	0.02	0.00	20.02	0.03	0.01	23.25
La	52.20	17.46	33.44	52.14	13.03	24.98	38.80	18.40	4.02	21.84	10.07	4.22	41.97
Ce	125.98	37.29	29.60	126.61	29.52	23.31	104.00	42.47	12.54	29.54	21.57	9.27	42.96
Th	15.84	3.99	25.19	20.37	3.30	16.19	14.00	7.02	2.05	29.22	2.72	0.91	33.61
Cr	63.83	11.43	17.91	**92.41**	15.98	17.29	**38.20**	40.53	12.72	31.40	10.78	2.55	23.65
Hf	10.19	1.21	11.93	11.98	0.92	7.71	11.60	4.10	1.21	29.62	2.03	0.84	41.18
Cs	**6.22**	0.63	10.14	2.27	1.21	53.28	1.36	1.02	0.73	72.08	0.38	0.11	28.34
Tb	1.26	0.33	26.46	1.48	0.30	20.27	1.18	0.46	0.19	40.10	0.22	0.10	43.29
Sc	12.24	2.34	19.08	16.56	2.59	15.63	**8.91**	6.81	1.68	24.61	2.03	0.46	22.84
Rb	**55.24**	7.75	14.03	14.63	6.23	42.56	**7.95**	10.42	3.59	34.50	2.76	1.78	64.55
Fe	2.89	0.33	11.29	**4.61**	0.61	13.16	**1.97**	2.07	0.80	38.74	0.53	0.08	15.69
Ta	2.05	0.51	25.00	1.66	0.28	16.95	1.24	0.68	0.13	19.43	0.25	0.06	25.15
Co	10.34	2.19	21.14	15.28	4.34	28.37	**4.81**	7.25	4.69	64.77	1.35	0.42	31.43
Eu	1.88	0.50	26.58	1.90	0.36	18.89	1.43	0.64	0.18	28.26	0.34	0.11	32.93

deposits (Boom soils) and includes the inland deposit, the adjacent creek-side deposit and the sample taken from the north bank at the Mouth of Dawson Creek. Group B consists of the Yalbac clay samples taken from directly below the ground's surface on the west side of the Lagoon, at the on-site and milpa locations. Group C comprises all of the 'Harbour clays', as well as the sample taken from 100m upstream on Dawson Creek (DCB), but can be considered as principally corresponding to clays that form in association with horizons of weathering limestone on the west side of the site side of the lagoon. Group D corresponds to creek deposits, including the samples taken from the Bar Mouth location on the New River and the unnamed creek north of Dawson Creek.

Table 5.4 gives the average concentration values and standard deviations (actual and in %) for the four chemical groups and the Barber Creek sample (the full set of chemical data is given in Appendix V). Groups A and B the Barber Creek sample are characterized by comparatively high concentrations for the rare earth elements (Sm, Lu, Yb, La, Ce, Th and Cr), the alkali metals (Na, Rb and Cs) as well as Fe, Sc, and Co and by low Ca values. In comparison, in Groups C and D, Ca levels are high and all other elemental concentrations are significantly lower that in Groups A and B. Groups A and B are discriminated due to differences in the amount of Cr, Cs Rb and Fe they contain. Cs and Rb concentrations are clearly higher in Group A and Cr and Fe levels are clearly higher in Group B. Groups C and D are separated by discrete differences in Sm, Lu, Th, Cr, Sc, Rb and Fe concentrations, these being significantly lower in Group D. The Barber Creek sample, which is the only chemical outlier in the data set, is discriminated from Groups A and B by lower levels of Na, Rb, Cr, Fe, Sc and Co; and from Groups C and D, by higher Hf and rare earth element concentrations, as well as by significantly lower levels of Ca.

Further inspection of the structure of the modern clay data set with principal components (PCA) analysis provides additional insight into the relationships between the different chemical groups and subgroups identified, as well as the elements most responsible for the differences among them. PCA extracts 2 components, which together account for about 82% of the total variation within the data set. Component 1 is loaded mainly by the rare earth elements Lu, Yb, Sm, La, Eu, Ce and Th, and also Sc and Co, whilst component 2 is characterized by high loadings for the alkali metals Na, Rb as well as Fe (Table 5.5).

Figure 5.3 shows a plot of components 1 and 2 with Varimax rotation. As can be observed, the four chemical groups identified through cluster analysis are discriminated, occupying different areas of the plot. The 'looseness' of these clusters reflects the considerable internal variation in elemental concentrations that exists. The position of Groups A (Boom clays) and C (Yalbac clays associated with weathering limestone) at opposite ends of the Y axis indicates that their discrimination is mainly dependent on component 2, which relates to Fe and the alkali metals Na and Rb. Since thin section analysis indicated that Boom clays (Group A) contain inclusions of potassium feldspar whereas the 'Harbour clays' do not, it might be inferred that the discrimination of the Boom clays from the Yalbac clays, and especially the clays comprising Group C, in part reflects mineralogical differences in their coarse fractions (aplastic inclusions). In contrast, the positioning of Group B (upper horizons of Yalbac clays) at right side of the plot indicates that its discrimination is dependent upon component A, which is characterised by high loadings for the rare earth elements. In this case, therefore, the discrimination relates primarily to chemical differences associated with clay minerals, and thus the fine fraction. In the case of Group D (creek clays), the discrimination clearly depends on both components, which is in keeping with the distinctive compositional characteristics of the creek clays, as observed on the microscopic level.

The clear chemical discrimination of Boom clays (Group A) from Yalbac clays (Groups B and C), might be expected as they derive from a different geological zone. However, when considering that Groups B and C comprise Yalbac clays that are formed within different geological contexts within the same geological zone, their chemical discrimination is quite significant. In addition, PCA also provides some evidence that differences in the environmental context in which clays are formed can lead to chemical distinctions. For example, the samples labelled DCM and DCB, which occur within the plot as chemical outliers of Groups A and C, respectively, derive from creek deposits within the area of Sibal soils that are subject to different environmental conditions and consequently, formation processes. DCM represents the

Table 5.5: Component matrix showing loadings of the components plotted in Figure 5.3.

	Component	
	1	2
Fe	.925	.261
Sc	.904	.377
Cr	.903	.322
Yb	.902	.413
Lu	.893	.428
Co	.882	.282
Th	.881	.447
Hf	.850	.466
Sm	.842	.491
Eu	.838	.502
La	.830	.471
Ce	.821	.489
Cs	.388	.895
Na	.317	.885
Rb	.386	.853

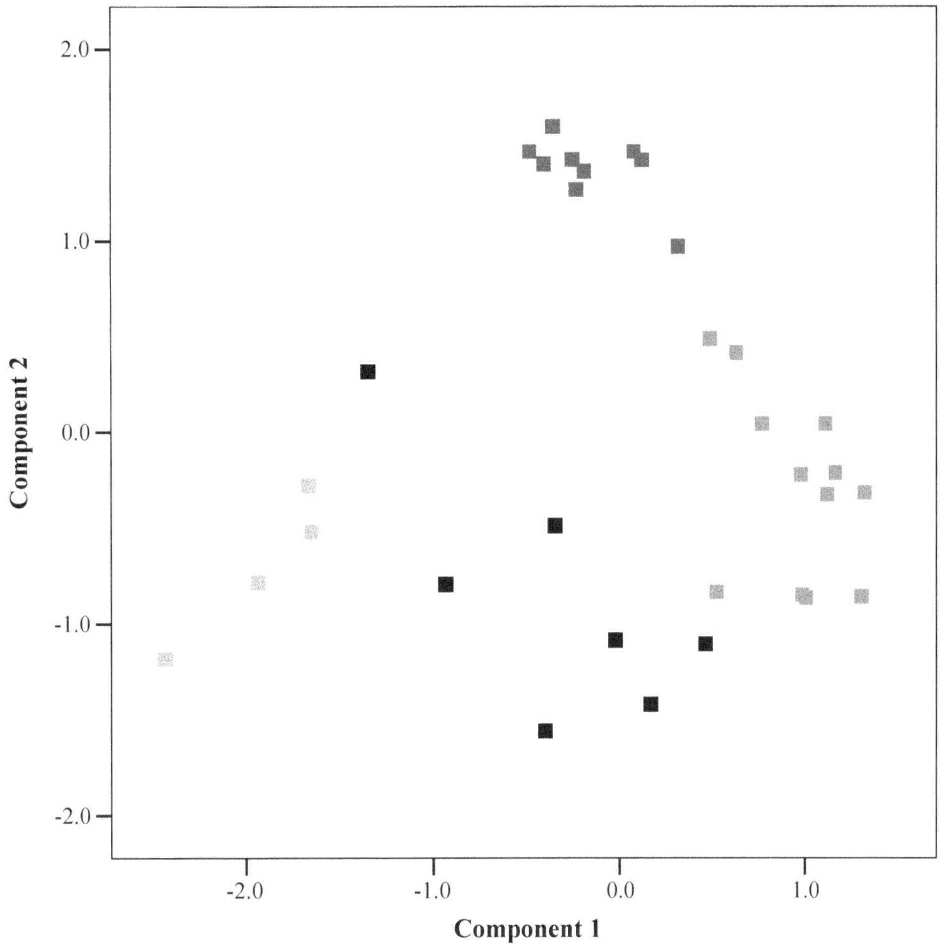

Groups

- A (clay samples from the east side of the New River Lagoon - Boom soils)
- B (Yalbac clay samples from directly below the ground's surface)
- BC (Barber Creek clay sample)
- C (Yalbac clay samples associated with horizons of weathering limestone and Dawson Creek sample [DCB])
- D (clay samples taken from creek deposits - Sibal soils)

Figure 5.3: Plot of components 1 and 2 of the clay data set with Varimax rotation based on Sm, Lu, Yb, Na, La, Ce, Th, Cr, Hf, Cs, Sc, Rb, Fe, Co and Eu. Chemical groups identified through cluster analysis are differentiated with coloured symbols (see colour version on accompanying CD).

only clay sample taken from the mouth of Dawson Creek, where as DCB, whilst also associated with Dawson Creek, derives from a deposit located 100m upstream. These samples, although chemically similar to the other clays that occur within their associated chemical groups are clearly differentiated both from the clays they are most chemically similar to, and perhaps more significantly, from other creek deposits consisting of Sibal soils.

5.3. The Local Area as a Potential Setting for Pottery Manufacture and Technological Investigation

The most important resources for pottery manufacture are clay, raw materials for tempering and fuel for firing (Rice 1987c:115). Accordingly, at first glance, it would appear that the Lamanai area would provide a particularly advantageous setting for pottery manufacture since clay is abundant, potential tempering materials such as limestone, shell and sand are widely available, and the fuel supply offered by the surrounding forest is virtually limitless. One potential drawback, however, is the nature of the clays available in the immediate vicinity of the site, which primarily consist of smectite and in some cases contain a considerable quantity of large rock fragments (principally limestone). Smectites are characterized by small particles and high expandability through base exchange, and as result, are very plastic and 'sticky'. While high plasticity might be a beneficial quality at certain stages of the manufacturing process, the tendency of these clays to absorb water means that they are prone to high shrinkage and often crack during the drying stage. Likewise, when fired, they are prone to cracking and explosion due to rapid heating and removal of chemically bound water from the clay body. (Rice 1987c:48-49, 86-88). Observations made during the firing experiment underline these potential difficulties (Appendix IV). The presence of large limestone fragments in the Yalbac clays is another potential problem as the fragments appear to promote the development of flaws during both the drying and firing stages (Appendix IV). In addition, the results of the firing experiment suggest that when using these clays, firing temperatures would have to be carefully monitored to ensure that they remain below at least 850°C, in order to avoid calcite decomposition and the consequent defects and damage it causes. The experimental results also show that some of the river and creek clays are equally problematic for the same reasons due to their high calcium carbonate content. In comparison, the Boom clays appear to be less problematic, developing fewer flaws during the drying and firing stages. Nonetheless, the sandy texture of these clays might have been undesirable for fine ware production, affecting the quality of the surface finish. In addition, their location across the lagoon poses logistical problems for their procurement.

When attempting to reconstruct local pottery making activities, the potential drawbacks associated with the use of particular local clays are important considerations as these might have been contributing factors in the choices made by local potters in the way of raw material selection and paste preparation strategies and techniques. Since many of the local clays tend to develop flaws during the drying and firing stages, it would appear that some initial processing would have been necessary to remove larger rock fragments or other unwanted constituents. Alternatively, clays displaying certain physical characteristics might have been actively sought out or avoided. In addition, many of the undesirable qualities in clays such as the high shrinkage potential of smectites, can be counteracted through the addition of temper or other paste preparation techniques.

The study of local clay resources has shown that there is considerable variation in their physical, compositional and textural properties, as well as their natural and fired colour. This variation is of particular significance if one considers that the selection of clays for pottery manufacture might have been in some way based on visual or tactile criteria. The petrographic and chemical analyses of the clays have shown that there are important mineralogical and compositional differences among clays that form in different geological and environmental contexts within the local area. It is this variability that permits the comparison of clays and ceramic bodies as a basis for suggesting particular clays as raw materials for local pottery production. The specific choices made by local potters in this regard are explored and discussed in Chapter 8.

CHAPTER 6

ARCHAEOLOGICAL CONTEXTS OF THE TERMINAL CLASSIC TO EARLY POSTCLASSIC CERAMIC ASSEMBLAGE AND ITS BROADER CULTURAL ASSOCIATIONS

As was seen in Chapter 2, the archaeological record at Lamanai suggests that the Terminal Classic to Early Postclassic period was a time of 'continued vibrance' in community life. It was a period of growth which saw the construction of new ceremonial and residential structures and the expansion of the settlement southward. It was also a time of innovation in which new ideas were rapidly introduced and integrated into existing traditions. Construction activities were marked by the emergence of previously unseen architectural features and design element in both new and existing structures. Large refuse deposits in association with communally built ceremonial structures and adjacent public areas make their first appearance and significant changes occur in regard to archaeological manifestations of offertory and funerary practices. The continuing use of Classic period ceremonial structures and public areas, however, is suggestive of continuity in day-to-day aspects of community life as well as strong links to established traditions. This chapter describes the archaeological contexts of the ceramics included in this study, considering the larger cultural meaning of the different kinds of deposits from which the ceramics derive.

6.1. ARCHAEOLOGICAL CONTEXTS OF CERAMIC MATERIALS INCLUDED IN THE STUDY

The ceramic assemblage that forms the basis of this study includes fragmentary ceramic materials (sherd assemblages) recovered from two large primary refuse deposits, as well as reconstructed vessels deriving a variety of other archaeological contexts. Both of the refuse deposits are directly associated with ceremonial structures (N10-9 and N10-27) within the central precinct that clearly were sites of ritual activity. The reconstructed vessels derive from various primary contexts mostly within and in association with various structures in the settlement that had ceremonial, residential or administrative functions. Vessels from caches, burials and isolated finds within structures N10-9 and N10-27 form part of this body of material.

Structure N10-9 and the Associated Midden Assemblage

Structure N10-9 is a large, terraced, pyramidal platform (ca. 20m high) located at the southern end of the central precinct of the site (Map 6.1, Plate VII.30). The structure features a chambered building set on a landing athwart the lower central stair and thus conforms to the Lamanai Building Type that emerges in the Late Classic period. The earliest evidence of construction is Early Classic in date. N10-9 was subsequently modified at various times, with the last major modifications taking place some time during the Early to Late Postclassic (Pendergast 1981a). Owing to its prominent role within the plaza group of which it is the main focus, structure N10-9 was likely an important setting for ritual activities, perhaps on a community level, from Early Classic through to early Late Postclassic times.

At a time when portions of N10-9 were still in use, deposition of a midden (refuse deposit) began at the building's northeast corner. Although there is no indication of the nature or the time required for accumulation of the deposit, the event clearly spanned sufficient time, or took place on a sufficient scale, to result in a metre-deep mass of debris which not only engulfed the east base of N10-9 and the south side of N10-7 (the nearest neighbour of N10-9), but flowed right over the platform edge and spread along the rear face of the buildings situated directly to the east of the plaza area (N10-2) (Pendergast 1981a:44).

The sherd assemblage analysed as part of the present study derives from an area of the deposit situated at the east base and northeast corner of the structure N10-9 and comprises material recovered during three separate excavations. Since the reasons for these excavations varied greatly, so do the excavation techniques that were employed, as well as the specific content of the recovered material. Despite these inherent dissimilarities among the bodies of material recovered on any one occasion, it can be assumed that the general content of the ceramic assemblages is fairly consistent, since on all three occasions even the most fragmentary material was kept.

The most substantial excavations of the midden deposit took place between 1974 and 1980. These excavations began with the aim of determining the construction history of structure N10-9 as well as the adjacent plaza area, but once the midden was discovered, it was separated from the architecture excavations. The excavations focussed on the midden involved screening and wet-screening of the midden fill and resulted in removal of approximately half of the original deposit. The materials recovered included large amounts of faunal remains, a large proportion of which was crocodile, a wide range of bone and stone artifacts, and roughly 150,000 pottery sherds (approximately 900 diagnostics were kept for subsequent study).

Subsequent excavation of the midden did not occur until 1999, when a small portion of the deposit located just to the northeast of structure N10-9 was cleared away in an effort to improve access to the plaza area from the main road to the site. Owing to the somewhat urgent nature of the clearing activities, much of the midden fill was set

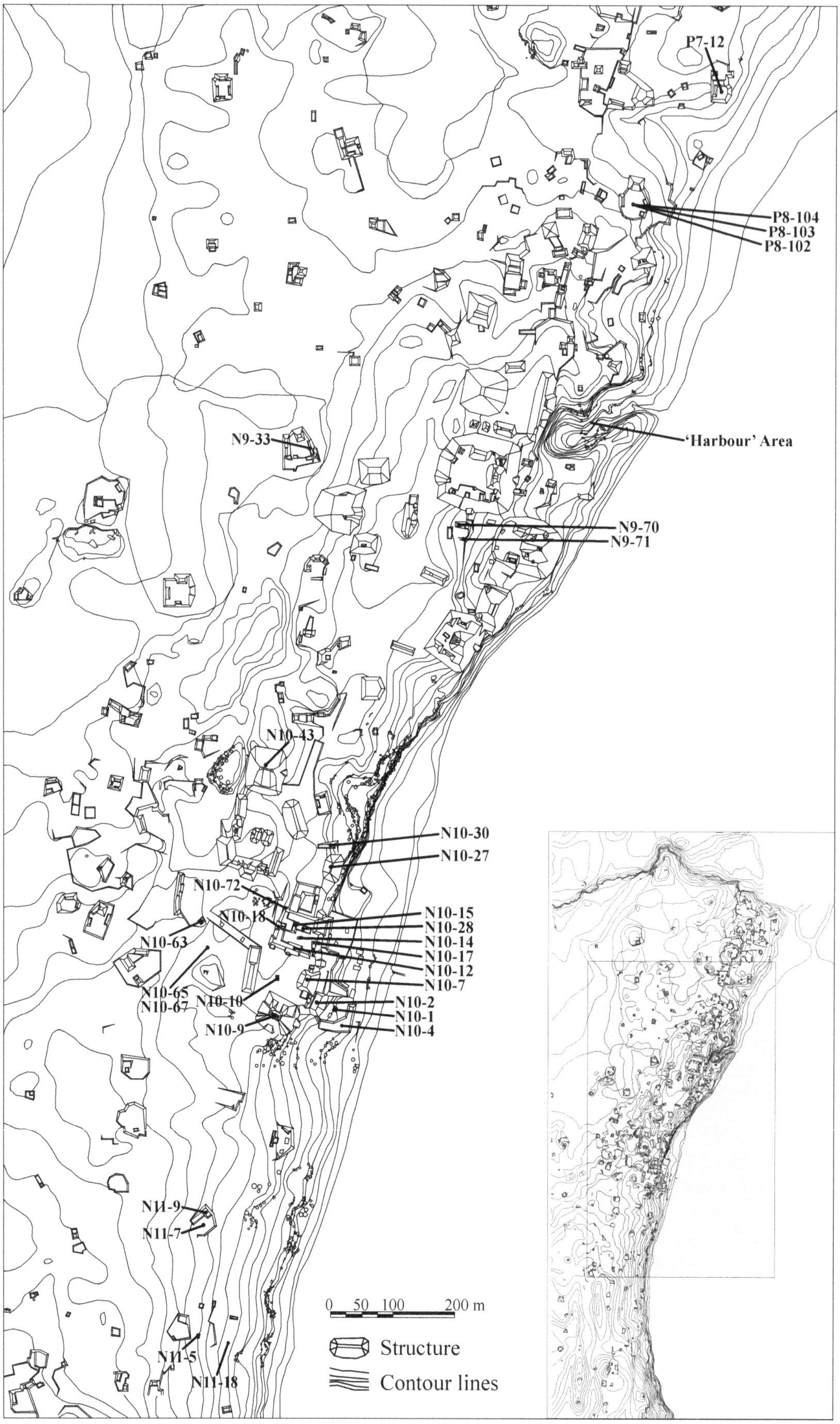

Map 6.1: Map of Lamanai showing locations of structures with which the midden assemblages are associated and from which the reconstructed vessels derive.

aside for screening during subsequent field seasons, and the artifacts were picked out by hand as the fill was removed. These efforts resulted in the recovery of approximately 1,000 sherds as well as numerous other artifacts.

The following field season, a controlled excavation of a small section of the midden (two trenches roughly 1mx5m in size) was undertaken by the Lamanai Archaeological Project Field School (Plates VII.31-32). This effort resulted in the recovery of 4560 sherds including many complete profiles. At this time, it was observed that the deposit was generally uniform throughout, lacking an obvious stratigraphy and comprising dark brown to black soil, with a distinctively greasy texture, intermixed with artifacts.

Structure N10-27 and the Associated Midden Assemblage

Structure N10-27, which is situated on the eastern side of the central precinct, is a moderately-sized, terraced, pyramidal platform of the Lamanai Building Type, featuring a chambered building on a landing straddling the lower central stair (Map 6.1, Plate VII.33). The earliest phase of this structure dates to the Late Preclassic and nothing is known of its Early Classic life span. However, during the Late Classic period, it underwent a series of modifications, before it was abandoned or fell into disuse in the Terminal Classic period.

The ceremonial nature of the structure is clearly indicated by the occurrence of a large carved limestone monument (Stela 9), which was uncovered at the foot of the central stairway, where it appears to have moved, after being toppled from its original position within the chambered building (Pendergast 1988) (Plate VII.34). Evidence of burning within the room and the nature of fracture pattern on the broken end of the stela, suggest that the breakup of the monument resulted from a fire that was lit at its base (Pendergast 1988:1-4) The carving on the stela comprises an iconographic representation of a male figure dressed in 'royal regalia' and a hieroglyphic inscription (Closs 1988; Reents-Budet 1988) Intentional defacement of the left eye of the individual, which appears to have occurred when the stela was broken and moved, may represent the ritual 'killing' of the monument or perhaps the power of the individual depicted (Reents-Budet 1988: 25-31). Based on the content of the stela inscription it seems clear that the monument was erected in honour of the male figure depicted. Likewise, the placement of the stela within the central room of the chambered building strongly suggests the structure itself was dedicated to him. As part of the dedicatory activities associated with the monument's erection, a multiple interment of five juveniles was placed in a pit beneath the stela. This offering would seem to be of special significance, since comparable examples at other sites in the Maya Lowlands are uncommon (Pendergast 1988:5-6).

Not long after the upper portion of the stela was toppled as a result of shattering its base, the deposition of a midden, which is concentrated on the building's south and west sides, began. Since the lower terrace faces of the structure and plaza surfaces had experienced no deterioration or exposure prior to when the refuse accumulation began, it would appear that at least parts of the structure and the surrounding plaza were still being actively maintained up until that point, although collapse of the end walls of the chambered building had already taken place (Graham 2004; Pendergast 1988). The presence of Terminal Classic pottery sherds between this collapse and the plaza floor suggests that the structure (and the building) started to fall into disuse some time during the 9th century (Graham 2004). The close proximity of structure N10-27 to the Ottawa Group elite residential and administrative complex situated just to the south, as well as the identical pottery styles recovered from these areas, suggests that refuse deposition at N10-27 possibly resulted from ritual activity associated with or centred in the Ottawa Group (Graham 2004:16) (see Map 2.4).

Graham (2004:15) reports that, in the area of heaviest accumulation on the southwest side of N10-27, midden deposition was "spotty and resulted in a highly idiosyncratic pattern of preservation". Accumulation appears to have first occurred against the lowermost terraces and stairside outset faces, and deposition continued outward from the structure to ultimately cover the central stair, as well as the toppled portion of the stela. The deposit is deepest in areas where it abuts the structure (over 1m), becomes increasingly shallow with further distance away from the structure and is thinnest at its margins.

The sherd assemblage that was included in the present study comprises fourteen lots of material (31,529 sherds) that derive from various different localities within the main area of refuse accumulation abutting the south front terraces (Figure 6.1). Lots LA701 and LA717 derive from a trench dug by Pendergast as part of excavations conducted in the early 1980's, which were aimed at investigating the construction history of N10-27.

The lower trench lot, LA717, includes refuse recovered from directly above the plaza floor, which was interspersed with collapse debris, while the upper lot, LA701, comprises material from the upper level of the deposit, in which collapse debris was largely absent. Other lots of material that were recovered from the deepest area of the midden include LA1346, 1347and 1348; LA1350, 1351 and 1352 and LA1354, 1355 and 1356. These lots derive from excavation units dug in 25cm levels at the southwest corner of N10-27 as part of the excavations undertaken by Graham from 1997 to 2002. The remaining lots of material (LA1114, LA1115 and LA1119), also recovered as part of Graham's investigations, derive from areas in front of the structure, where extensive horizontal exposure was undertaken (Plate VII.35). Since the midden tapers off in this area, these lots are associated with areas of lesser accumulation and the margins of the deposit. A main focus of Graham's excavations has been investigating the pattern of

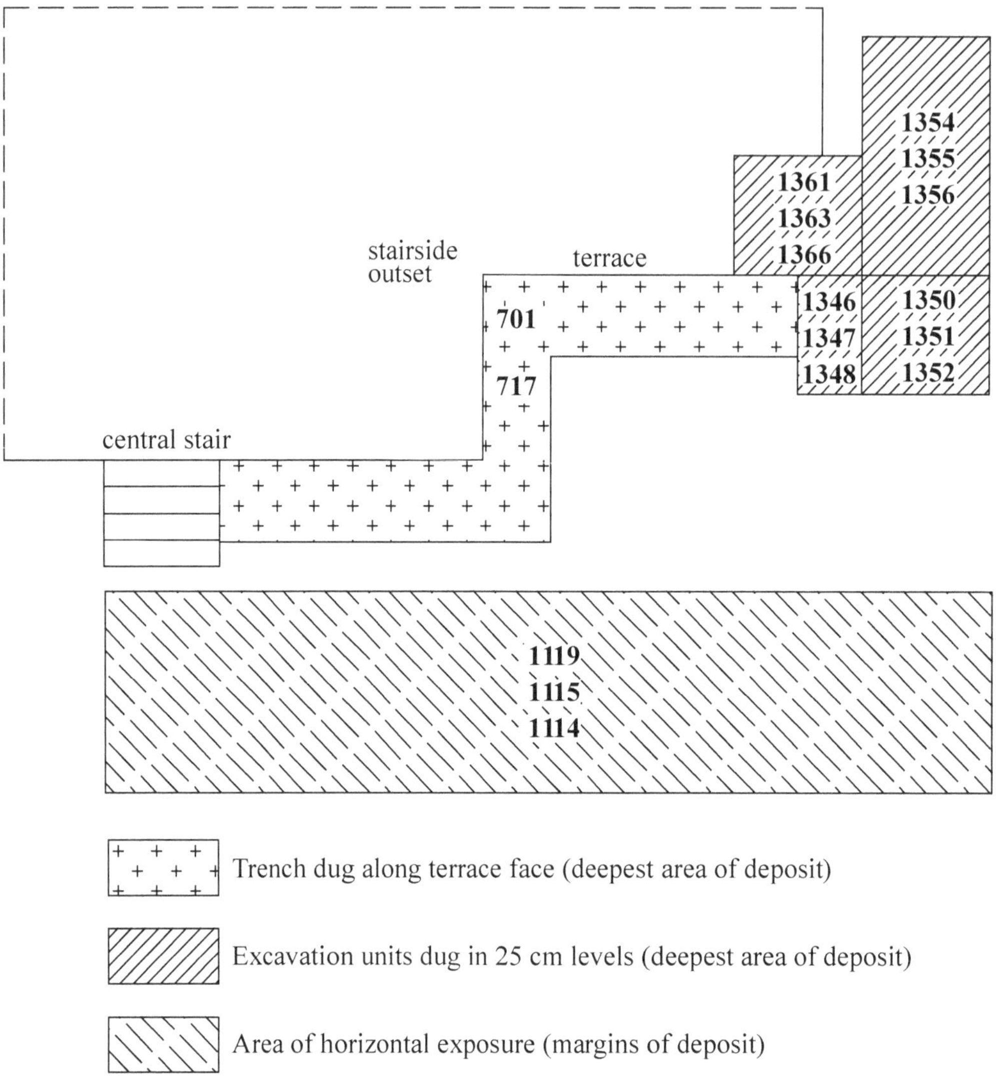

Figure 6.1: Schematic drawing of south front of structure N10-27 showing excavated areas of the refuse deposit and stratigraphic relationships among lots.

deposition within the midden deposit, and accordingly has permitted a close examination of the stratigraphic relationships (vertical and horizontal) among the different kinds of vessels comprising the midden assemblage. The results of this examination are presented in Chapter 7, along with a discussion of patterns of continuity and change in ceramic styles over the period in which the midden accumulated.

Ritual and Ceremonial Activities Associated with Ceremonial Structures

On a very general level, the kinds of activities associated with refuse accumulations can be inferred based on their contents, including the kinds of ceramic vessels they contain. Maya archaeologists generally recognize at least five functional classes of vessels, which are distinguished based on the differing functional significance of the vessel forms, as well as the comparative quality and presence/absence of decorative treatments (e.g. Fry 1981, 1979; Reents-Budet 1994). These classes include serving vessels, water jars, utilitarian vessels, musical instruments and incense-burning paraphernalia. Serving vessels include vases, bowls, dishes and plates that are comparatively well made and have high quality decorative treatments. Among these wares, figure-painted polychrome vessels featuring detailed pictorial scenes and/or functional hieroglyphic inscriptions are considered to have played an important role in elite household rituals and feasting (Reents-Budet 1994). The remainder of the serving vessels, which includes less elaborately decorated polychrome vessels as well as bichrome and monochrome slipped vessels, sometimes with additional decoration, are thought to have served as table wares and items for display and use in a range of social and ritual contexts. Large dishes and bowls may have had additional utilitarian functions. Although also a serving vessel, slipped narrow mouth jars are generally thought to have functioned specifically as containers for water and other liquids. Utilitarian vessels generally include unslipped s

Figure 6.2: Scene on a polychrome vase showing a meeting of dignitaries at which food and drink are consumed (Reents-Budet 1994:76). (see colour version on accompanying CD)

dishes and bowls and large, wide mouth jars that are most often comparatively poorly made, including the decorative treatment accorded (when present), and that have comparatively coarse-textured pastes. These vesselundoubtedly facilitated a range of tasks, including those related to the preparation and storage of food. Ceramic musical instruments include various forms of drums and whistles, whilst vessels thought to have functioned primarily as containers for burning incense include such forms as incense burners or colanders, frying pan or ladle censers and other censer forms.

The pictorial scenes depicted on polychrome vessels and other media provide additional evidence of the kinds of activities for which ceremonial structures served as a backdrop or setting. The central role that the consumption of food and drink played in a variety of historical and religious events is confirmed by the numerous representations of pottery vessels in the historical and mythological scenes painted on figure painted polychrome vessels (Figure 6.2). In addition, the scenes and events depicted on carved stone monuments sometimes feature dancers and musicians playing drums, suggesting that at least some of the events that took place in the vicinity of ceremonial structures involved music and dancing (Reents-Budet 1994:85). Within the context of particular religious rites, plates and dishes may have also served as containers for sacrificial and dedicatory offerings. Plates containing sacrificial blood, strips of cloth or paper spattered with human blood, and other ritual objects, sometimes occur in scenes depicted on stone monuments, painted murals and polychrome vessels (Reents-Budet 1994:80). The carbon deposits that occur on the interior bases of composite silhouette and rounded dishes and chalices at Lamanai provide further evidence that substances or materials, perhaps in the form of offerings, were sometimes burnt in these vessels.

As reported in Chapter 7, all of the functional classes of ceramics described above are represented in the midden deposits at N10-9 and N10-27, however, a disproportionately large volume of serving vessels and large storage jars occurs in each case. This situation, together with the fragmentation pattern of the ceramics, which tends towards large fragments and nearly complete vessel profiles, suggests that these accumulations cannot be considered as representing standard domestic refuse; a contention that is supported by the faunal data from the N10-9 midden (see Chapter 2). The basic composition of the ceramic component of the deposits, in terms of the range of functional classes of ceramics represented, as well as their relative frequencies, suggests that associated events and activities involved the consumption of food and drink, music and dancing, and the burning of incense and perhaps other materials or substances. In addition, the large quantity of storage jars contained in the deposits might be perceived as suggesting that, at least in some cases, these events and activities involved large gatherings of people. Alternatively, their occurrence might simply reflect additional functions of the ceremonial structures and associated perishable buildings, which may have served, in part, as storehouses or 'drop off points' for tribute, including materials and products stored in large jars. As was discussed in Chapter 2, the first occurrence of large refuse accumulations in association with ceremonial structures during the Terminal Classic would appear to signify a significant shift in attitude toward the maintenance and upkeep of areas of the site that served as important settings for ceremonial and ritual activity. However, when the nature of the ceramic component of these deposits is considered, it would seem that the appearance of these deposits may also reflect a important change in attitude regarding the appropriate treatment of vessels after their use in particular contexts of ceremonial and ritual activity, or alternatively, a significant change in religious practice in general.

Reconstructed Vessels Deriving from Other Cultural Contexts

The reconstructed vessels that were included in this study derive from a variety of different contexts within numerous structures located throughout the central precinct and the immediate areas surrounding it (Map 6.1, Table 6.1). The structures themselves have a variety of different functions, as sites of ceremonial and religious activities, as residences, and as buildings and building complexes where elites, including public and religious officials, resided, and from which they conducted their public and administrative duties. The majority of the vessels derive from primary archaeological contexts within these structures, such as burials, caches and other offerings, that reflect a range of ritual and ceremonial practices associated with different habitation and communally built structures within the community. Fewer vessels derive from refuse deposits found in direct association with various structures, and most of these deposits appear to be primary depositions or accumulations. While the history of the contents of such deposits is nearly always indiscernible, the fact that the entire vessel was present in most cases, suggests, at the very least, that the deposition of the broken vessels was not followed by subsequent removal to another location, or conversely, that the vessels were not first deposited elsewhere. In two cases, the vessels derive from primary contexts for which broader cultural associations are even more ambiguous (see 'Isolated Contexts in Areas Adjacent to the Lagoon' in Table 6.1). One of the vessels comes from a burial and the other from an isolated find, and both discoveries were made during construction and related activities in the excavation camp, which at the time of the discoveries, was situated adjacent to the

Table 6.1: Structures and archaeological contexts of the reconstructed vessels.

Structure Number	Type of Structure	Contexts
Ceremonial Structures		
N10-1	very small platform (uncertain use)	1 burial
N10-7	large ceremonial structure	3 burials 1 cache
N10-10	small platform	1 isolated find in stair core
N10-9	temple	1 burial 2 caches in architectural features 1 cache of smashed vessels over central stair associated refuse deposit (1 vessel)
N10-43	temple	2 caches 2 isolated finds in architectural features
N10-27	temple	6 isolated finds in architectural features associated refuse deposit (2 vessels)
Residential/Administrative Structure		
N10-2		6 burials 4 isolated finds in architectural features
Residential/administrative Complex north of N10-9 plaza		
N10-12		1 cache
N10-14		2 burials 2 caches
N10-15		1 burial 4 caches 7 isolated finds in architectural features levine pit (7 vessels) associated refuse deposit (1 vessel)
N10-17		1 burial 1 cache
N10-18		2 isolated finds in architectural features 1 isolated find in courtyard associated refuse deposit (8 vessels)
N10-28		1 burial
court yard area		2 isolated finds

Table 6.1: Structures and Archaeological Contexts of the Reconstructed Vessels (continued).

Structure Number	Type of Structure	Contexts
Residential Structures		
N9-33		1 burial
N9-70		1 burial
N9-71		3 isolated finds in architectural features
N10-4		11 burials
N10-66		3 burials, 1 cache
N10-67		1 burial
N11-5		1 burial 1 isolated find in an architectural feature
N11-7		1 isolated find in an architectural feature
N11-9		1 burial
N11-18		pit feature (1 vessel)
P7-12		1 burial
P8-102		5 burials associated refuse deposit (1 vessel)
P8-103		associated refuse deposit (2 vessels)
P8-104	residential structure/ special use	1 cache 1 burial
Structure of uncertain use		
N10-30		burial (post abandonment interment)
Isolated Contexts in Areas Adjacent to the Lagoon		
Isolated find in camp garbage pit#3, depth ca. 55cm		
Miscellaneous Burial 8		

lagoon, directly west of the Ottawa Group. Although explicit contextual information for these vessels on a broader level is lacking, it is clear that neither of the discoveries derive from structures, suggesting that such deposits were not confined to these particular contexts, and also, at least in the case of the burial, that these discoveries reflect a facet of ritual activity that is perhaps less well know owing a continuing focus on the excavation of architectural remains. A summary of the structural and depositional contexts of the reconstructed vessels is presented in Table 6.1 and the vessels are presented and described individually, together with the contextual information, in Appendix VIII. More detailed descriptions and illustrations appear in Appendix IX.

Behavioural Associations of Offerings and Burials

While burials, caches and other offerings provide a wealth of evidence about the specifics of ritual and ceremonial practices, the broader context of these discoveries – the particular kind of structure or location (for example a cave) in which they occur – provides additional insight into the objectives and beliefs that lay behind these ritual acts. When placed in buildings and other structures, these discoveries take on an additional temporal commemorative significance, reflecting beliefs surrounding the construction and maintenance (in both a physical and spiritual sense) of public and private architecture.

OFFERINGS

Coe (1965:462) defines a cache as a 'hidden offering' placed in the fill of a structure or structural modifications and additions, often occurring on the central, front-rear axis of the structure. These offerings presumably have a dedicatory or commemorative function – to a personage, a deity, an event a chronological cycle or other purpose – their objective relating to the construction itself. Alternatively, Pendergast (1998:61) suggests that in cases where caches were placed on the central axis of a ceremonial structure, in building additions or as part of modifications, part of the intention may have been to amplify already existing power, since the central axis forms the 'lifeline' of the structure and the "main avenue of communication with a deity". At Lamanai, caches occur in habitation structures as well as communally built structures that served ceremonial or other functions, suggesting that these offerings may have had the same function, dedicatory or otherwise, within these different contexts (Plate VII.36). Residential caches, however, tend to be more variable in regard to their position and makeup, perhaps signifying greater freedom in the rites performed in these family contexts as compared to those associated with communally built structures dedicated to public welfare (Pendergast 1998:61-62). Although cache composition and deposition patterns are generally more consistent in communally built structures at Lamanai, it would appear that "there was clearly no hard fast rule regarding placement of offerings along the primary axis of new construction" (Pendergast 1998:57). Pendergast (1998:57) observes that "a building renewal might contain an offering although its predecessor had none, and the presence of such an offering certainly did not dictate similar deposition in succeeding modification". This evidence not only raises the important point that over the course of a structure's construction history caches were interred rather inconsistently, but also suggests that placement of hidden offerings cannot be considered as a mandatory part of construction activities.

Other offerings in structures appear to mark the termination of use of particular areas of the structure or the structure itself, whether it is on the verge of abandonment or being buried and replaced by a new one. In these cases objects are unhidden, often left in a whole or fragmentary state on the surface of the obsolete structure or superficially intruded into it, or in interior rooms (Coe 1965). At Lamanai, many of the reconstructed vessels recovered from isolated contexts within rooms, on structure surfaces and within building cores and other architectural elements would appear to be offerings of this type. Termination rituals most often involve the intentional breakage of material items such as pottery, or other destructive behaviour. The intention of these acts was possibly to "terminate contact with the powerful conduits to the underworld", and when they occurred in conjunction with construction activities, to desanctify the previous structure and to enclose its sacred portals (Mock 1998:10). Alternatively, material items may be broken as part of rituals with a more dedicatory objective. In this case, the intention can be perceived as an attempt to rid objects of their power once the ritual act has been completed.

BURIALS

When considering the Maya area as a whole, the burial patterns associated with habitation and communally built structures vary greatly across time and space (see reviews by Ruz 1965 and Welsh 1988). This apparent lack of patterning would seem to suggest that the beliefs, reasons and motives surrounding these interments were equally variable and highly complex. One thing that is certain, however, is that not all Maya were buried in these locations, suggesting that this was a distinction accorded only to certain or selected members of the community. In some cases, the burials appear to have had a commemorative or dedicatory function, marking the inauguration or termination of the construction of civil or ceremonial buildings, a calendrical event or cycle or the erection of a monument (as is the case with the burial below the stela at structure N10-27). In addition, in some instances the individuals laid to rest were sacrificial victims. Mock (1998:7) suggests that the deeper meaning behind interments placed in the foundations of a structure was to bring it 'to life'. A universal characteristic of Maya burial practice up until at least the Terminal Classic period is that of placing with the corpse, food or drink contained in pottery vessels. Ceramic items and other material culture were sometimes included as well, and are often perceived as the belongings of the individual or objects characterizing his or her office, status, age or sex (Ruz 1965).

THE FUNCTION OF POTTERY IN OFFERINGS AND BURIALS

The use of pottery vessels as offering containers, in funerary and other contexts, is a practice of great longevity in the Maya lowlands, beginning in the Middle Preclassic and continuing through to the contact period and beyond. These containers may have served as a "metaphor for *pibs*, 'earth ovens', where like food, the gods were transformed or conjured" (Mock 1998:7). D. Chase (1988:98) suggests, however, that during the Postclassic period, emphasis shifted away from the container and towards the anthropomorphic character of the vessel itself – particularly in regard to the effigy vessels and incense figurines that emerge during this period – since they embodied a standardized religious symbolism that was intended to unite the Maya populace. As was discussed in Chapter 2, at Lamanai, a marked change in offertory and funerary practices involving ceramic vessels occurs during the Terminal Classic, when the practice of smashing vessels *en masse* as offerings (for example the large cache recovered from N10-9) and scattering the broken pieces of smashed vessels over the dead as part of funerary rites (Plate VII.37) replaces the earlier pattern involving the interment of whole vessels. This change in the treatment of pottery vessels within contexts of ritual and ceremonial activity would appear to reflect an increase in the importance of termination offerings starting in the Terminal Classic period, while the importance of dedicatory offerings was decreasing.

6.2. DATING OF TERMINAL CLASSIC TO EARLY POSTCLASSIC CERAMICS AT LAMANAI

The basic chronological sequence of pottery styles at Lamanai was established by Pendergast (1981a, 1982) and Graham (1987). Terminal Classic pottery types were identified through comparison to the San Jose IV and San Jose V assemblages (Thompson 1939: Plates 15, 26, 18, 20 and 21) at the site of San Jose, situated 31km south west of Lamanai, and Spanish Lookout phase and New Town phase ceramics at the site of Barton Ramie in the Belize Valley (Gifford 1976:Fig. 148, 149 and 199). The Early Postclassic pottery styles at Lamanai, which are poorly represented or entirely absent at other sites, were previously termed Middle Postclassic based on similarities with the Hocaba ceramic complex (A.D. 1200-1300) (Smith 1971) at Mayapan; traits such as the use of incised and gouged decoration and the occurrence of serpent motifs, segmented basal flanges and hollow oven and globular vessel supports (feet) (Pendergast 1975:38, 1981a:48, 1982). In addition, it was observed that these Postclassic styles overlay and chronologically succeeded a discrete group of styles that appeared to represent manufacture from Terminal Classic to Early Postclassic times (Graham 1987). The terminology for the Terminal Classic to Late Postclassic sequence has since been adjusted in closer accord with terms as they are applied to specific time spans elsewhere, and also because the dating of the appearance of these styles (eleventh to twelfth century) is more secure than their end (Graham 2004). Radiocarbon dates from structure N10-2, which contained a large number of burials, indicate that production of Early Postclassic vessel styles was well established by A.D. 1140 (Pendergast 1981a:48). Building on the basic chronological sequence of pottery types established by Graham and Pendergast, aspects of continuity and change in pottery styles over the Terminal Classic to Early Postclassic period are examined in detail in the succeeding chapter, through a detailed analysis of the stylistic and stratigraphic relationships among the different vessel types comprising the ceramic assemblage that is the focus here.

CHAPTER 7
STYLISTIC ASPECTS OF THE CERAMICS: THE TERMINAL CLASSIC TO EARLY POSTCLASSIC ASSEMBLAGE

This chapter presents the results of the macroscopic analysis of the sherd assemblages and whole and reconstructed vessels described in the previous chapter. This analysis builds upon the work of Pendergast and Graham, who have presented the ceramic sequence corresponding to the Terminal Classic to Middle Postclassic periods of occupation at Lamanai based on the evidence from complete vessels in previous publications (e.g. Graham 1987, Pendergast 1982). The present study expands upon this body of knowledge by examining the composition of large sherd assemblages according to the nature, stratigraphic relationships and relative frequency of the vessels they contain, producing new detail regarding the ceramic sequence. The descriptions of the stylistic groups presented here are based on a detailed consideration of surface treatment, morphological and paste attributes, although aspects of variation surface treatment and morphology are the main focus. Variation in paste attributes is described in the detailed group descriptions (Appendix II), and is examined on the microscopic and chemical levels in subsequent chapters (Chapters 8 and 9).

7.1. General Composition of the Assemblage

The ceramic assemblage that forms the basis of this study comprises 32,428 sherds and 379 whole vessels. Approximately 8% of the assemblage was suitable for detailed study, a body of material that includes 2,330 sherds and 152 whole and reconstructed vessels (Table 7.1)

Whole Vessel Component

The whole vessel component of the assemblage includes complete and partially reconstructed vessels, as well as vessel forms represented by a complete or nearly complete profile. Of the whole vessels that were studied in detail, 96% (N=146) are fine ware vessels and only 3.9% (N=6) are coarse ware (Table 7.2). The preponderance of fine ware vessels within the whole vessel component of the assemblage stands in direct contrast to the near equal frequency of fine ware and coarse ware vessels in the sherd assemblages (see below). Since the majority of the whole vessels derive from specialized deposits relating to mortuary and offertory practices, it would seem that fine ware vessels (serving vessels and tableware) are more strongly associated with this kind of ritual and ceremonial activity.

Sherd Component

Out of the 32,428 sherds that were analysed, 899 derive from the midden associated with structure N10-9 and 31,529 derive from the midden associated with structure N10-27. As discussed in the previous chapter, the N10-27 midden comprises 14 lots and the N10-9 midden comprises only one lot, excavated over three different field seasons. Table 7.1 gives a breakdown of the total number of sherds recovered from each lot within the two middens. Within the N10-9 midden assemblage, 26% of the sherds derive from fine ware vessels and 74% derive from coarse ware vessels. Similarly, within the N10-27 midden, fine ware represents 21% of the material and coarse ware represents 79%. Accordingly, in contrast to ceramics deriving from specialized deposits, the majority of the material comprising the midden assemblages derives from undecorated, coarse-textured vessels with largely utilitarian functions.

Diagnostic sherds comprise 7.9% (N=2,559) of the material from both midden assemblages and a minimum of 2,330 individual vessels are represented (Table 7.1): 663 vessels in the N10-9 midden and 1,667 from N10-27 (Tables 7.3 and 7.4). Of the diagnostic sherds recovered from the N10-9 midden, 52.5% (N=348) derive from fine ware vessels and 47.5% (N=315) are from coarse ware vessels. This breakdown is similar to the midden associated with structure N10-27, where 49% (N=817) of the vessels are coarse ware and 51% (N=850) are fine ware. In terms of their basic composition, therefore, the two midden assemblages are practically identical, containing near equal amounts of serving vessels and other types of fine ware and utilitarian vessels, the vast majority of which (approximately 90%) are storage jars (see Table 7.5). This basic similarity between the two middens in terms of the proportion of serving vessels to utilitarian forms suggests that the types of human activity that gave rise to these accumulations may have been quite similar in nature. The high frequency of utilitarian vessels would appear to be of particular significance since it presents a sharp contrast to totals reported for domestic middens at other sites in the southern lowlands. For instance, Fry (1980:10) reports that unslipped utilitarian vessels comprise only 18-21% of domestic midden assemblages at Tikal. The high percentage of coarse ware vessels in the Lamanai assemblage, therefore, suggests an important difference between the nature of the ceramic refuse comprising the Lamanai middens, which is associated with activities taking place in proximity to ceremonial structures, and that generally found in association with domestic structures.

7.2. Description of Ceramic Groups

The ceramic descriptions that follow are organized according to *ware* (i.e. *fine ware* and *coarse ware*). The descriptions presented for each group of fine ware and coarse ware vessels is based on the combined information

Table 7.1: Sherd counts and minimum numbers of vessels for all lots.

CONTEXT	Lot Number	Sherd Count	Minimum Number of Vessels
Midden at Structure N10-9	LA 187	899	663
Midden at Structure N10-27	LA 701	15404	671
	LA 717	3826	278
	LA 1114	8091	346
	LA 1115	216	30
	LA 1119	296	20
	LA 1346	227	20
	LA 1347	260	10
	LA 1348	161	17
	LA 1350	189	24
	LA 1351	491	45
	LA 1352	183	30
	LA 1354	165	2
	LA 1355	630	89
	LA 1356	1361	58
	other lots	29	27
Sub Total of Lots from N10-27		31529	1667
Total Sherd Count		32428	2330
Whole Vessels - Various Contexts			
All vessels			379
Vessels studied in detail			152
Sub Total			152
Vessel Count (Grand Total)			**2482**

obtained through the detailed study of whole vessels on the one hand, and the sherd assemblages on the other. When viewed independently, neither of these two bodies of material provide a sufficiently detailed and complete picture of the Terminal Classic to Early Postclassic ceramic assemblage as a whole, or the variation within particular vessel groups. For example, although the identification of many of the vessel forms represented in the sherd assemblages could be achieved through direct comparison of vessel fragments with the whole and nearly complete vessels, there were a variety forms, particularly different kinds of jars, for which whole or nearly complete examples were entirely absent. Hence, the characterization of the morphology and surface treatment of these forms had to pieced together based on the study of fragments deriving from various anatomical areas of what appeared to be the same kind of vessel. Likewise, some of the vessel forms and surface treatments represented among the whole and nearly complete vessels appear to be absent in the sherd assemblages. Since the sherds and whole vessels provide different and complementary information about Terminal Classic to Early Postclassic ceramics at the site, the final descriptions of the vessel groups are based on a consideration of both bodies of data, and by doing so provide a more comprehensive characterization of ceramics at the site during this time period in terms of the following criteria:

- specific details of the morphology and surface treatment of different kinds of vessels.

- variation in various aspects of morphology and surface treatment within groups of related vessels defined according to particular attributes they share in common.
- the relative frequency of different kinds of vessels within the assemblage as a whole or comprising particular vessel groups.

For the sake of space, the descriptions that follow are summarized versions of the more detailed descriptions that appear in Appendix II.

Table 7.2: Surface treatment summarized by count and percent for whole vessels.

SURFACE TREATMENT	Type Totals	Type Totals (%)	% of Type within ware
Polychrome Slipped and Painted	10	6.6	6.8
Black on Red Slipped and Painted	6	3.9	4.1
Red-Orange-Black Resist	2	1.3	1.4
Monochrome Black	15	9.9	10.3
Monochrome Red Slipped	27	17.8	18.5
Monochrome Orange Slipped	37	24.3	25.3
Monochrome Red to Orange Slipped (interior-slipped)	2	1.3	1.4
Monochrome Orange Slipped and Incised	38	25.0	26.0
Buff/Cream Slipped - Plain or Incised	2	1.3	1.4
Monochrome Brown Slipped	2	1.3	1.4
Monochrome Red Slipped and Incised	1	0.7	0.7
Monochrome Red Slipped, Stuccoed and Painted	1	0.7	0.7
Monochrome Red Slipped and Model-Carved	1	0.7	0.7
Monochrome Red Slipped and Groove-Incised	2	1.3	1.4
Total Fine Ware	146	96.1	100.0
Unslipped-Striated	1	0.7	16.7
Unslipped/Smoothed-Buff	1	0.7	16.7
Unslipped-Utilitarian	2	1.3	33.3
Unslipped with Appliqué	2	1.3	33.3
Total Coarse Ware	**6**	3.9	100.0
GRAND TOTAL	**152**	100.0	

Table 7.3: Surface treatment summarized by count and percent for LA187.

	Whole Assemblage					LA187			
SURFACE TREATMENT	LA 187	N10-27 All Lots	Type Totals	Type Totals (%)	% of Type within ware	% within ware (in LA187)	% within LA 187 (Grand Total)	% within total ware	% within Whole Assemblage
Polychrome Slipped and Painted	0	64	64	2.7	5.3	0	0	0	0
Black on Red Slipped and Painted	0	2	2	0.1	0.2	0	0	0	0
Red-Orange-Black Resist	0	23	23	1	1.9	0	0	0	0
Monochrome Black	1	21	22	0.9	1.8	0.3	0.2	0.1	0
Monochrome Red Slipped	6	80	86	3.7	7.2	1.7	0.9	0.5	0.3
Monochrome Orange Slipped	198	309	507	21.8	42.3	56.9	29.9	16.5	8.5
Monochrome Red to Orange	0	114	114	4.9	9.5	0	0	0	0
Monochrome Red-Orange-too eroded	3	28	31	1.3	2.6	0.9	0.5	0.3	0.1
Red-Orange-Resist-too eroded	1	24	25	1.1	2.1	0.3	0.2	0.1	0
Monochrome Orange Slipped and Incised	135	160	295	12.7	24.6	38.8	20.4	11.3	5.8
Monochrome Orange Slipped and Groove-Incised	0	3	3	0.1	0.3	0	0	0	0
Monochrome Orange to Red Slipped - Notched and Incised	3	4	7	0.3	0.6	0.9	0.5	0.3	0.1
Monochrome Orange with Appliqué	1	13	14	0.6	1.2	0.3	0.2	0.1	0
Buff/Cream Slipped - Plain or Incised	0	2	2	0.1	0.2	0	0	0	0
Monochrome Grey (Plumbate)	0	1	1	0	0.1	0	0	0	0
Monochrome Red Slipped and Notched	0	2	2	0.1	0.2	0	0	0	0
Fine Ware Total	**348**	**850**	**1198**	**51.4**	**100**	**100**	**52.5**	**29**	**14.9**
Slipped Rim/Lip - Smoothed Body	7	202	209	9	18.5	2.2	1.1	0.6	0.3
Red-Brown-Striated	0	17	17	0.7	1.5	0	0	0	0
Unslipped/Smoothed-Perforated	15	4	19	0.8	1.7	4.8	2.3	1.3	0.6
Unslipped/Smoothed-Buff	76	23	99	4.2	8.7	24.1	11.5	6.7	3.3
Unslipped Utilitarian	217	565	782	33.6	69.1	68.9	32.7	19.2	9.3
Washed/Smoothed	0	6	6	0.3	0.5	0	0	0	0
Coarse Ware total	**315**	**817**	**1132**	**48.6**	**100**	**100**	**47.5**	**27.8**	**13.5**
GRAND TOTAL by context	**663**	**1667**	**2330**	**100**					

Table 7.4: Surface treatment summarized by count and percent for LAN10-27 (all lots).

	Whole Assemblage					N10-27 - All Lots			
SURFACE TREATMENT	LA 187	N10-27 All Lots	Type Totals	Type Totals (%)	% of Type within ware	% within ware in N10-27 (all lots)	% within N10-27 all lots (Grand Total)	% within total ware	% within Whole Assemblage
Polychrome Slipped and Painted	0	64	64	2.7	5.3	7.5	3.8	5.3	2.7
Black on Red Slipped and Painted	0	2	2	0.1	0.2	0.2	0.1	0.2	0.1
Red-Orange-Black Resist	0	23	23	1	1.9	2.7	1.4	1.9	1
Monochrome Black	1	21	22	0.9	1.8	2.5	1.3	1.8	0.9
Monochrome Red Slipped	6	80	86	3.7	7.2	9.4	4.8	6.7	3.4
Monochrome Orange Slipped	198	309	507	21.8	42.3	36.4	18.5	25.8	13.3
Monochrome Red to Orange	0	114	114	4.9	9.5	13.4	6.8	9.5	4.9
Monochrome Red-Orange-too eroded	3	28	31	1.3	2.6	3.3	1.7	2.3	1.2
Red-Orange-Resist-too eroded	1	24	25	1.1	2.1	2.8	1.4	2	1
Monochrome Orange Slipped and Incised	135	160	295	12.7	24.6	18.8	9.6	13.4	6.9
Monochrome Orange Slipped and Grooved-Incised	0	3	3	0.1	0.3	0.4	0.2	0.3	0.1
Monochrome Orange to Red Slipped - Notched and Incised	3	4	7	0.3	0.6	0.5	0.2	0.3	0.2
Monochrome Orange with Appliqué	1	13	14	0.6	1.2	1.5	0.8	1.1	0.6
Buff/Cream Slipped - Plain or Incised	0	2	2	0.1	0.2	0.2	0.1	0.2	0.1
Monochrome Grey (Plumbate ware)	0	1	1	0	0.1	0.1	0.1	0.1	0
Monochrome Red Slipped and Notched	0	2	2	0.1	0.2	0.2	0.1	0.2	0.1
Fine Ware Total	348	850	1198	51.4	100	100	51	71	36.5
Slipped Rim/Lip - Smoothed Body	7	202	209	9	18.5	24.7	12.1	17.8	8.7
Red-Brown-Striated	0	17	17	0.7	1.5	2.1	1	1.5	0.7
Unslipped/Smoothed-Perf0rated	15	4	19	0.8	1.7	0.5	0.2	0.4	0.2
Unslipped/Smoothed-Buff	76	23	99	4.2	8.7	2.8	1.4	2	1
Unslipped Utilitarian	217	565	782	33.6	69.1	69.2	33.9	49.9	24.2
Washed/Smoothed	0	6	6	0.3	0.5	0.7	0.4	0.5	0.3
Coarse Ware total	315	817	1132	48.6	100	100	49	72.2	35.1
GRAND TOTAL by context	663	1667	2330	100					

Table 7.5: Surface treatment and form summarized by count and percent for the two sherd assemblages.

SURFACE TREATMENT		FORM	LA 187		N10-27 All Lots		Whole Assemblage			
			Count	%	Count	%	Totals by Surface Treatment	Totals by Surface Treatment (%)	Total by Form	% Form within Surface Treatment
Polychrome Slipped and Painted		vases	0	0	5	0.3			5	7.8
		bowls	0	0	16	1			16	25
		tripod bowls	0	0	2	0.1			2	3.1
		thick-walled, open vessels (basin?)	0	0	3	0.2			3	4.7
	(interior-slipped)	dishes	0	0	38	2.3			38	59.4
	Total		**0**	**0**	**64**	**3.8**	**64**	**2.7**		
Black on Red Slipped and Painted	**Total**	**massive bowls**	**0**	**0**	**2**	**0.1**	**2**	**0.1**		
Red-Orange-Black Resist	**Total**	**composite silhouette dishes**	**0**	**0**	**23**	**1.4**	**23**	**1**		
Monochrome Black (all decorative treatments)		vases	0	0	4	0.2			4	18.2
		bowls	1	0.2	17	1			18	81.8
	Total		**1**	**0.2**	**21**	**1.3**	**22**	**0.9**		
Monochrome Red Slipped		vases	0	0	1	0.1			1	1.2
		bowls	0	0	22	1.3			22	25.6
		tripod bowls	0	0	3	0.2			3	3.5
		jars	2	0.3	11	0.7			13	15.1
		composite silhouette dishes	2	0.3	17	1			19	22.1
		chalices*	0	0	3	0.2			3	3.5
	(interior-slipped)	rounded bowls	0	0	18	1.1			18	20.9
		rounded dishes	2	2.3	5	0.3			7	8.1
	Total		**6**	**2.9**	**80**	**4.8**	**86**	**3.7**		
Monochrome Orange Slipped		composite vessels	2	0.3	0	0			2	0.4
		sieve	1	0.2	0	0			1	0.2
		vases	0	0	2	0.1			2	0.4
		bowls	128	19.3	145	8.7			273	53.8
		tripod bowls	42	6.3	38	2.3			80	15.8
		jars	21	3.2	52	3.1			73	14.4

Table 7.5: Surface treatment and form summarized by count and percent for the two sherd assemblages (continued).

SURFACE TREATMENT		FORM	LA 187		N10-27 All Lots		Whole Assemblage			
			Count	%	Count	%	Totals by Surface Treatment	Totals by Surface Treatment (%)	Total by Form	% Form within Surface Treatment
Monochrome Orange Slipped		drums	3	0.5	18	1.1			21	4.1
		plate/lid	1	0.2	0	0			1	0.2
		composite silhouette dishes	0	0	3	0.2			3	0.6
		chalices**	0	0	23	1.4			23	4.5
	(interior-slipped)	rounded bowls	0	0	1	0.1			1	0.2
		rounded dishes	0	0	27	1.6			27	5.3
	Total		**198**	**29.9**	**309**	**18.5**	**507**	**21.8**		
Monochrome Red to Orange		composite silhouette dishes	0	0	31	1.9			31	27.2
		jars	0	0	16	1			16	14
	(interior-slipped)	rounded bowls	0	0	49	2.9			49	43
		rounded dishes	0	0	18	1.1			18	15.8
	Total		**0**	**0**	**114**	**6.8**	**114**	**4.9**		
Monochrome Red-Orange-too eroded		bowls	0	0	2	0.1			2	6.5
		jars	3	0.5	22	1.3			25	80.6
	(interior-slipped)	rounded dishes	0	0	4	0.2			4	12.9
	Total		**3**	**0.5**	**28**	**1.7**	**31**	**1.3**		
Red-Orange-Resist-too eroded	**Total**	**composite silhouette dishes**	**1**	**0.2**	**24**	**1.4**	**25**	**1.1**		
Monochrome Orange Slipped and Incised		bowls	51	7.7	83	5			134	45.4
		chile grinders	34	5.1	3	0.2			37	12.5
		tripod bowls with segmented basal flange	12	1.8	3	0.2			15	5.1
		pedestal-based jars	9	1.4	7	0.4			16	5.4
		jars	7	1.1	2	0.1			9	3.1
		drums	1	0.2	4	0.2			5	1.7
		frying pans	2	0.3	2	0.1			4	1.4
		whistles	2	0.3	1	0.1			3	1
		chalices***	17	2.6	55	3.3			72	24.4
	Total		**135**	**20.4**	**160**	**9.6**	**295**	**12.7**		

Table 7.5: Surface treatment and form summarized by count and percent for the two sherd assemblages (continued).

SURFACE TREATMENT		FORM	LA 187		N10-27 All Lots		Whole Assemblage			
			Count	%	Count	%	Totals by Surface Treatment	Totals by Surface Treatment (%)	Total by Form	% Form within Surface Treatment
Monochrome Orange Slipped and Groove-Incised		drum	0	0	1	0.1			1	33.3
		vases	0	0	2	0.1			2	66.7
	Total		0	0	3	0.2	3	0.1		
Monochrome Orange to Red Slipped - Notched and Incised	Total	composite silhouette bowls	3	0.5	4	0.2	7	0.3		
Total Monochrome Orange with Appliqué	Total	bowls	1	0.2	13	0.8	14	0.6		
Total Buff/Cream Slipped - Plain or Incised	Total	bowls	0	0	2	0.12	2	0.1		
Total Monochrome Grey (Plumbate ware)	Total	pedestal base	0	0	1	0.06	1	0		
Total Monochrome Red Slipped and Notched	Total	large bowls	0	0	2	0.12	2	0.1		
Total Fine Ware			**348**	**54.5**	**850**	**51**	**1198**	**51.4**		
Slipped Rim/Lip - Smoothed Body	Total	jars	7	1.1	202	12.1	209	9		
Red-Brown-Striated	Total	jars	0	0	17	1	17	0.7		
Unslipped/Smoothed-Perforated	Total	Incense burners	15	2.3	4	0.2	19	0.8		
Unslipped/Smoothed-Buff		jars	58	8.7	18	1.1			76	76.8
		plate/lid	18	2.7	0	0			18	18.2
		bowls	0	0	4	0.2			4	4
		broad rimmed bowl	0	0	1	0.1			1	0.1
	Total		76	11.5	23	1.4	99	4.2		
Unslipped Utilitarian		jars	182	27.5	486	29.2			668	85.4
		bowls	2	0.3	20	1.2			22	2.8
		dishes	6	0.9	0	0			6	0.8
		plates/lids	23	3.5	57	3.4			80	10.2

Table 7.5: Surface treatment and form summarized by count and percent for the two sherd assemblages (continued).

SURFACE TREATMENT		FORM	LA 187		N10-27 All Lots		Whole Assemblage			
			Count	%	Count	%	Totals by Surface Treatment	Totals by Surface Treatment (%)	Total by Form	% Form within Surface Treatment
Unslipped Utilitarian		comals/large thick-walled plates	3	0.5	0	0			3	0.4
		censer base, basal apron	1	0.2	0	0			1	0.1
		large ring base	0	0	1	0.1			1	0.1
		insloping, thick-walled jar		0	1	0.1			1	0.1
	Total		**217**	**32.7**	**565**	**33.9**	**782**	**33.6**		
Total Washed/smoothed	**Total**	**jars**	**0**	**0**	**6**	**0.4**	**6**	**0.3**		
Total Coarse Ware			**315**	**47.5**	**817**	**49**	**1132**	**48.6**		
GRAND TOTAL by cpntext			**663**		**1667**		**2330**			

Notes: * Counts are based on the total number of dish fragments
** Counts are based on the total number of plain orange-slipped pedestal vases
*** Counts are based on the total number of dish fragments minus the total number of plain orange-slipped pedestal vases

Fine Ware – General Summary

Surface Treatment - The fine ware component of the assemblage constitutes 146 whole vessels and 1,198 sherds (Table 7.6). A total of sixteen different surface treatments were recognized, nine of which are quite rare within the assemblage as a whole, being represented by less than five vessels in each case. The majority of the fine ware vessels (approximately 65%) are either orange-slipped or orange-slipped with incised decoration. Red-slipped vessels are the next most abundant, followed by polychrome vessels and monochrome black vessels with plain or decorated surfaces. Comparatively less frequent are vessels with resist decoration, orange-slipped vessels with an appliqué impressed fillet and red-slipped vessels with black painted decoration. Rare surface treatments within the assemblage include brown-slipped vessels, buff- or cream-slipped vessels with plain or incised surfaces (two of these are Slate ware), orange-to-red-slipped vessel with notched and incised decoration, orange-slipped vessels with groove-incised decoration and red-slipped vessels with painted stucco, incised, model-carved, groove-incised or notched decoration. The one monochrome grey-slipped sherd within the assemblage likely derives from a Plumbate ware vessel. The rare surface treatment types tend to occur either in the sherd assemblages or among the whole vessels but not in both. The exception to this is Buff/Cream Slipped – Plain or Incised group which was encountered in both sets of material. It also should be noted that the surface treatment of two of the vessels within this group is consistent with that characteristic of 'Slate ware'.

There is considerable variation in the general 'quality' of the surface treatments on fine ware vessels, both within and between the different surface treatment groups, in terms of the relative hardness, glossiness and consistency of the slip (thickness and colour) on finished surfaces. These qualitative differences between the surface treatments seen on individual vessels and sherds appear to have had a direct effect on the ability of finished surfaces to survive within different postdepositional environments. Perhaps not surprisingly, vessels and sherds with hard, glossy, or somewhat waxy surface finishes tend to be the best preserved within the assemblage. In contrast, vessels with softer and less glossy surface treatments are often heavily eroded, and the slip is so easily removed, that when washing them with a toothbrush, it was often impossible to remove the dirt without also removing the slip. Since most of the sherd lots contain both well preserved and heavily eroded examples of the same surface treatment, it would seem that the differential preservation of vessel surfaces within the assemblage is perhaps less due to postdepostional environments and processes than to fundamental differences in techniques employed to achieve particular surface treatments. These differences, which are reflected both within and between surface treatments, are explored in detail in Chapter 10.

As was discussed in the section of the methodology chapter dealing with the recording of macroscopic vessel

Table 7.6: Surface treatment summarized by count and percent for whole vessels and sherd lots.

	Whole Vessels			Sherds			Whole Assemblage		
SURFACE TREATMENT	Type Totals	Type Totals (%)	% of Type within ware	Type Totals	Type Totals (%)	% of Type within ware	Type Totals	Type Totals (%)	% of Type within ware
Polychrome Slipped and Painted	10	6.6	6.8	64	2.7	5.3	74	3	5.5
Black on Red Slipped and Painted	6	3.9	4.1	2	0.1	0.2	8	0.3	0.6
Red-Orange-Black Resist	2	1.3	1.4	23	1	1.9	25	1	1.9
Monochrome Black	15	9.9	10.3	22	0.9	1.8	37	1.5	2.8
Monochrome Red Slipped	27	17.8	18.5	86	3.7	7.2	113	4.6	8.4
Monochrome Orange Slipped	37	24.3	25.3	507	21.8	42.3	544	21.9	40.5
Monochrome Red to Orange Slipped (interior-slipped)	2	1.3	1.4	114	4.9	9.5	116	4.7	8.6
Monochrome Red-Orange-too eroded	0	0	0	31	1.3	2.6	31	1.2	2.3
Red-Orange-Resist-too eroded	0	0	0	25	1.1	2.1	25	1	1.9
Monochrome Orange Slipped and Incised	38	25	26	295	12.7	24.6	333	13.4	24.8
Monochrome Orange Slipped and Groove-Incised	0	0	0	3	0.1	0.3	3	0.1	0.2
Monochrome Orange to Red Slipped - Notched and Incised	0	0	0	7	0.3	0.6	7	0.3	0.5
Monochrome Orange with Appliqué	0	0	0	14	0.6	1.2	14	0.6	1
Buff/Cream Slipped - Plain or Incised	2	1.3	1.4	2	0.1	0.2	4	0.2	0.3
Monochrome Grey (Plumbate ware)	0	0	0	1	0	0.1	1	0	0.1
Monochrome Brown Slipped	2	1.3	1.4	0	0	0	2	0.1	0.1
Monochrome Red Slipped and Incised	1	0.7	0.7	0	0	0	1	0	0.1
Monochrome Red Slipped, Stuccoed and Painted	1	0.7	0.7	0	0	0	1	0	0.1
Monochrome Red Slipped and Model-Carved	1	0.7	0.7	0	0	0	1	0	0.1
Monochrome Red Slipped and Groove-Incised	2	1.3	1.4	0	0	0	2	0.1	0.1
Monochrome Red Slipped and Notched	0	0	0	2	0.1	0.2	2	0.1	0.1
Total Fine Ware	**146**	**96.1**	**100**	**1198**	**51.4**	**100**	**1344**	**54.1**	**100**
Slipped Rim/Lip - Smoothed Body	0	0	0	209	9	18.5	209	8.4	18.4
Red-Brown-Striated	0	0	0	17	0.7	1.5	17	0.7	1.5
Unslipped/Smoothed-Perforated	0	0	0	19	0.8	1.7	19	0.8	1.7
Unslipped/Smoothed-Buff	1	0.7	16.7	99	4.2	8.7	100	4	8.8
Unslipped-Utilitarian	2	1.3	33.3	782	33.6	69.1	784	31.6	68.9
Washed/Smoothed	0	0	0	6	0.3	0.5	6	0.2	0.5
Unslipped with Appliqué	2	1.3	33.3	0	0	0	2	0.1	0.2
Unslipped-Striated	1	0.7	16.7	0	0	0	1	0	0.1
Total Coarse Ware	**6**	**3.9**	**100**	**1132**	**48.6**	**100**	**1138**	**45.9**	**100**
GRAND TOTAL	**152**	**100**	**200**	**2330**	**100**	**200**	**2482**	**100**	**200**

Table 7.7: Surface treatment and form summarized by count and percent for whole vessels.

SURFACE TREATMENT		FORM	Count	%	% form within Surface Treatment
Polychrome Slipped and Painted		cylindrical vase	1	0.7	10
		composite silhouette bowls with a ring base	4	2.6	40
	(interior-slipped)	rounded dishes	4	2.6	40
		rounded dish with an incipient ring base	1	0.7	10
		total dishes	5	3.3	50
	Total		**10**	**6.6**	**100**
Black on Red Slipped and Painted		massive bowl	1	0.7	16.7
		rounded dishes	5	3.3	83.3
	Total		**6**	**3.9**	**100**
Red-Orange-Black Resist		composite silhouette dishes with a ring base	1	0.7	50
		out-curving dish	1	0.7	50
	Total	**total dishes**	**2**	**1.3**	**100**
Monochrome Black	(all decorative treatments)	cylindrical vases	4	2.6	26.7
		barrel shaped vases	1	0.7	6.7
		total vases	5	3.3	33.3
		Flaring bowls	2	1.3	13.3
	(plain)	in-curving bowl	1	0.7	6.7
		rounded bowl	1	0.7	6.7
	(all decorative treatments)	deep bowls	6	3.9	40
		total bowls	10	6.6	66.7
	Total		**15**	**9.9**	**100**
Monochrome Red Slipped		rounded bowls	2	1.3	7.4
		composite silhouette bowls	2	1.3	7.4
		composite silhouette bowls with a ring base	3	2	11.1
		Flaring bowl	1	0.7	3.7
		total bowls	8	5.3	29.6
		out-curving tripod bowls	2	1.3	7.4
		flaring tripod bowl	1	0.7	3.7
		total tripod bowls	3	2	11.1
		Jars	2	1.3	7.4
		composite silhouette dishes with a ring base	3	2	11.1

Table 7.7: Surface treatment and form summarized by count and percent for whole vessels (continued).

SURFACE TREATMENT		FORM	Count	%	% form within Surface Treatment
Monochrome Red Slipped	(Interior-slipped)	rounded bowls	3	2	11.1
		composite silhouette bowl	2	1.3	7.4
		total bowls	5	3.3	18.5
		rounded dishes	5	3.3	18.5
		composite silhouette dish	1	0.7	3.7
		total dishes	6	3.9	22.2
	Total		**27**	**17.8**	**100**
Monochrome Orange Slipped		composite vessel	1	0.7	2.7
		cylindrical tripod vase	1	0.7	2.7
		out-curving bowls	2	1.3	5.4
		rounded bowls	4	2.6	10.8
		composite silhouette bowl	1	0.7	2.7
		composite silhouette bowl with a ring base	1	0.7	2.7
		composite silhouette bowl with a pedestal base	1	0.7	2.7
		rounded bowl with strap handles	1	0.7	2.7
		total bowls	10	6.6	27
		flaring tripod bowl	1	0.7	2.7
		out-curving tripod bowls	5	3.3	13.5
		out-curving tripod bowl with slab feet	1	0.7	2.7
		rounded tripod bowls	5	3.3	13.5
		total tripod bowls	12	7.9	32.4
		out-curving dish	1	0.7	2.7
		composite silhouette dishes with a ring base	2	1.3	5.4
		Chalice	1	0.7	2.7
	(painted with Maya Blue)	Chalice	1	0.7	2.7
		out-curving tripod dish	1	0.7	2.7
		rounded tripod dish	1	0.7	2.7
		total dishes	7	4.6	18.9
		Jars	2	1.3	5.4
		jar with strap handles	1	0.7	2.7
		total jars	3	2	8.1

Table 7.7: Surface treatment and form summarized by count and percent for whole vessels (continued).

SURFACE TREATMENT		FORM	Count	%	% form within Surface Treatment
Monochrome Orange Slipped		Double chamber drum	2	1.3	5.4
		single ovoid chamber drum	1	0.7	2.7
		total drums	3	2	8.1
	Total		**37**	**24.3**	**100**
Monochrome Red to Orange Slipped (interior-slipped)	**Total**	rounded dishes	**2**	**1.3**	**100**
Monochrome Orange Slipped and Incised		out-curving bowls	6	3.9	15.8
		rounded bowls	4	2.6	10.5
		composite silhouete bowl	1	0.7	2.6
		total bowls	11	7.2	28.9
		chile grinders	3	2	7.9
		tripod bowls with a segmented basal flange	7	4.6	18.4
		tetrapod bowls	2	1.3	5.3
		tetrapod or tripod stand with a segmented basal flange	1	0.7	2.6
	(notched and appliqué impressed fillet)	composite silhouette dish	1	0.7	2.6
		out-curving tripod dish	1	0.7	2.6
		Chalice	2	1.3	5.3
	(piercework)	Chalice	3	2	7.9
		total dishes	7	4.6	18.4
		pedestal-based jar (censer)	1	0.7	2.6
		Jars	3	2	7.9
	(appliqué bird head on shoulder)	Jar, necked, bird effigy on shoulder	1	0.7	2.6
		total jars	5	3.3	13.2
		bell chamber drum	1	0.7	2.6
	(piercework and appliqué bird head on chamber)	bell chamber drum	1	0.7	2.6
		total drums	2	1.3	5.3
	Total		**38**	**25**	**100**

Table 7.7: Surface treatment and form summarized by count and percent for whole vessels (continued)..

SURFACE TREATMENT		FORM	Count	%	% form within Surface Treatment
Buff/Cream Slipped - Plain or Incised		rounded bowl	1	0.7	50
		rounded dish	1	0.7	50
	Total		**2**	**1.3**	**100**
Monochrome Brown Slipped		tripod bowl	1	0.7	50
		torch'	1	0.7	50
	Total		**2**	**1.3**	**100**
Monochrome Red Slipped and Incised	**Total**	cylindrical vase	**1**	**0.7**	**100**
Monochrome Red Slipped, Stoccoed and Painted	**Total**	cylindrical vase	**1**	**0.7**	**100**
Monochrome Red Slipped and Model-Carved	**Total**	Flaring bowl	**1**	**0.7**	**100**
Monochrome Red Slipped and Groove-Incised	**Total**	Bottles (in-sloping vases)	**2**	**1.3**	**100**
Total Fine Ware			**146**	**96.1**	
Unslipped-Striated	**Total**	Jar	**1**	**0.7**	**100**
Unslipped/Smoothed-Buff	**Total**	bowl, broad flaring rim	**1**	**0.7**	**100**
Unslipped Utilitarian		comal (thick-walled plate)	1	0.7	50
	(appliqué spikes)	pedestal-based censer	1	0.7	50
	Total		**2**	**1.3**	**100**
Unslipped with Appliqué	**Total**	jars with strap handles	**2**	**1.3**	**100**
Total Coarse Ware			**6**	3.9	
GRAND TOTAL			**152**		

attributes, slip colour was sometimes impossible to access with an acceptable level of accuracy for fine ware vessels with red to orange slips or resist decoration. In such cases, the slip was either too severely eroded or the colour of it was observed to vary across the vessel's surface. Vessels for which colour assessments were problematic comprise approximately 13% of all fine ware, which includes 2 whole vessels and 172 fragments (Table 7.6). Principal forms include composite silhouette dishes (N=56), rounded bowls and dishes on which the slip is confined to the rim area of the exterior surface ('interior-slipped') (N=71), and jars (N=41) (Tables 7.5 and 7.7).

Fine ware vessels are always slipped on at least one surface and approximately 40% have some sort of decoration. Decoration is most often achieved through techniques involving the alteration of a vessel's surface, such as incising, notching or stamping. Comparatively less often, it is accomplished through additions to the vessels surface via appliqués, the application of paint, Maya blue or stucco, or the manipulation of slip colour using a resist technique. The most elaborate decoration occurs on vessels with painted decoration (Polychrome Slipped and Painted and Black on Red Slipped and Painted), the Monochrome Orange Slipped and Incised vessels, and Monochrome Black vases, which most often feature a combination of decorative techniques. Regardless of the technique(s) utilized, decorative design appears to follow some very basic conventions in terms of the organization and positioning of decorative elements. Orange-slipped vessels with an impressed fillet appliqué (Monochrome Orange Slipped with Appliqué) and the Monochrome Black deep bowls with groove-incising are the only examples of highly standardized decorative design. Within the other surface treatment groups, the specific content of the decoration on individual vessels is most often unique.

Vessel Form - The fine ware component of the assemblage is dominated by bowl, dish, vase and jar forms (Tables 7.5 and 7.7). Bowls, including tripod bowls, are most common, followed by dishes and jars. Vases are comparatively less frequent and rare forms include: composite vessels, sieves, drums, plate/lids, frying pan censers, whistles, 'bottles' and a 'torch'. A wide range of different bowl and tripod bowl forms were encountered, although rounded bowls are most common. In contrast, there are only three principal forms of dishes – rounded dishes, composite silhouette dishes with a ring base and 'chalices'. Among the jars, although most forms appear to be globular to slightly maliform in shape, there is considerable variation in the morphology of the rim to neck area. Vases have a more consistent morphology and most often are cylindrical in form. Many of the dish and bowl forms, such as rounded and flaring bowls and rounded and composite silhouette dishes, were encountered in three or more of the main surface treatment groups. Jars, however, are strictly red-slipped, or orange-slipped with a plain or incised exterior surface. Most of the comparatively rare forms within the assemblage, including pedestal-based jars, drums, frying pan censers, composite vessels, sieves, plate/lids and whistles are restricted to the Monochrome Orange Slipped and/or Monochrome Orange Slipped and Incised groups. Similarly, many of the vessels with rare surface treatments have distinct or unique morphologies

Forming Techniques - Definitive macroscopic evidence of forming techniques employed in the manufacture of fine ware vessels is often entirely absent, especially in regard to bowl and dish forms. It would appear that such evidence was completely obliterated during the process of smoothing, burnishing or polishing vessel surfaces in preparation for the application of a slip or in order to achieve the desired surface finish. Many of the forms with angular junctions in their profiles such as composite silhouette bowls and dishes, however, might have been achieved most successfully by using some sort of mould to form at least part of the vessel body. Vessel forms in which changes in the curvature of the interior and exterior profile form particularly sharp angles, such as the chalices, exhibit a very high level of consistency in terms of the details and proportions of their morphology and wall thickness (which is quite thin), both individually and as a form category. In such cases, the use of a mould would be of great assistance in achieving standardized and consistent morphologies when making thin-walled vessels with such a complex curvature.

The only vessel forms or anatomical areas of a vessel that display clear evidence of the forming techniques used in their manufacture are those in which the interior surface is only poorly smoothed and is unslipped, as is the case with pedestal bases, jars and vases. Pedestal bases appear to have been coil built, since 'relic coils', or undulations, are clearly visible on the interior surface. Similarly, the interior surfaces of jars often have percussion facets, suggesting they were shaped using a paddle and anvil technique. There is some evidence to suggest that, at least in some cases, vases were slab built. The two most complete vases, both monochrome black, have a slightly raised or thickened area on the interior surface that runs along the height of the vessel, perpendicular to the base. This inconsistency in the vessel wall is most likely the result of joining the edges of a slab of clay to create a cylindrical form.

Summarized Descriptions of Fine Ware Stylistic Groups

POLYCHROME SLIPPED AND PAINTED

Frequency: Polychrome Slipped and Painted vessels constitute one of the smaller groups within the assemblage, comprising just 5.5% of the fine ware and 3.0% of all vessels. A minimum number of 74 vessels is represented, including ten whole vessels and 64 rim fragments (Table 7.6). All of the fragmentary examples were recovered from the midden associated with structure N10-27 (Tables 7.3 and 7.4). The majority of vessels are rounded dishes (60% of fragments and 40% of whole vessels), followed by composite silhouette bowls (25% of fragments and 40% of whole vessels). There are comparatively fewer vases (one whole vessel and five

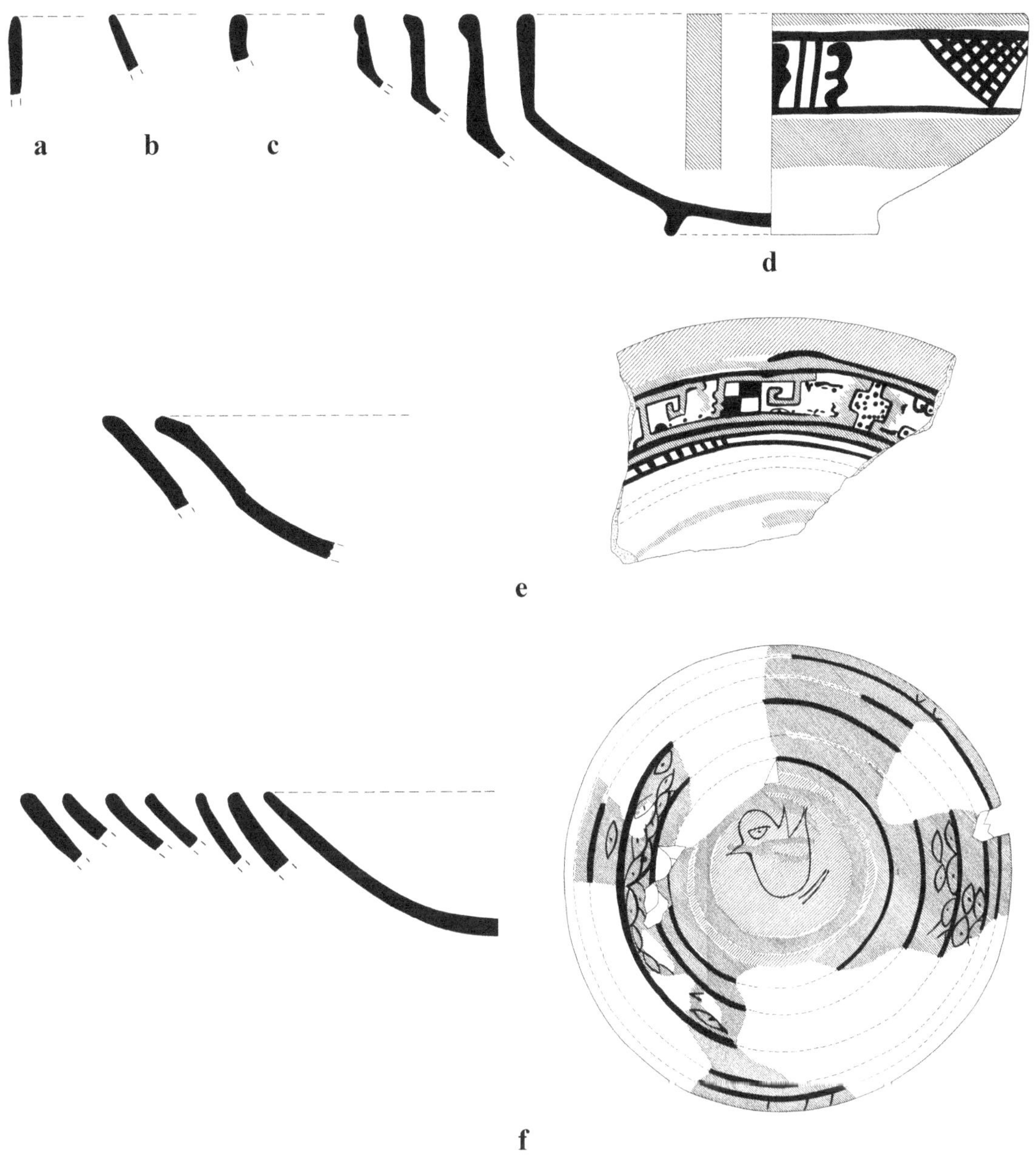

Figure 7.1: Polychrome Slipped and Painted vessel forms: a) vase; b) flaring bowl; c) rounded bowl; d) composite silhouette bowls; e) composite silhouette dishes; f) rounded dishes.

fragments), tripod bowls (two fragments) and large open vessels (three fragments), which are possibly basins (Tables 7.5 and 7.7).

Principal Identifying Modes: Vessels with slipped or polished surfaces that also display a painted decorative treatment comprising more than one colour (Figure 7.1, Plates VII.38-40).

BLACK ON RED SLIPPED AND PAINTED

Frequency: Black on Red Slipped and Painted vessels are comparatively rare within the assemblage, representing less than one percent (0.6%) of fine ware as well as the total number of vessels (0.3%). A minimum number of eight vessels was identified, including six whole vessels and 2 rim fragments (Table 7.6). Both of the fragmentary examples were recovered from the midden associated with structure N10-27 (Tables 7.3 and 7.4). Five out of the six whole vessels are rounded dishes (Table 7.7). The remaining whole vessel is a massive bowl and both rim fragments derive from morphologically similar bowls (Tables 7.5 and 7.7).

Principal Identifying Modes: Rounded dishes and massive bowls with a red slip and black painted decoration (Figure 7.2, Plate VII.41).

RED-ORANGE-BLACK RESIST

Frequency: Vessels with a resist surface treatment comprise one of the least well represented groups within the assemblage, constituting just 1.9% of the fine ware and 1.0% of all vessels. A minimum number of 25 vessels is represented, including two whole vessels and 23 rim fragments (Table 7.6). All of the fragmentary

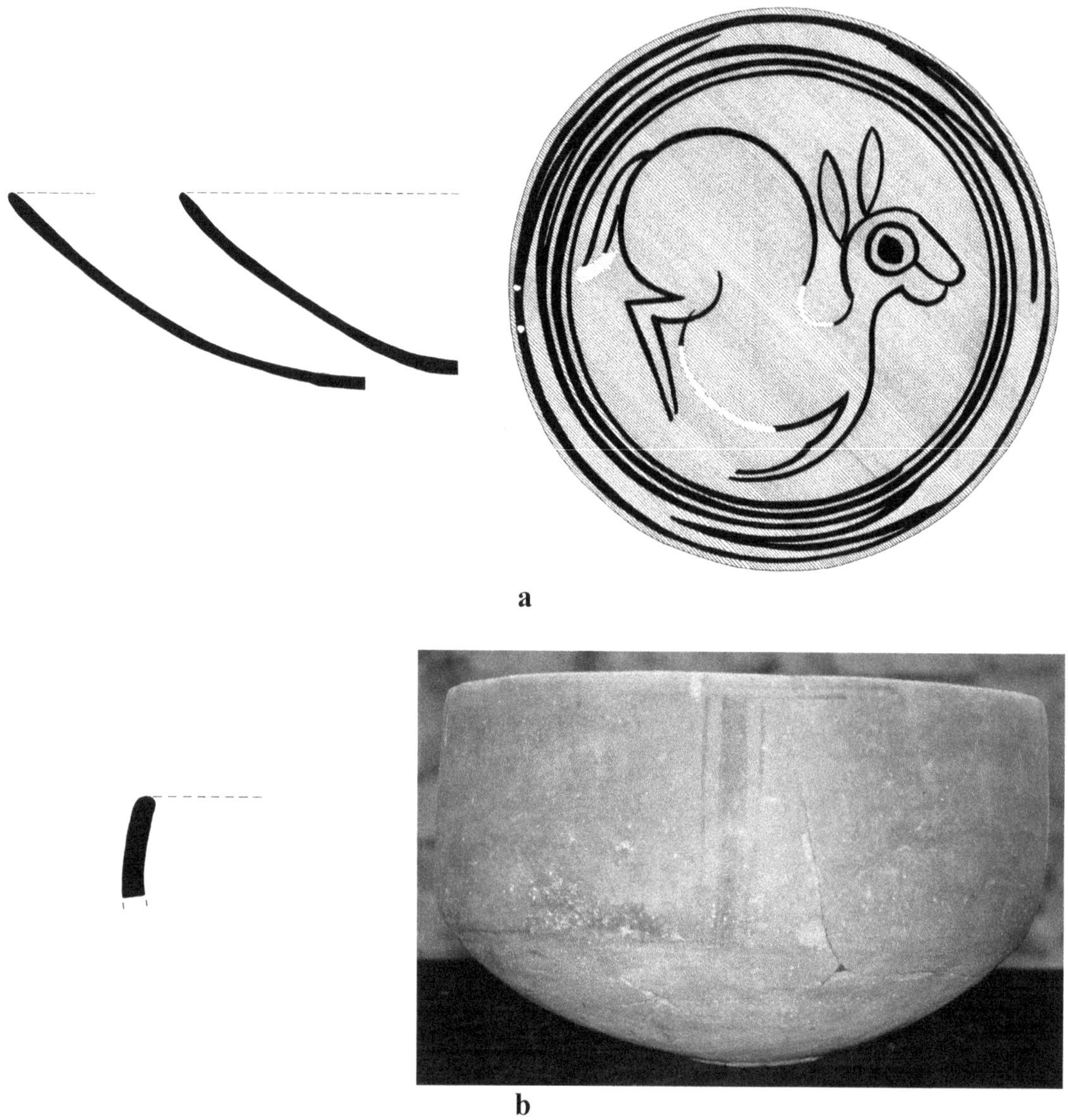

Figure 7.2: Black on Red Slipped and Painted vessel forms: a) rounded dishes; b) massive bowl (see colour version on accompanying CD).

examples were recovered from the midden associated with structure N10-27 (Tables 7.3 and 7.4). One of the whole vessels is an out-curving dish and the other is a composite silhouette dish with a ring base. All of the vessels fragments derive from composite silhouette dishes (Tables 7.5 and 7.7).

Principal Identifying Modes: Bichrome, red and black dishes on which the slip has been purposely manipulated during firing through a 'resist' technique to produce comparatively large, bold, decorations. The dominant colour on the vessel (colour of exterior and majority of interior) can be either deep red to orange or black and the decoration, which appears as irregular-shaped or amorphous blotches, occur in the opposite colour - i.e. they are red on a predominantly black vessel and black on a predominantly red vessel (Figure 7.3, Plate VII.42).

MONOCHROME BLACK

Frequency: Monochrome Black vessels constitute another of the least well represented groups within the assemblage, comprising just 2.8% of the fine ware and 1.5% of all vessels. A minimum number of 37 vessels is represented, including 15 whole vessels and 22 rim fragments (Table 7.6). All but one of the fragmentary examples were recovered from the midden associated with structure N10-27 (Tables 7.3 and 7.4). The majority of vessels are bowls (81.8% of fragments [N=17] and 66.7% of whole vessels[N=10]), flaring bowls and deep bowls being the most common forms. Comparatively fewer vessels are vases (five whole vessel and four fragments) (Tables 7.5 and 7.7).

Principal Identifying Modes: Monochrome black vessels, principally bowls and vases, with or without a secondary decorative treatment (addition or alteration to the vessel's

Figure 7.3: Red-Orange-Black Resist vessel forms: a) out-curving dish; b) composite silhouette dishes.

surface). Secondary decorative treatments include incising, gouging, groove-incising, appliqué, stamping, false gardooning, impressing and punctuation, and on vases, these different treatments occur together in varying combinations (Figure 7.4, Plates VII.43-44).

MONOCHROME RED SLIPPED

Frequency: Red Slipped vessels constitute one of the smaller groups within the assemblage, comprising 8.4% of the fine ware and 4.6% of all vessels. A minimum number of 113 vessels is represented, including 27 whole vessels and 86 rim fragments (Table 7.6). Most of the fragmentary examples (N=80) were recovered from the midden associated with structure N10-27, where as only six fragments derive from the midden at structure N10-9 (Tables 7.3 and 7.4). Among the Red Slipped vessels, bowls are most common (46.5% of fragments and 48.1% of whole vessels), followed by dishes (33.7% of fragments and 33.3% of whole vessels) and there are a variety of forms represented in both of these shape classes. There are comparatively fewer jars (two whole vessels and 13 fragments) and tripod bowls (three whole vessels and three fragments), and only one fragment from a vase (Tables 7.5 and 7.7).

Principal Identifying Modes: Vessels with a red to dark red slip (Figures 7.5 and 7.6, Plates VII.45-50).

MONOCHROME ORANGE SLIPPED

Frequency: Orange Slipped vessels form the largest fine ware group within the assemblage, comprising 40.5% of the fine ware and 21.9% of all vessels. A minimum number of 544 vessels is represented, including 37 whole vessels and 507 rim fragments (Table 7.6). This group is well represented in both of the midden assemblages. A total of 309 fragmentary examples were recovered from the midden associated with structure N10-27 and 198 fragments derive from the midden at structure N10-9 (Tables 7.3 and 7.4). The majority of Orange Slipped vessels are either bowls or tripod bowls (53.8% and 15.8% of fragments, and 27.0% and 37.9% of whole vessels, respectively), and a variety of different forms are represented. There are comparatively fewer dishes (13.5% of whole vessels and 9.8% of fragments), jars (8.1% of whole vessels and 14.4% of fragments) and durms (8.1% of whole vessels and 4.1% of fragments). Rare forms include composite vessels (one whole vessel and two fragments), vases (two fragments), a tripod vase (whole vessel) and a sieve (fragment) (Tables 7.5 and 7.7). The principal dish forms are rounded dishes,

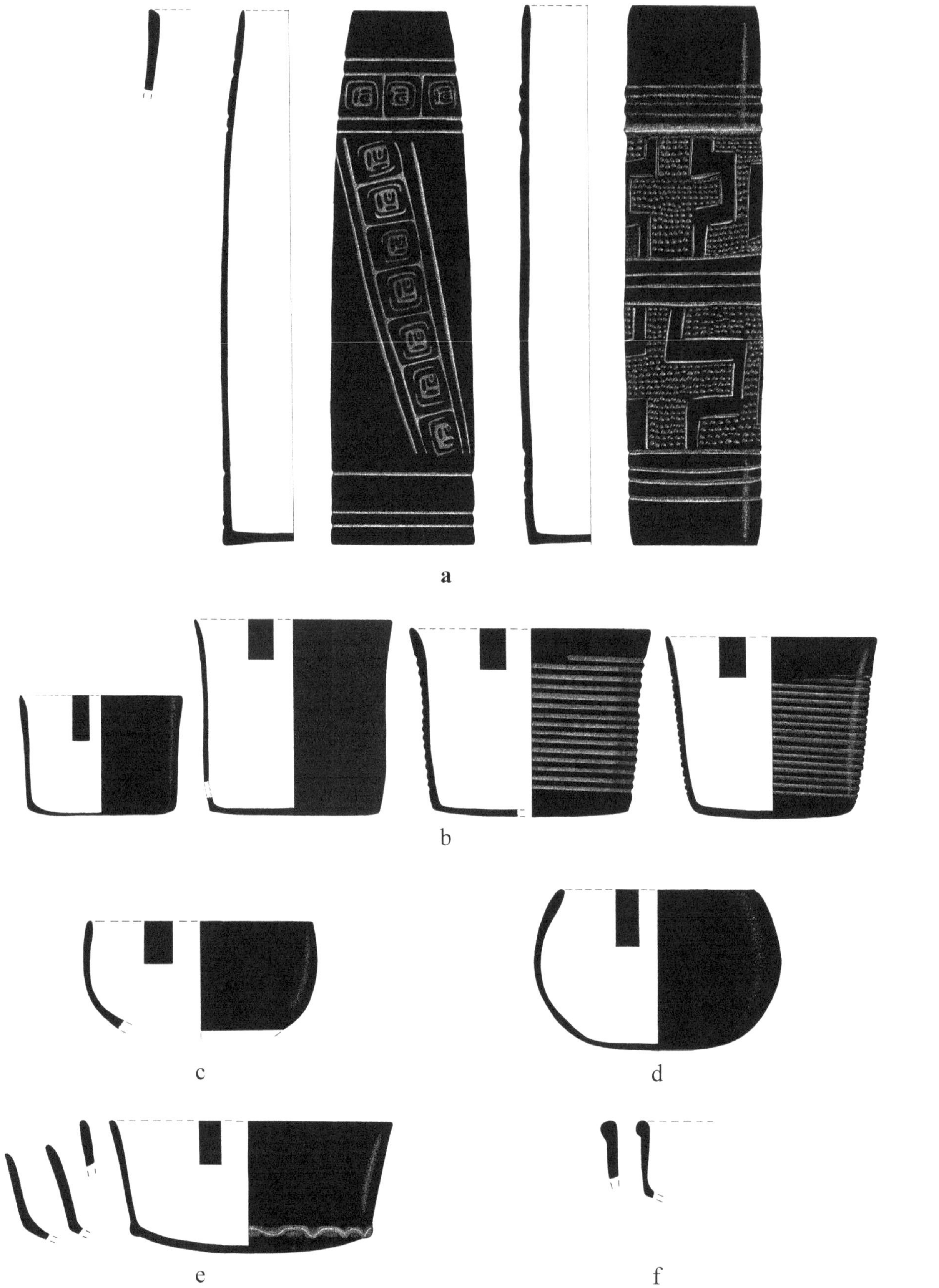

Figure 7.4: Monochrome Black vessel forms: a) vases; b) deep bowls; c) rounded bowl; d) in-curving bowl; e) flaring bowls; f) composite silhouette bowls.

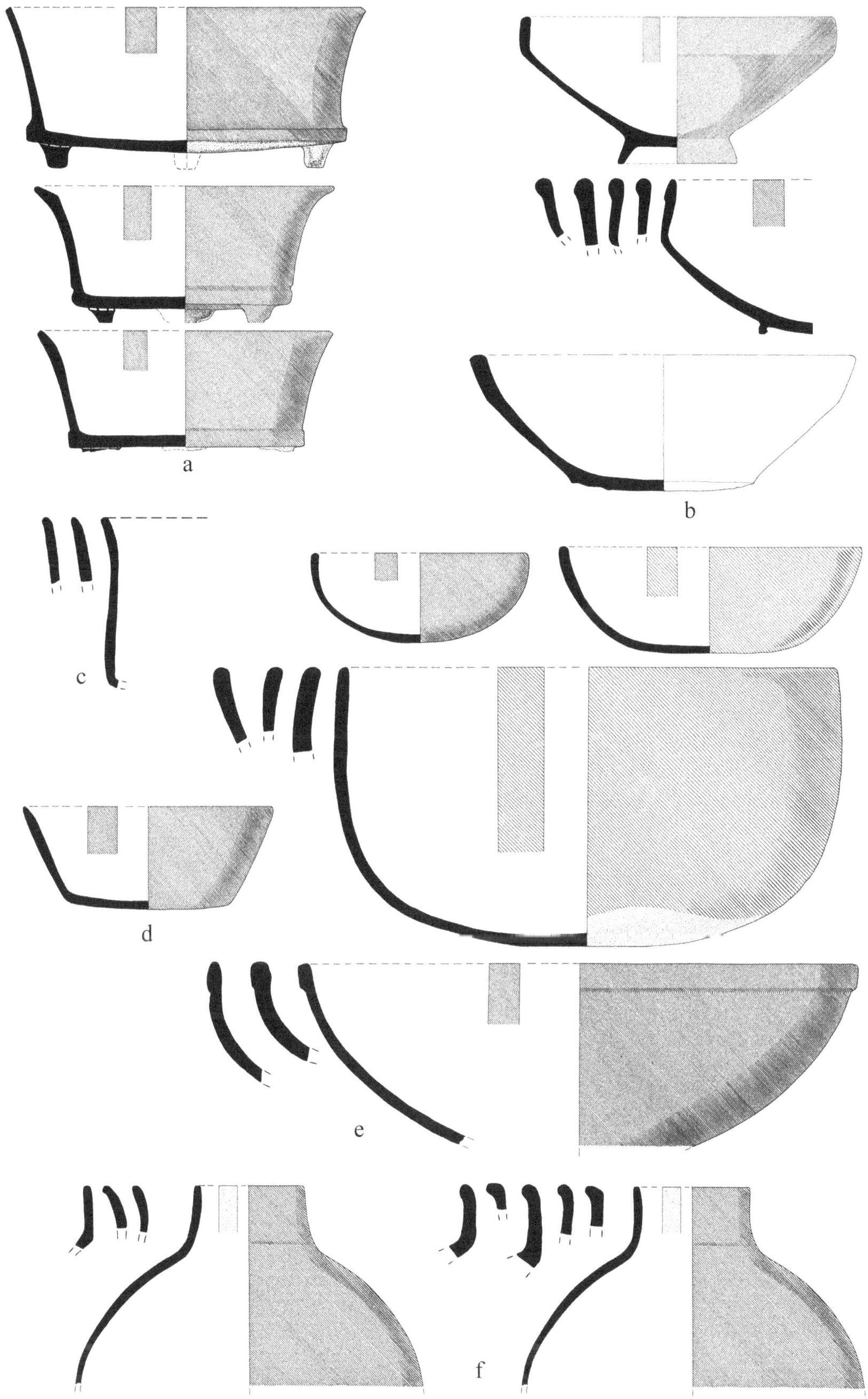

Figure 7.5: Monochrome Red Slipped vessel forms: a) flaring and out-curving tripod bowls; b) composite silhouette bowls; c) flaring to out-curving bowls/deep bowls; d) flaring bowls; e) rounded bowls; f) jars with different rim forms.

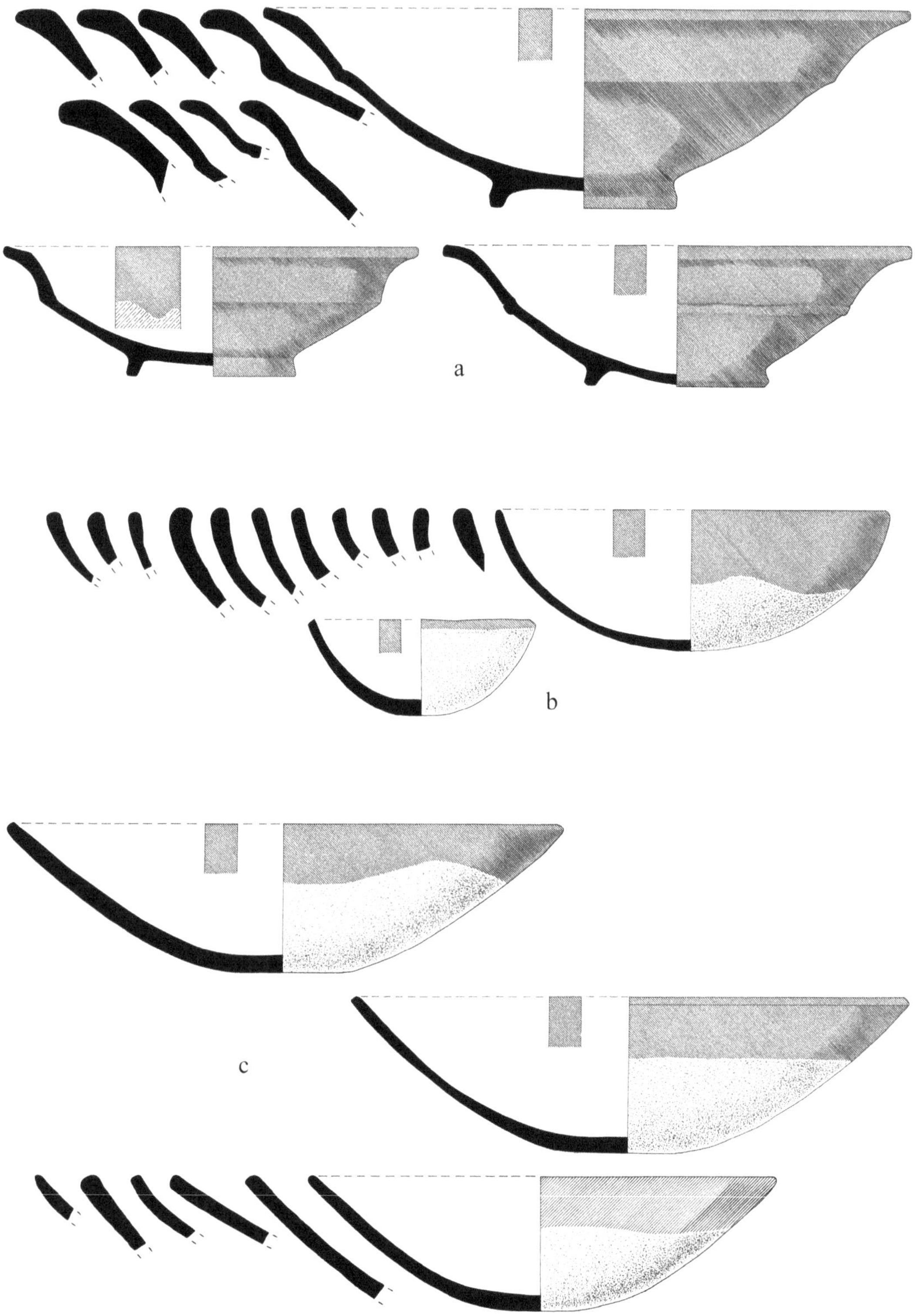

Figure 7.6: Monochrome Red Slipped and 'interior-slipped'(slip is confined to rim area on exterior surface) vessel forms: a) Monochrome Red composite silhouette dishes; b) red to orange 'interior-slipped' rounded bowls; c) red to orange 'interior-slipped' rounded dishes.

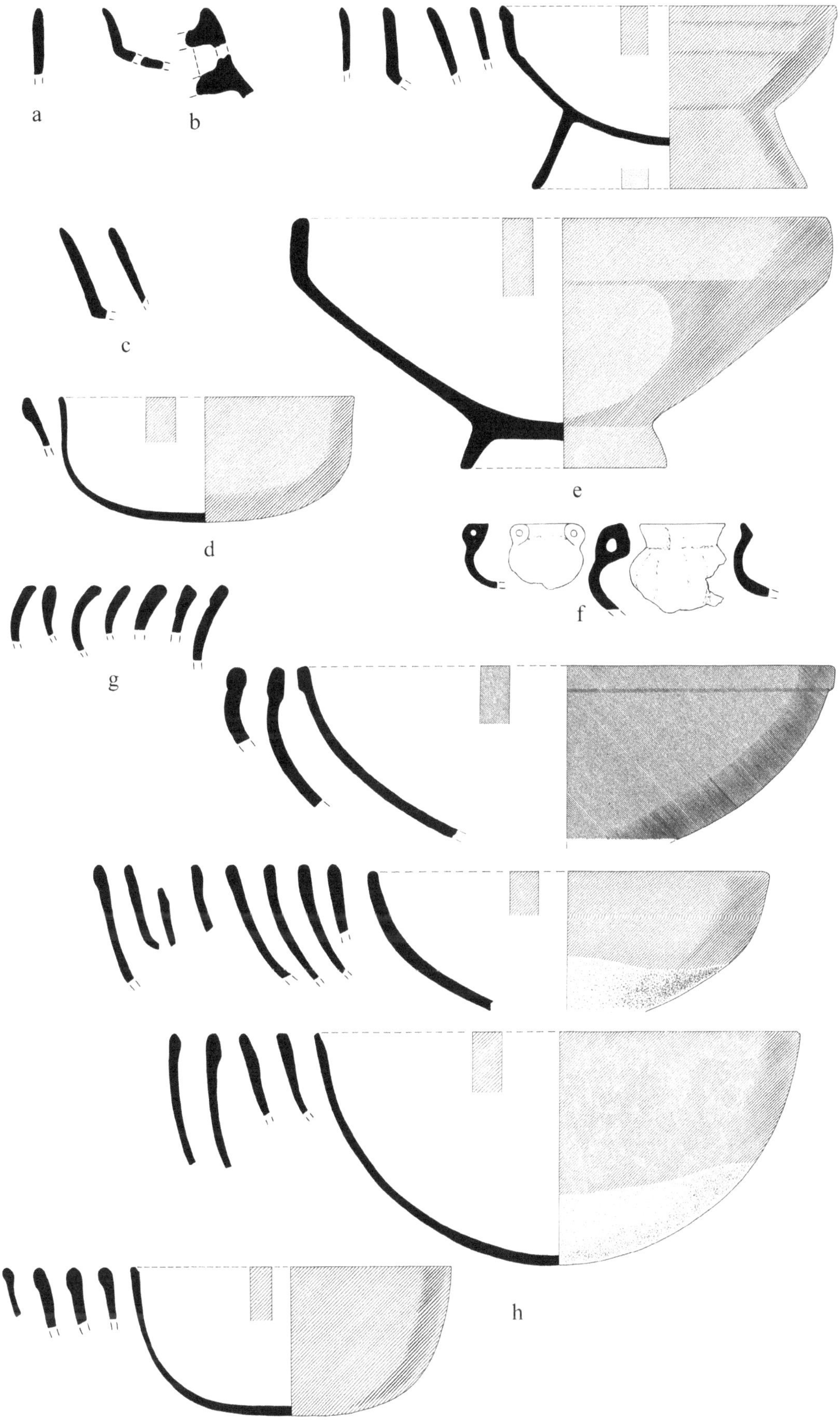

Figure 7.7: Monochrome Orange Slipped vessel forms: a) vase; b) sieve; c) flaring bowls; d) out-curving bowl; e) composite silhouette bowls; f) miniature jars; g) in-curving bowls with different rim forms; h) rounded bowls with different rim forms.

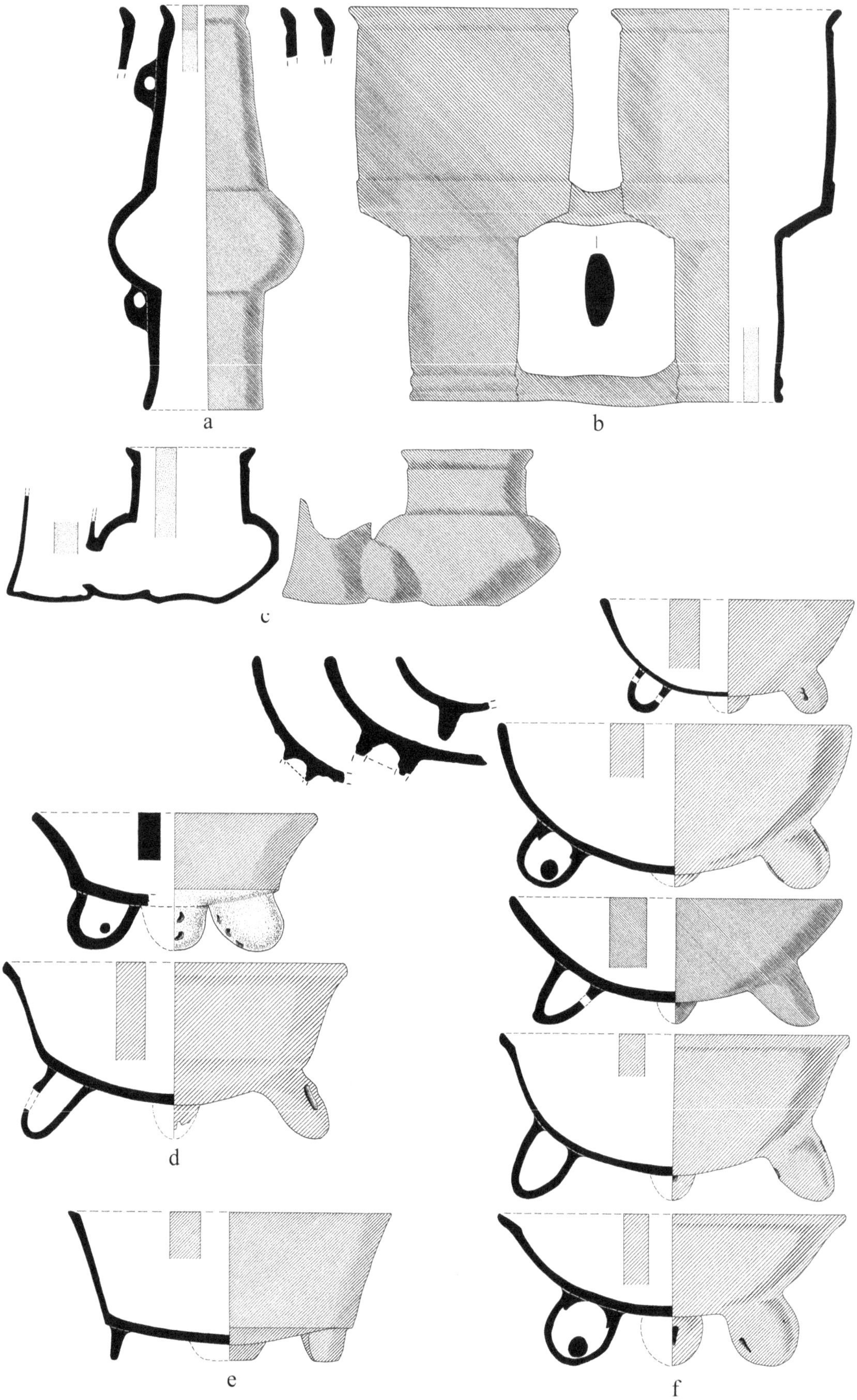

Figure 7.8: Monochrome Orange Slipped vessel forms: a) ovoid chamber drum; b) double chamber drum; c) composite vessel; d) out-curving tripod bowls; e) flaring tripod bowl; f) rounded tripod bowls.

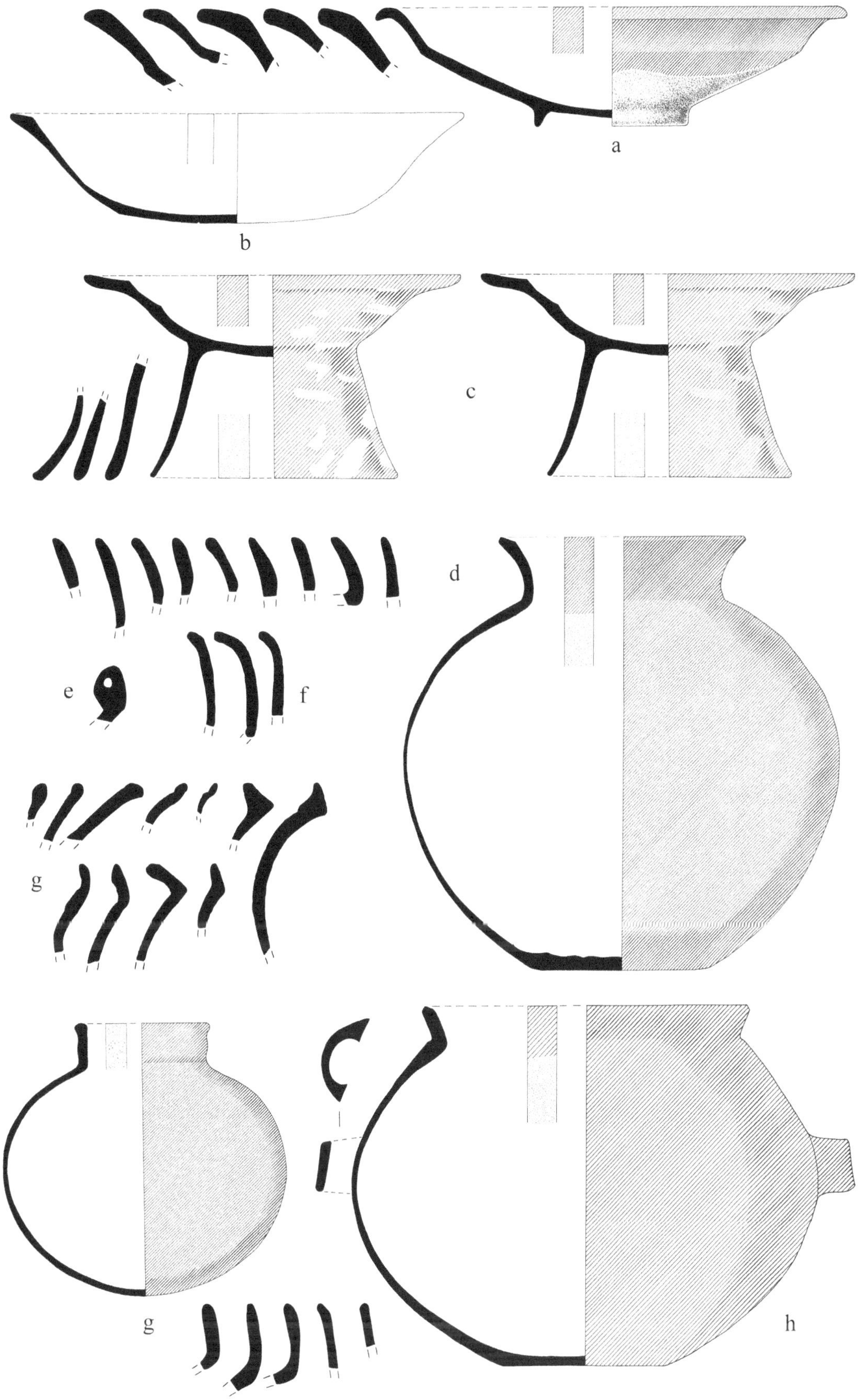

Figure 7.9: Monochrome Orange Slipped vessel forms: a) composite silhouette dishes; b) out-curving dish; c) 'chalices' or pedestal-based dishes; d) jars with out-curving necks; e) jar with a loop handle; f) jars with high out-curving necks; g) 'neckless' jars and jars with low flaring to slightly out-curving necks; h) jar with strap handles.

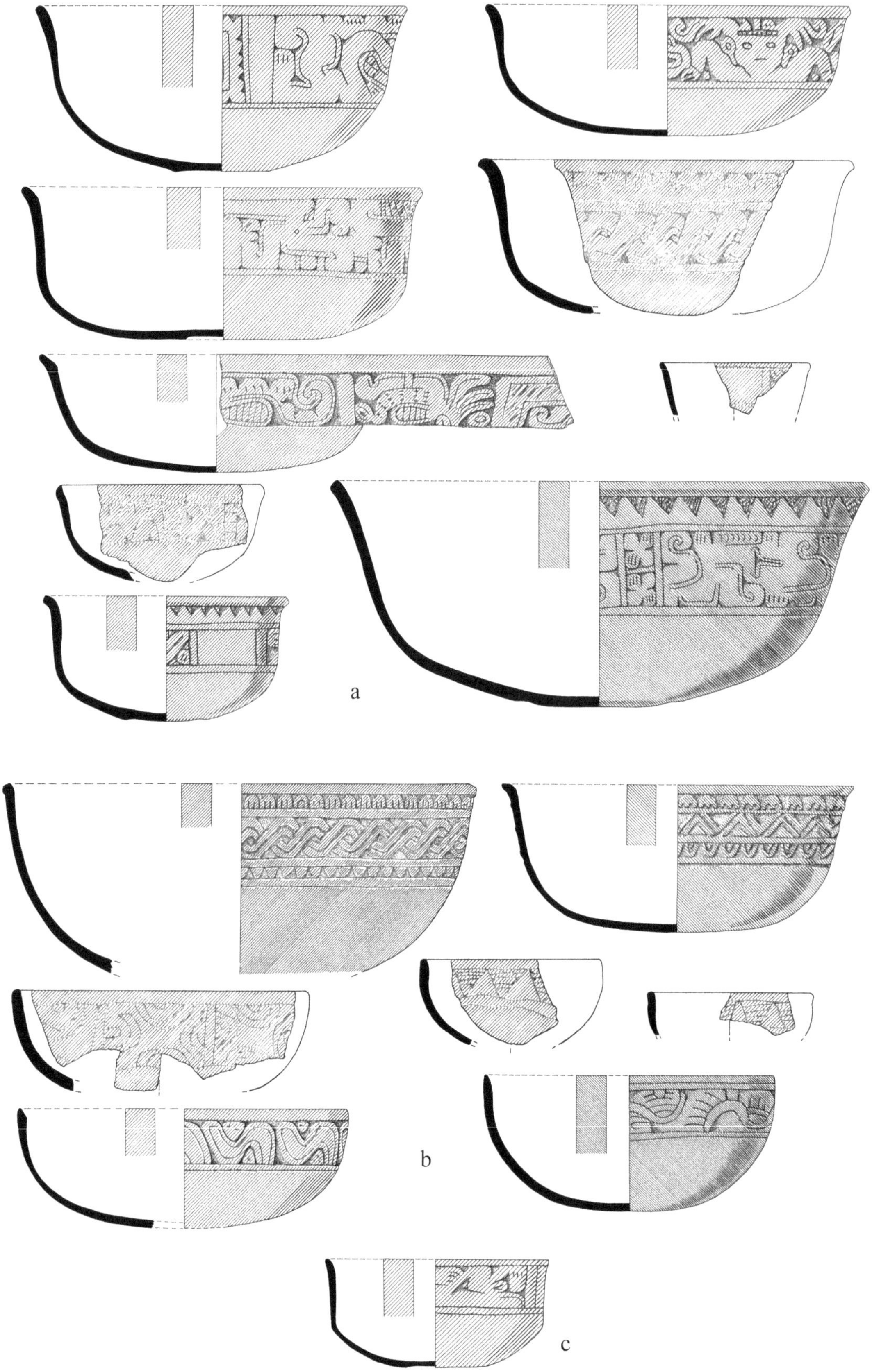

Figure 7.10: Monochrome Orange Slipped and Incised bowl forms: a) out-curving bowls; b) rounded bowls c) composite silhouette bowl.

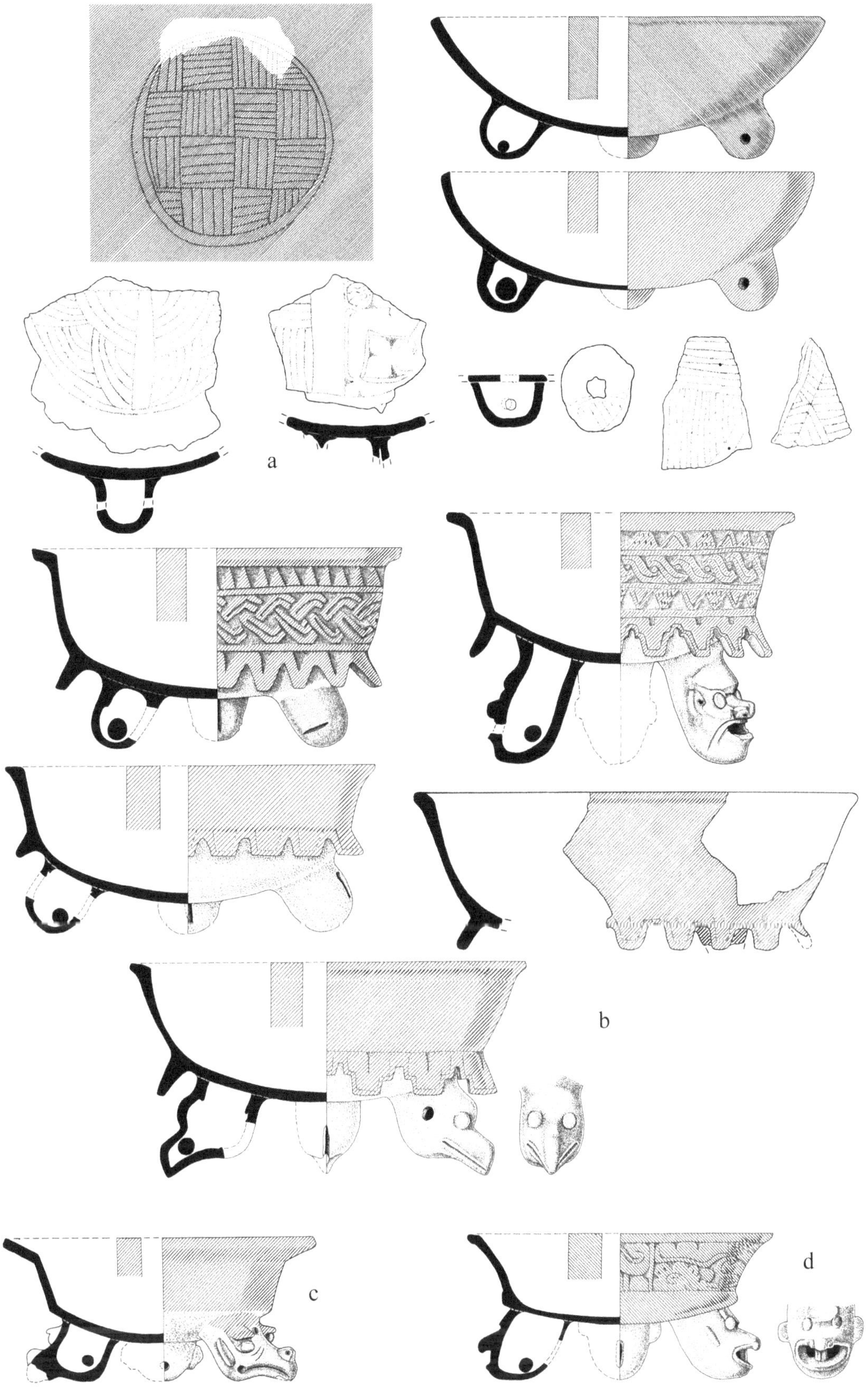

Figure 7.11: Monochrome Orange Slipped and Incised vessel forms (Note: vessel c) is from the Monochrome Orange Slipped group) : a) chile grinders; b) tripod bowls with a segmented basal flnge; c) Monochrome Orange Slipped flaring tripod bowl; d) out-curving tripod bowl.

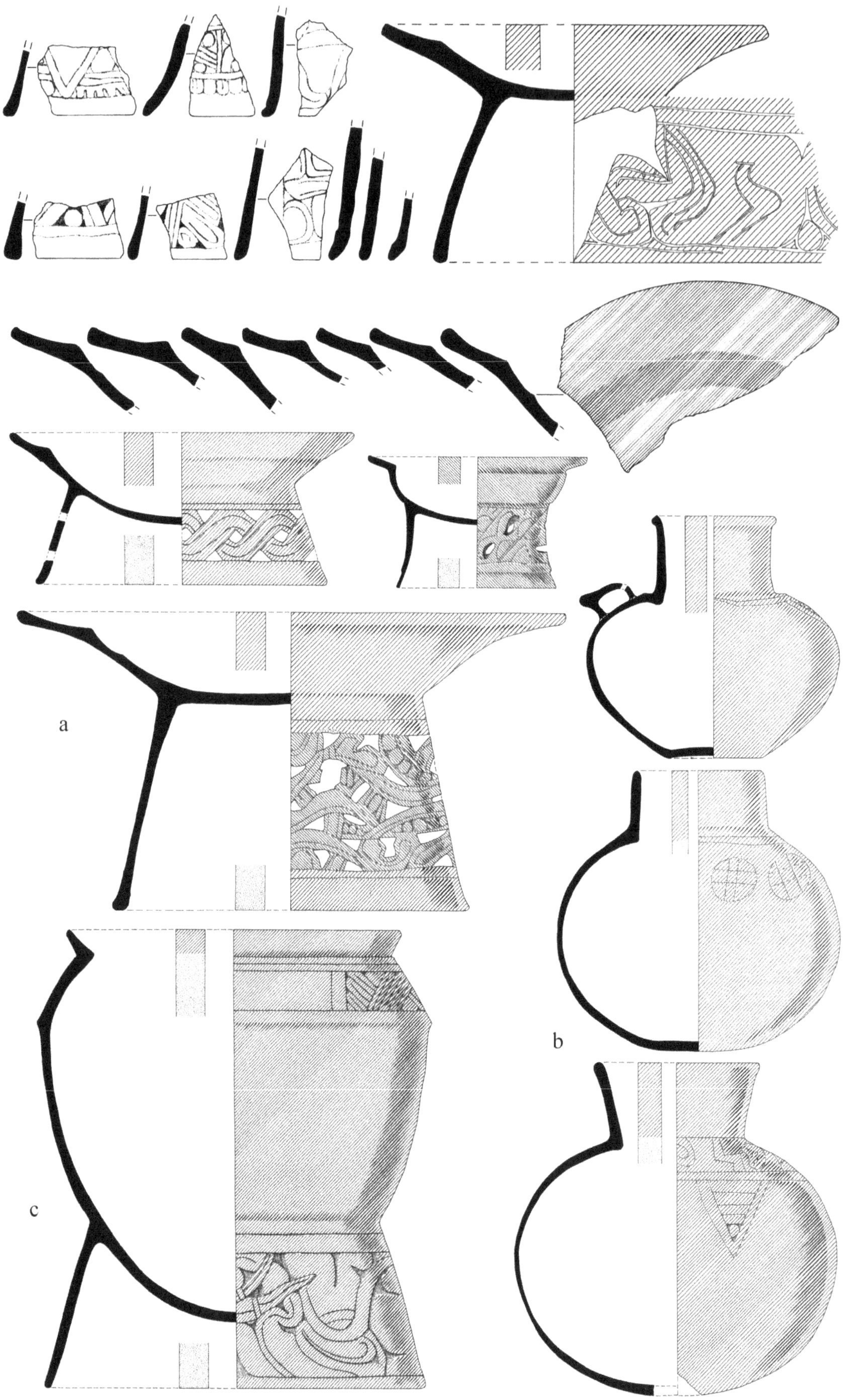

Figure 7.12: Monochrome Orange Slipped and Incised vessel forms: a) 'chalices' or pedestal-based dishes with incision or incision and piercework; b) jars; c) 'censer' or pedestal-based jar.

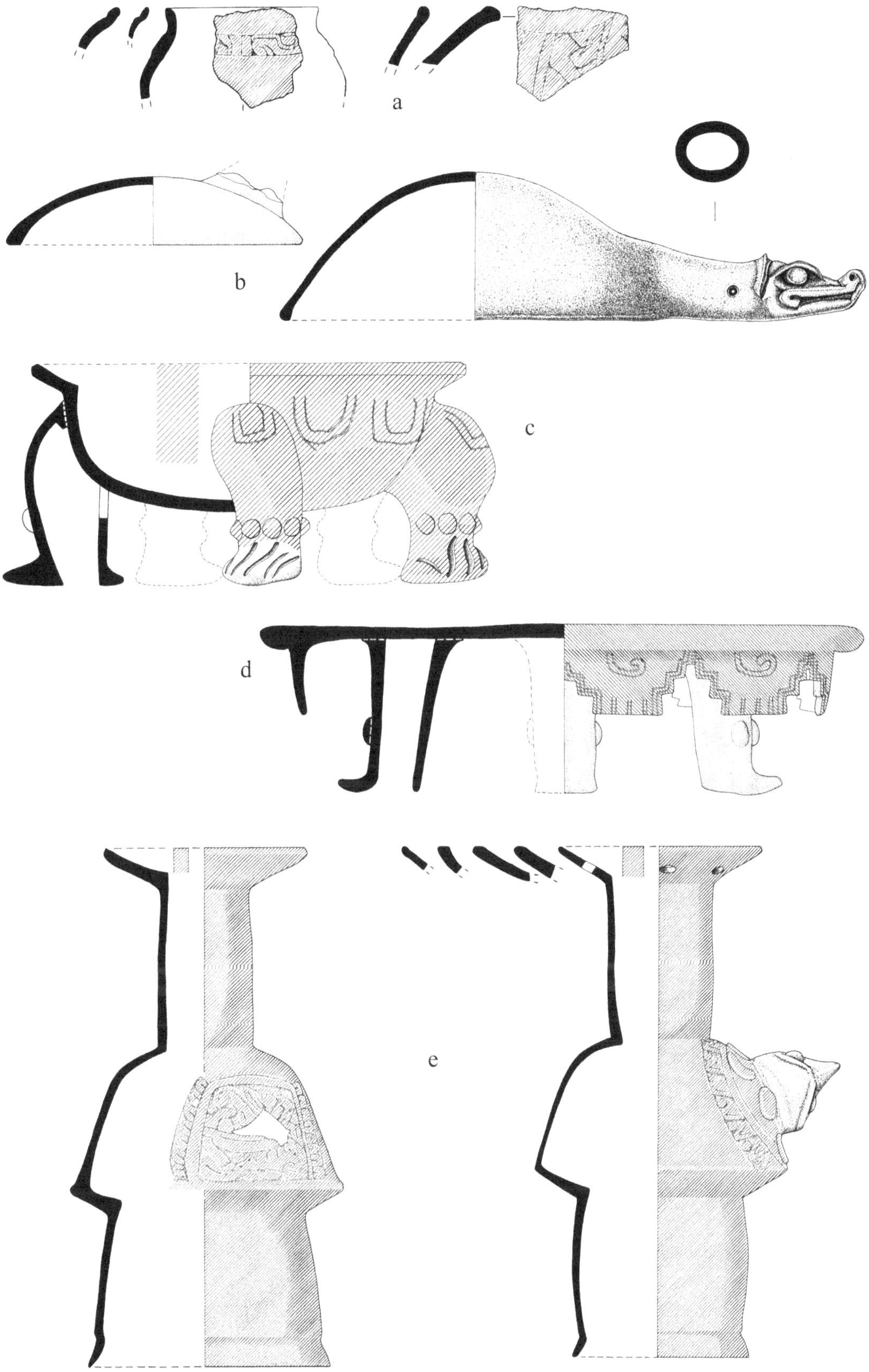

Figure 7.13: Monochrome Orange Slipped and Incised vessel forms: a) 'neckless' jars and jars with very low flaring rims; b) frying pan censers; c) tetrapod bowl; d) stand; e) bell chamber drum.

Figure 7.14: Monochrome Orange Slipped and Groove Incised vessel forms: a) vase; b) drum.

chalices and composite silhouette dishes, whereas jar morphology is more variable.

Principal Identifying Modes: Vessels with an orange to red-orange or brown-orange slip (Figures 7.7 to 7.9 and 7.11c, Plates VII.51-55).

MONOCHROME ORANGE SLIPPED AND INCISED

Frequency: Orange Slipped and Incised vessels form the second largest fine ware group within the assemblage, comprising 24.8% of the fine ware and 13.4% of all vessels. A minimum number of 333 vessels is represented, including 38 whole vessels and 295 fragments (Table 7.6). This group also is well represented in both of the midden assemblages, with 160 fragmentary examples were recovered from the midden associated with structure N10-27 and 135 recovered from the midden at structure N10-9 (Tables 7.3 and 7.4). Approximately half of the Orange Slipped and Incised vessels are either bowls or tripod bowls (45.4% and 17.6% of fragments, and 28.9% and 26.3% of whole vessels, respectively), with out-curving and rounded forms being the most common. Chalices are the next most abundant (24.4% of fragments and 13.2% of whole vessels), followed by jars (8.4% of fragments and 13.2% of whole vessels), half of which are pedestal-based jars (N=17). Comparatively fewer vessels are drums (1.7% of fragments and 5.3% of whole vessels; N=7), frying pans (four fragments) or whistles (three fragments). Rare forms include tetrapod bowls (two whole vessels) and a tripod or tetrapod stand (whole vessel) (Tables 7.5 and 7.7).

Principal Identifying Modes: Vessel with an orange to red-orange slip and post slip- pre firing incised and gouged decoration (Figures 7.10 to 7.13, Plates VII.53-54,56-59).

MONOCHROME ORANGE SLIPPED AND GROOVE-INCISED

Frequency: Orange-slipped vessels with groove-incised decoration are rare within the assemblage, representing 0.1% of all ceramics and 0.2% of all fine ware vessels (Table 7.6). The group comprises just three fragmentary examples, all of which were recovered from the midden associated with structure N10-27 (Table 7.4). Two of the fragments derive from vases and one is from a drum (Table 7.5).

Principal Identifying Modes: Thin-walled, ash tempered vases with an orange slip on the exterior surface only and multiple grooves encircling the body. The orange slip covers the grooves, which are oriented either horizontally or vertically (Figure 7.14, Plate VII.40).

MONOCHROME ORANGE TO RED SLIPPED - NOTCHED AND INCISED

Frequency: Orange- or red-slipped vessels with notched and incised decoration are also rare within the assemblage, comprising just 0.3% of all ceramics and 0.5% of all fine ware vessels (Table 7.6). A minimum number of seven vessels is represented, all of which are fragments. Four of the fragments were recovered from the midden associated with structure N10-27 and three derive from the midden at structure N10-9 (Tables 7.3 and 7.4). All of these vessels are composite silhouette bowls (Table 7.6).

Principal Identifying Modes: Orange- or red-slipped, composite silhouette bowls with a bolstered rim and impressed and incised, or just impressed, decoration (Figure 7.15, Plate VII.60).

MONOCHROME ORANGE WITH APPLIQUÉ

Frequency: Orange-slipped vessels with an appliqué impressed fillet form another of the rare groups within the assemblage, constituting just 0.6% of all ceramics and approximately 1.0% of all fine ware vessels (Table 7.6). A minimum number of 14 vessels are represented and all of these are fragments that derive from rounded bowls (Table 7.5). All but one of these bowl fragments were recovered from the midden associated with structure N10-27 (Tables 7.3 and 7.4). Four of the fragments are rim sherds and the remaining ten are body sherds.

Principal Identifying Modes: Orange-slipped, rounded bowls with an impressed fillet encircling the body below the rim area (Figure 7.16, Plate VII.61).

RARE SURFACE TREATMENTS

Buff/Cream Slipped – Plain on Incised (includes two possible Slate ware vessels): This group is represented by two rims sherds and two whole vessels (LA673/5 and LA717/1). Both of the rim sherds derive from rounded bowls, as does vessel LA717/1. The other whole vessel is a rounded dish. The whole vessels are illustrated and described in detail in Appendix IX, Figures IX.73,84. All of these vessels have a buff or cream coloured slip on the interior and the exterior surface and either a glossy or 'soapy/slatey' surface finish. Burnishing marks are visible in each case. Vessel LA 717/1 has a band on incised decoration on the exterior surface of the body, and the other vessels have plain surfaces.

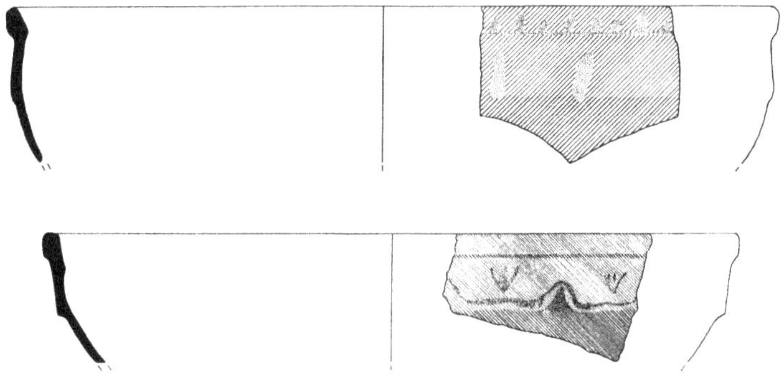

Figure 7.15: Monochrome Orange to Red Slipped – Notched and Incised bowls.

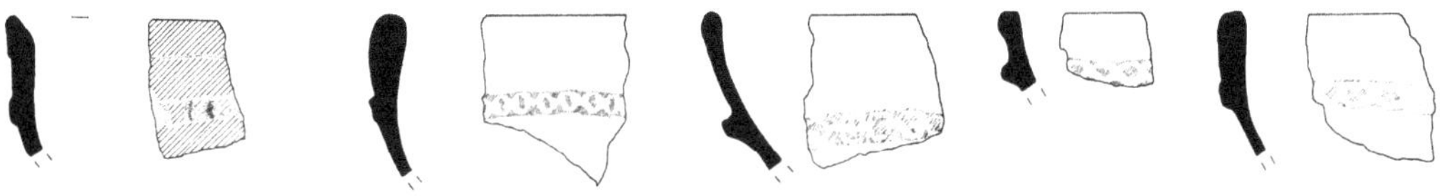

Figure 7.16: Monochrome Orange with Appliqué bowls.

Figure 7.17: Monochrome Red Slipped and Notched massive bowls.

Monochrome Grey Slipped (Plumbate ware) – Only a single sherd was encountered with this surface treatment. This sherd appears to derive from a small, low pedestal base, possibly from a vase form.

Monochrome Brown Slipped – There are two vessels with this surface treatment, LA662/2 and LA630/8. Vessel LA 662/2 is a tripod bowl and LA630/8 can best be described as a 'torch'. These vessels are described in detail in Appendix IX, Figures IX.71,63g respectively.

Monochrome Red Slipped and Incised – This cylindrical vase (LA630/1) has incised decoration that is highly distinctive and unique within the assemblage as a whole. The decorative design comprises repeating panels of incised decoration. Two of the panels feature large, conventionalized, 'glyph-like' motifs, which are rendered through a combination of incising and gouging. The other two panels are in-filled with lines of opposing oblique incisions. See Figure IX.63a in Appendix IX for a detailed description and illustration.

Monochrome Red Slipped, Stuccoed and Painted – This surface treatment was encountered on only one vessel, LA 630/3, which is a cylindrical vase. The vessel is illustrated and described in detail in Appendix IX, Figure IX.63c.

Monochrome Red Slipped and Model-Carved – This bowl (LA115/33) is the only vessel within the assemblage with model-carved decoration. See Figure IX.16 in Appendix IX for a detailed description and illustration.

Monochrome Red Slipped and Groove Incised – There are two vessels, LA656/1 and LA515/1, with this surface treatment. The vessels are conformal and their shape, which is similar to a 'bottle', can best be described as an in-sloping vase with a sharp basal angle. These vessels are described in detail in Appendix IX, Figures IX.68a,53a.

Monochrome Red Slipped and Notched – This group includes three rim sherds and one whole vessel (LA656/6). Two of the rim sherds derive from large rounded bowls with bolstered rims and rounded lips (Figure 7.17, Plate IX.62) and the third is from a composite silhouette dish with a ring base. All of these vessels have red slip on their interior and exterior surfaces and a line of regular-spaced notches encircle the body on the exterior surface. On the composite silhouette dish the notching occurs on the angular junction between the upper and lower body. In each case, the notching was done before the vessels were slipped. The whole vessel is a ring-based composite silhouette bowl. This vessel also has a bolstered rim and a rounded lip, but the red slip on the exterior surface is restricted to the upper portion of the body (see Appendix IX, Figure IX.68e for a detailed description and illustration).

Coarse Ware – General Summary

Surface Treatment and Fabric - The coarse ware component of the assemblage constitutes six whole vessels and 1,132 sherds (Table 7.6). Eight different groups were recognized based on surface treatment and fabric characteristics. Three of the groups are quite rare within the assemblage as a whole, being represented by six vessels or less. The majority of coarse ware vessels (approximately 70%) have dark coloured (grey, brown, dark grey to black), crudely smoothed surfaces and coarse-textured, sandy fabrics (Unslipped-Utilitarian group). Coarse-textured, calcite-tempered jars with red-slipped rims (Slipped Rim/Lip-Smoothed Body group) are the next most abundant, followed by coarse-textured, calcite-tempered vessels with light brown, 'white-flecked' exterior surfaces (Unslipped/Smoothed –Buff group). Comparatively less frequent are coarse-textured, calcite-tempered jars with reddish brown to pinkish red surfaces and striated bodies (Red-Brown-Striated group) and small, medium to fine textured 'incense burners' (Unslipped/Smoothed-Perforated group). Rare coarse ware vessels within the assemblage include sandy textured jars with a red to orange wash on their exterior surface, sandy textured jars with grooved strap handles and an appliqué impressed fillet on their upper shoulder, and a calcite-tempered jar with a heavily striated body and a bolstered rim.

Apart from the jars that have a red to orange wash, a red-slipped rim or an impressed fillet, vessels generally have plain unslipped surfaces. The vessels in most groups are generally 'well made' in that a considerable amount of time was invested in smoothing and finishing their surfaces. The Unslipped-Utilitarian and Washed/Smoothed vessels, however, have more crudely finished surfaces and drag marks are visible where inclusions were pulled across the surface as the vessel was hastily smoothed.

Vessel Forms and Morphology - The vast majority of coarse ware vessels are jars (Tables 7.5 and 7.7). Comparatively less frequent are plates/lids, followed by bowls, incense burners and dishes. Rare forms include a broad rimmed bowl, comals, pedestal-based censers, large basins and an insloping jar with a flat base. Bowls, dishes and plates are generally rounded in form and their morphology is fairly consistent between groups. In contrast, there is a lot of variation in the morphology of jars, specifically in regard to the rim to neck area of these vessels, and this variation occurs both within and between vessel groups. In general, vessel forms within the groups that have calcite-tempered fabrics tend to be fairly uniform in terms their basic morphology and proportions. The same can be said of the two strap handle jars with an impressed fillet. Among the two groups comprising vessels with sandy fabrics, however, the morphology of the jars is highly variable and unstandardized, particularly in relation to the rim to neck area.

Forming Techniques - As with the fine ware jars, the coarse ware jars often have percussion facets on their interior surfaces, indicating they were shaped using a paddle and anvil technique. Macroscopic evidence of the techniques employed in the manufacture of the open forms (bowls, dishes and plates), however, is less than clear. Most fragments of these forms are small, and thus any clues as to the forming techniques that were used are obscured. It is equally possible, however, that such evidence was obliterated when the surfaces of these vessels were smoothed.

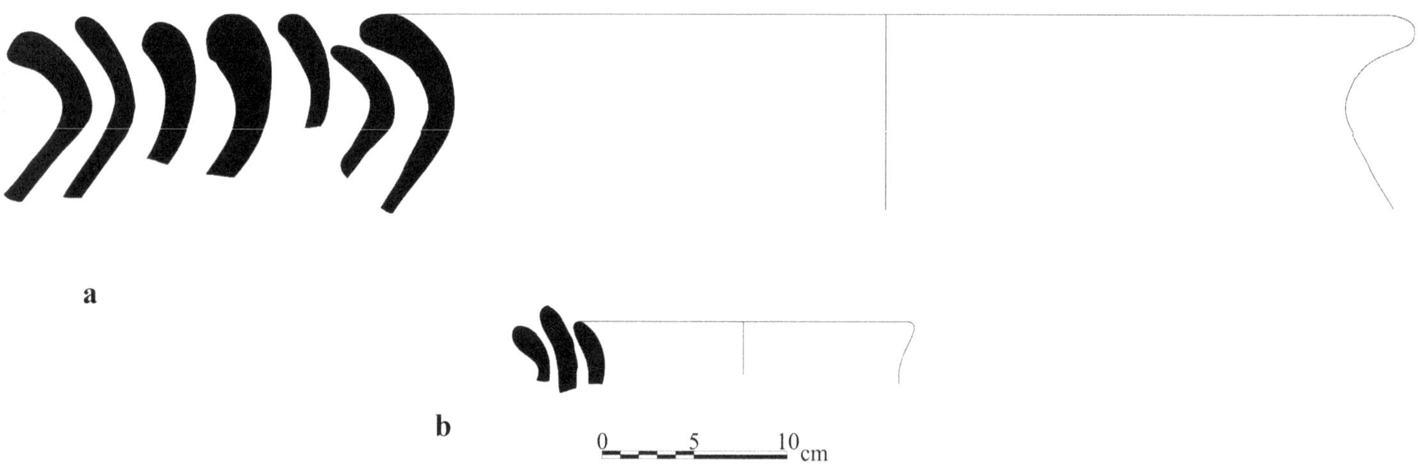

Figure 7.18: Slipped Rim/Lip-Smoothed Body jars: a) vessels with slipped rims; b) vessels with slipped lips.

Forming Techniques - As with the fine ware jars, the coarse ware jars often have percussion facets on their interior surfaces, indicating they were shaped using a paddle and anvil technique. Macroscopic evidence of the techniques employed in the manufacture of the open forms (bowls, dishes and plates), however, is less than clear. Most fragments of these forms are small, and thus any clues as to the forming techniques that were used are obscured. It is equally possible, however, that such evidence was obliterated when the surfaces of these vessels were smoothed.

Description of Coarse Ware Stylistic Groups

SLIPPED RIM/LIP-SMOOTHED BODY

Frequency: This group comprises 8.4% (N=209) of all ceramics and 18.4% of the coarse ware component of the assemblage (Table 7.6). A total of 202 diagnostic sherds were recovered at N10-27 and only seven were recovered from N10-9 (Tables 7.3 and 7.4). These totals include both rim and neck sherds and all but one of the vessels are jars.

Principal Identifying Modes: Coarse-textured, calcite-tempered jars with out-curving or out-flaring rims that are red-slipped to either just below the lip or to the base of the neck. Bodies are either well smoothed or have fine striations (
Figure, Plates IX.63-64).

RED-BROWN – STRIATED

Frequency: A total of 17 diagnostic sherds (rim and neck sherds) comprise this group, all recovered from the N10-27 midden (Table 7.4). This group represents 0.7% of the whole assemblage and constitutes 1.5% of the coarse ware (Table 7.6). All vessels are jars.

Principal Identifying Modes: Coarse-textured, calcite-tempered, unslipped jars with tall, slightly out-curving rims and striated bodies. Particularly distinctive of these jars are the pointed ridge that encircles the rim, below the lip area, and the red-brown to pinkish colour of the surface (Figure 7.18, Plate IX.65).

UNSLIPPED/SMOOTHED – PERFORATED

Frequency: This group comprises 0.8% (N=19) of the whole assemblage and 1.7% of all coarse ware (Table 7.6). A total of four sherds were recovered from the midden associated with N10-27, where as 15 derive from the midden at structure N10-9 (Tables 7.3 and 7.4). All vessels are incense burners

Principal Identifying Modes: Small globular vessels with a restricted orifice and small, circular, irregular-spaced perforations on the body (Figure 7.19, Plate IX.66).

UNSLIPPED/SMOOTHED – BUFF

Frequency: This group comprises 4.0% (N=100) of the whole assemblage and 8.83% of all coarse ware (Table 7.6) and includes only one whole vessel. Of the 99 vessel fragments that were recovered, 23 derive from the N10-27 midden and 76 derive from the midden at N10-9 (Tables 7.3 and 7.4). Approximately 77% (N=76) of the vessels in this group are jars, where as 18.2% (N=18) are plates/lids and 4.0% (N=4) are bowls (Table 7.6).

Principal Identifying Modes: Unslipped, thin walled jars and, less frequently, plates/lids and bowls with a light brown or buff coloured surface. These vessels tend to be calcite-tempered, although a very few examples appear to have grog-tempered fabrics which are identical in appearance to the grog-tempered fabrics that are typical

Figure 7.18: Red-brown Striated jar.

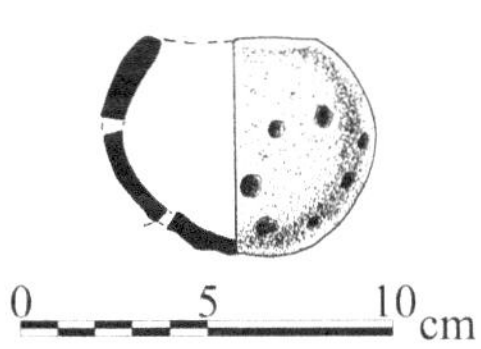

Figure 7.19: Unslipped/Smoothed - Perforated incense burner.

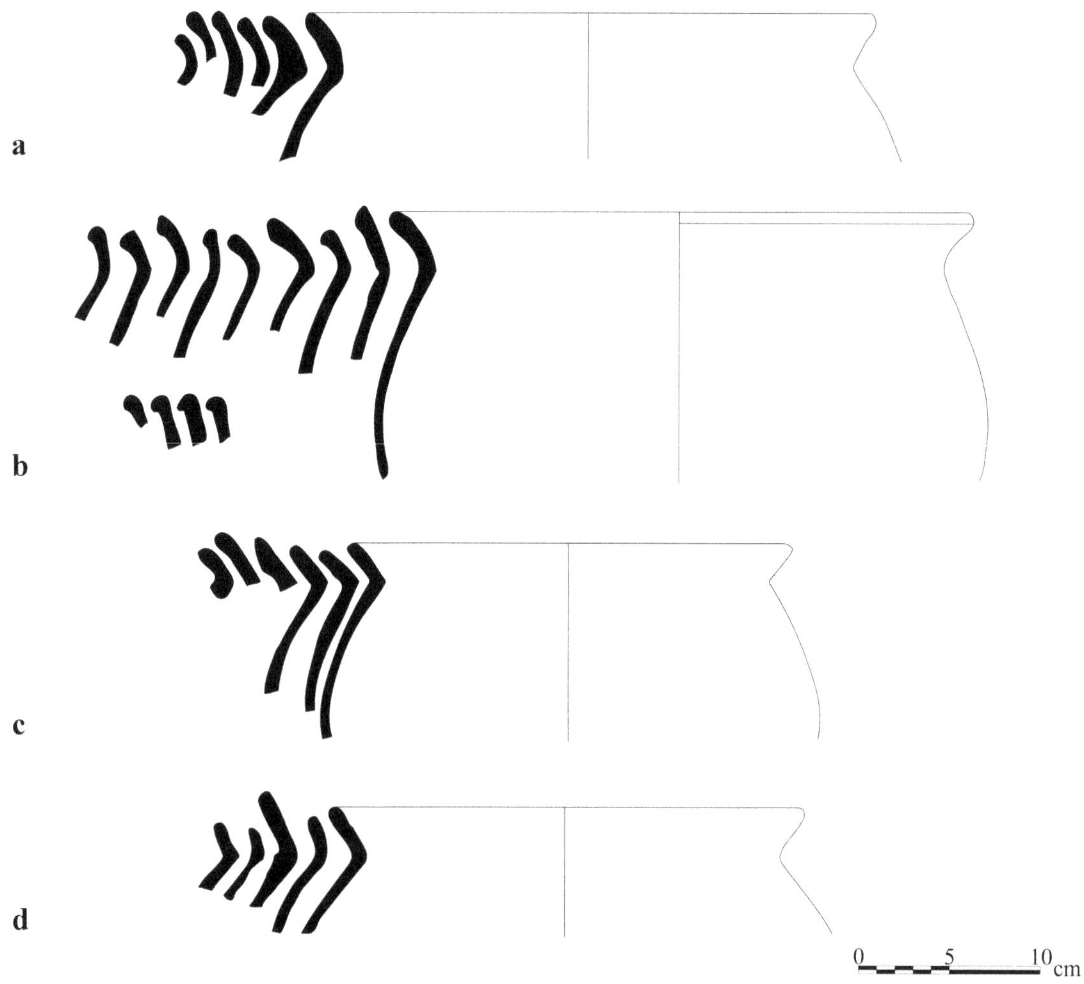

Figure 7.20: Unslipped/Smoothed-Buff jars: a) out-curving neck and exterior-thickened rim; b) out-curving neck and bolstered (top) or folded (bottom) rim; c) flaring neck (third from left has a collared rim); d) less well made jars.

of the Orange Slipped and Incised vessels discussed above. The forms that occur in the calcite- and grog-tempered fabrics, however, appear to be identical (Figure 7.20, Plates IX.67-68).

UNSLIPPED UTILITARIAN

Frequency: This group, which includes two whole vessels and 784 fragments, is the most ubiquitous within the overall assemblage, comprising 31.6% (N=784) of the whole assemblage and 68.9% of all coarse ware (Table 7.6). A total of 565 fragments of these vessels were recovered from the midden associated with N10-27, where as 217 were recovered from the midden at structure N10-9 midden (Tables 7.3 and 7.4). The majority of the vessels in this group are jars (85.4%, N=668). Plates/lids comprise 10.2% (N=80) and bowls represent 2.8% (N=22) (Table 7.5).

Principal Identifying Modes: These vessels are comparatively thick-walled, unslipped and have only crudely smoothed interior and exterior surfaces. The fabrics are coarse-textured, contain quartz sand and have a sandy look and feel. The vessel surfaces tend to be dark colours such as brown, grey and dark grey to black. Principal vessel forms include jars with flaring to slightly out-curving necks, direct rims and rounded lips, rounded bowls and dishes and plates/lids. Very rarely, the jars have loop handles straddling the neck (Figures 7.22 and 7.23, Plate IX.69).

WASHED/SMOOTHED

Frequency: This group is less well represented, comprising only 0.2% of all ceramics (N=6) and 0.5% of all coarse wares (Table 7.6). All fragments were recovered from the midden associated with N10-27 and all derive from jars (Tables 7.4 and 7.5).

Principal Identifying Modes: Thick-walled jars with crudely smoothed interior and exterior surfaces and a orange to red-orange wash on the exterior surface. Fabric-wise, these jars are identical to the Unslipped-Utilitarian jars in that the fabrics are coarse-textured, contain sand and have a sandy look and feel.

RARE COARSE WARE VESSELS

Unslipped-Striated – This vessel (LA487/1), which is a jar, appears to be unique within the assemblage. The body of the jar is globular and it has a high, vertical neck and a

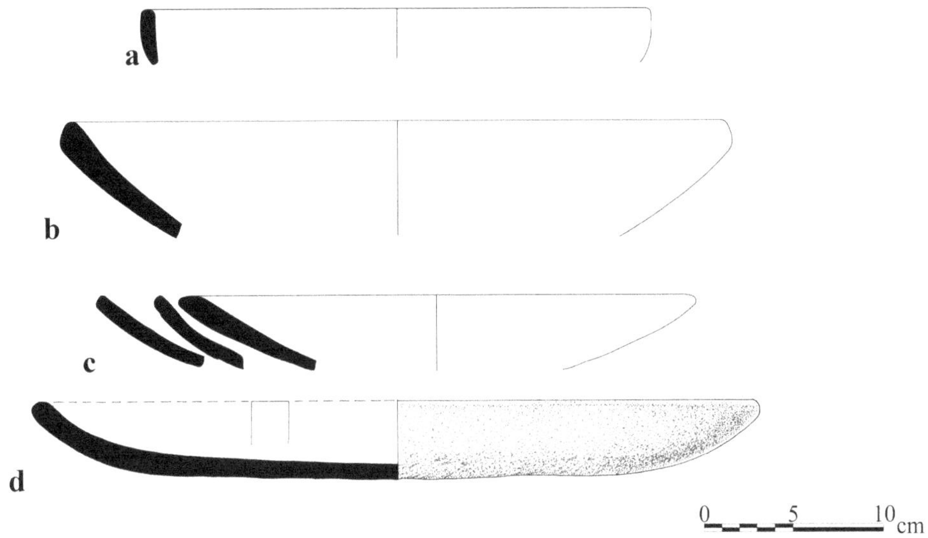

Figure 7.21: Unslipped Utilitarian vessels: a) bowl; b) dish; c) plates/lids; d) comal.

bolstered rim (see Figure IX.45 in Appendix IX for a detailed description and illustration). The interior and exterior surfaces of the vessel are unslipped. The exterior surface is well-smoothed from the rim to the shoulder and the body has deep striations. The interior surface is less well smoothed.

Unslipped with Appliqué – This group is represented by two partial vessels (LA469a and LA469b), both of which are jars. Necks are flaring, rims are direct and lips are bevelled-out. In both cases, vertical, grooved, strap handles straddle the neck. A similar strap handle was recovered from the midden associated with structure N10-27, and might also derive from this type of jar. The interior and exterior surface of the jars is unslipped and the undecorated areas of the exterior surface are well-smoothed. An appliqué impressed fillet encircles the upper shoulder of both jars and LA469a has incised decoration between the fillet and the neck. The jars are illustrated and described in detail in Figure IX.43 of Appendix IX.

7.3. Continuity and Change in Ceramic Styles: The Development of Vessel Forms and Decorative Techniques During the Terminal Classic to Early Postclassic Period

In the previous section and Appendix II, the content and composition of the Terminal Classic to Early Postclassic assemblage was described in detail. The fine ware and coarse ware components of the assemblage were characterized according to the different stylistic groups of vessels they comprise; documenting the relative frequency of each group, the vessels forms and surface treatments they entail, and any variation in this regard (Appendix II). This section places this information into a broader spatial and temporal framework. It looks at the relative frequencies of the vessels groups within different stratigraphic contexts, delineating their temporal associations and relationships, and investigates the nature of continuity and change in ceramic styles during the Terminal Classic to Early Postclassic period.

The Stratigraphic and Temporal Relationships of the Vessel Groups

The relative frequencies of the different fine ware and coarse ware vessel groups comprising each lot of ceramics from the midden deposits provide direct evidence of continuity and change in vessel forms and surface treatments over time, when the stratigraphic relationships between the lots are considered. As was discussed in Chapter 6, the lots of ceramics excavated from the midden associated with structure N10-27 include well known markers of the Terminal Classic period, as well as vessel styles that are typical of the Early Postclassic 'Buk' phase at Lamanai. The sherd assemblage from the midden at structure N10-9, however, is dominated by Early Postclassic vessel styles. Since the two midden assemblages overlap stylistically, in terms of the specific kinds of vessels they contain, they can be considered together as representing a continuous sequence of pottery styles that straddles the Terminal Classic and Early Postclassic periods. To facilitate comparison of the relative frequencies of vessel groups within different stratigraphic contexts, as well as within the two midden deposits, the lots of ceramics from the midden at structure N10-27 were organized into three groups that reflect the basic stratigraphy of this deposit. The deepest area of midden, which directly abuts the structure, is represented by the trench and the series of squares that were dug along, and at the corner of, the terrace face. The trench comprises two lots of material, which represent the upper (LA701) and lower (LA717) levels of this area of the deposit. The lots from the squares were organized into three groups – those lots

Figure 7.22: Unslipped Utilitarian jars.

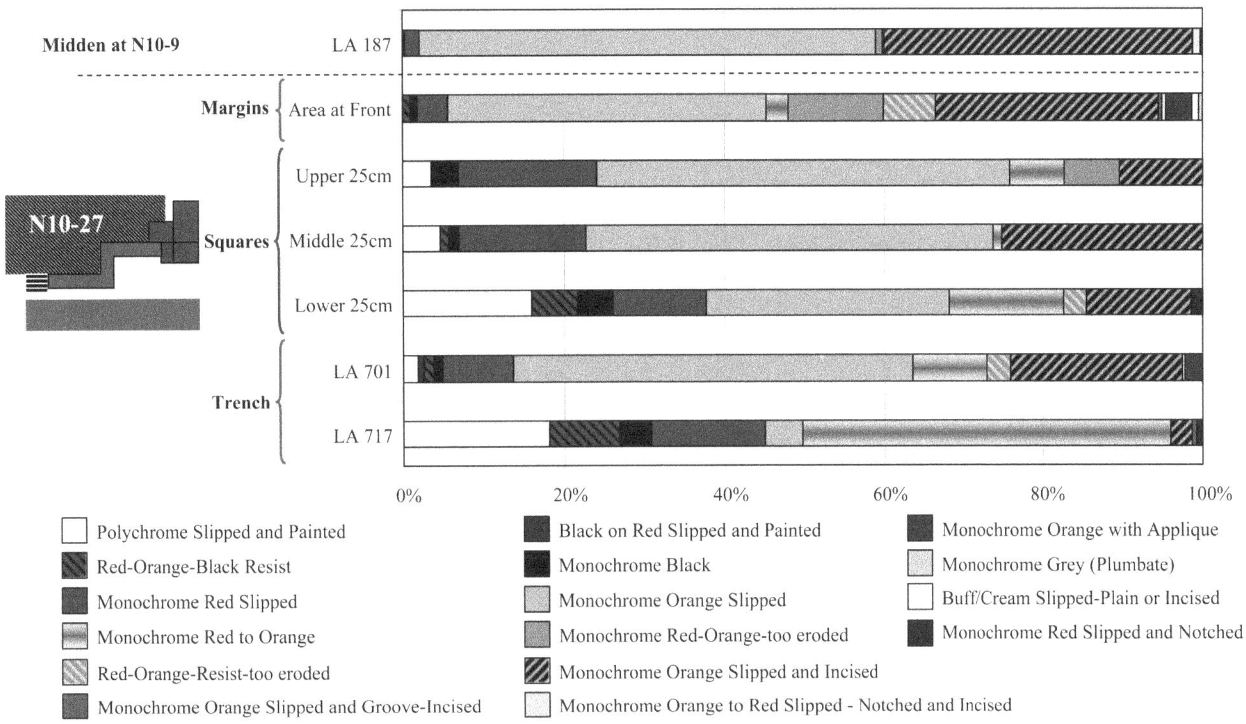

Figure 7.24: Frequency distribution of fine ware groups within different stratigraphic contexts (see colour version on accompanying CD).

deriving from the upper 25cm of the deposit (LA1346, LA1350 and LA1354), those from the middle 25cm (LA1347, LA1351 and LA1355), and those from the lower 25cm (LA1348, LA13521 and LA1355). The third group of lots (LA1114, LA1115 and LA1119) comprises material recovered from the area out in front of the structure, where the midden tapers off. Accordingly, this group of lots is associated with areas of lesser accumulation and the margins of the deposit. The ceramics from the midden associated with structure N10-9 comprise one lot of material (LA187). Since no obvious stratigraphy was observed within the deposit, this lot is considered to represent a single stratigraphic unit that can be directly compared to the lot groups from the midden at structure N10-27.

FINE WARE

The relative frequencies of the fine ware vessels groups comprising the different midden lots are presented in Figure. The most basic pattern that emerges through this comparison is that Polychrome, Black on Red, Monochrome Black, Red Slipped and Resist vessels are more strongly associated with the deepest areas of the midden at structure N10-27, whereas Orange Slipped and Orange Slipped and Incised vessels are more strongly associated with the upper levels and margins of the deposit, and clearly dominate the material from the midden at structure N10-9. This pattern seems to reflect a temporal trend in which Orange Slipped and Orange Slipped and Incised vessels increase over time and replace the decreasing Polychrome, Black on Red, Monochrome Black, Red Slipped and Resist vessels. This trend is most clearly seen in the relative frequencies of vessels groups within different areas of the midden at structure N10-27. Within the lowest level of the trench (LA717), which comprises material from the deepest area of the midden, Polychrome, Black on Red, Resist, Monochrome Black and Red Slipped vessels comprise approximately half of the ceramics, whereas Orange Slipped and Orange Slipped and Incised vessels are poorly represented. In the upper level of the trench (LA701), however, Orange Slipped and Orange Slipped and Incised vessels clearly predominate, whilst the other vessel groups together only form about 15% of the material. This complete shift in the vessels groups that form the majority of ceramics within the lower vs. the upper levels of the deepest area of the midden is mirrored in the frequency distributions associated with the squares that were dug in 25cm levels. The frequency distribution of the ceramics deriving from the lowest 25cm of the squares is very similar to that of the lowest level of the trench. The middle and upper levels, however, give distributions that are more similar to the upper level (LA701). Similarly, the margins of the midden also show a predominance of Orange Slipped and Orange Slipped and Incised vessels, whereas Polychrome vessels are entirely absent and Monochrome Black, Resist and Red Slipped vessels are only represented in very small amounts. The frequency distribution for LA187 – the lot deriving from the midden associated with structure N10-9 – shows an even greater prevalence of Orange Slipped and Orange Slipped and Incised vessels, accompanied by a complete absence of Polychrome and Resist vessels, and extremely small amounts of Monochrome Black and

Red Slipped vessels. The temporal associations that are revealed through the frequency distribution data, therefore, are that Polychrome, Black on Red, Monochrome Black, Red Slipped and Resist vessels are more strongly associated with the early part of the sequence, whereas Orange Slipped and Orange Slipped and Incised vessels typify the later part of the sequence.

Most of the vessel groups that are comparatively rare within the assemblage also show clear temporal associations. For example, the red-slipped bowls with notched decoration (Red Slipped and Notched) only occur in the lowest levels of the midden deposit at Structure N10-27, suggesting that this surface treatment is restricted to the early part of the sequence. In contrast, the red- and orange-slipped bowls with notched and incised decoration (Red to Orange Slipped – Notched and Incised) only occur in the upper level and margins of the deposit, as well as in the deposit at structure N10-9 – i.e. temporally later contexts – suggesting this is a later surface treatment. Similarly, the orange-slipped bowls with an appliqué impressed fillet (Monochrome Orange with Appliqué) also appear to represent a later decorative style, since these vessels are restricted to the same, later, stratigraphic contexts. Other rare vessels groups that tend to only occur in the most recent areas of accumulation within the midden at structure N10-27 include the single example of Plumbate ware and Buff/Cream Slipped vessels with plain surfaces or incised decoration (two examples are Slate ware). The latter vessel group, however, also includes a complete vessel that was recovered from the lower trench lot (LA717/1).

The temporal associations of the orange-slipped vessels with groove-incising (Monochrome Orange Slipped and Groove-Incised) are more ambiguous. These vessels only occur within the midden deposit at structure N10-27, but are found in both early and later areas of accumulation (the lower trench lot and margins of the deposit). The vessels that occur within the lowest levels of the midden, however, are vases – an early form within the sequence – whereas the vessel from the margins of the deposit is a drum, which is a later vessel form (see below).

Since each of the vessel groups comprises a particular suite of vessel forms, the frequency distribution data also provides evidence of continuity and change in vessel morphology over time. The focus here is on the most robust patterns in regard to the temporal associations of particular shape classes, as well as specific forms within the shape classes. Changes in specific discrete attributes such as base form or foot type are considered on a more general level.

As was discussed in the introductory section on the fine ware, the shape classes that are most prevalent include bowls, dishes, and jars, suggesting a certain level of continuity across the sequence, in terms of in the kinds of vessels that tend to be most frequent. Certain shape classes, however, tend to have specific temporal associations. For example vases, as a shape class, are more strongly associated with vessel groups that dominate the early end of the sequence, such as Polychrome, Monochrome Black and Red Slipped. There are only two vases in the Orange Slipped group, both of which derive from the lowest levels of the N10-27 midden deposit, and they are entirely absent in the Orange Slipped and Incised group. It would seem, therefore, that vases, as a shape class, tend to decrease in frequency and eventually disappear within the sequence. Other shape classes are more strongly associated with the later part of the sequence. For example, with one exception, drums only occur within the Orange Slipped and Orange Slipped and Incised groups and they increase in frequency in areas of more recent accumulation (upper levels and margins) within the deposit at structure N10-27, as well as in the midden at structure N10-9. Similarly, comparatively rare forms such as sieves and composite vessel, which exclusively occur in the Orange Slipped group, and frying pan censers and whistles, which only occur in the Orange Slipped and Incised group, are exclusively associated with the same, later, stratigraphic contexts. Accordingly, it would seems that these forms are introduced into the sequence as the frequency of Orange Slipped and Orange Slipped and Incised vessels increases.

Within the bowl class, forms that are more strongly associated with the early end of the ceramic sequence include flaring and out-curving bowls with flat bottoms and composite silhouette bowls with a flat or ring base. These forms are principally associated with the Polychrome, Monochrome Black and Red Slipped groups; are comparatively rare within the Orange Slipped group; and are entirely absent within the Orange Slipped and Incised group. The bowls within these later groups tend to have either rounded or out-curving sides and rounded bases, sometimes with a circular concavity or impression. Among the whole vessels, this circular concavity only occurs on Orange Slipped and Incised out-curving bowls, suggesting that this type of base may be restricted to these particular bowls. In regard to tripod bowls, there is a dramatic increase in their frequency in the upper levels and margins of the midden at structure N10-27 as well as in the deposit at structure N10-9. This form type is also most strongly represented within the Orange Slipped and Orange Slipped and Incised groups, suggesting that the increase in this form type is directly related to the proliferation of these surface treatments. Tripod bowls are comparatively infrequent among the vessel groups that are more typical of the early end of the sequence, and these vessels tend to have solid slab, or less often, nubbin feet. As tripod bowls increases in frequency within the Orange Slipped and Orange Slipped and Incised groups, feet tend to be hollow, and oven to globular or conical in shape. Effigy feet also occur within these vessel groups and they are most frequent in the midden at structure N10-9.

Within the dish class, forms that can be considered to be early within the ceramic sequence include rounded dishes and composite silhouette dishes with a flat or ring base. The Black on Red group only includes rounded dishes, and with only one exception, so does the Polychrome group. Rounded dishes occur in both the Red Slipped and Orange Slipped groups and there are also many examples upon which the colour the slip grades between red and orange. Within the Orange Slipped and Incised group, rounded dishes are entirely absent. Composite silhouette dishes with a ring base are predominantly associated with the Resist and Red Slipped groups. Orange Slipped examples are comparatively rare, and as with the rounded dishes, there are many of these vessels on which the slip colour varies. Composite silhouette dishes are far more frequent within the lowest levels of the midden at structure N10-27, suggesting that there is a considerable decrease in the frequency of this form over time. The frequency of rounded dishes also appears to decrease, although less dramatically than the composite silhouette dishes. 'Chalices', or pedestal-based dishes, appear to represent a later dish form within the sequence since they dramatically increase in frequency within later stratigraphic contexts – the upper levels and margins of the deposit at N10-27 and the midden at structure N10-9. Chalices are also almost exclusively associated with the Orange Slipped and Orange Slipped and Incised groups. Graham (1987) suggests that the chalices, as a form type, develop out of the ring-based composite silhouette dishes, since their basic morphology is very similar (see Figure 7.23). Individual examples of both form types sometimes have a low pedestal base, which is considered to be an intermediary form. Within the assemblage presented here, base fragments range from short ring bases to tall pedestal bases and gradational increase in height throughout the sequence seems to be indicated, as opposed to a discrete separation into two distinct base types. Some of the complete examples of composite silhouette bowls, however, also have a low pedestal base, as opposed to the more usual ring or flat base, suggesting that base heights tend to increase in general throughout the sequence, regardless of vessel form.

Fine ware jars appear to remain globular to slightly maliform in shape throughout the sequence. The morphology of the rim to neck area of these vessels is more variable, although this variation is generally continuous in nature and cross-cuts the different vessel groups. A few specific jar forms, however, appear to have more restricted temporal associations within the sequence, the best examples of which are the pedestal-based jars, the 'neckless' jars and jars with strap handles. The pedestal-based jars, which are confined to the Orange Slipped and Orange Slipped and Incised groups, are completely absent in the lowest levels of the midden at structure N10-27 and increase in frequency in later contexts. The 'neckless' jars show the same general pattern. It would appear, therefore, that these two jar forms are introduced into the sequence as the frequency of Orange Slipped and Orange Slipped and Incised vessels increases. Strap handle jars show a similar pattern in that fragments of orange-slipped strap handles only occur within later stratigraphic contexts – i.e. within the upper levels and margins of the midden at structure N10-27 and in the deposit at structure N10-9. The restricted occurrence of the handle fragments seems to suggest that these jars represent a form that emerges late in the ceramic sequence. The only jar form that appears to be restricted to the early end of the sequence has a vertical to slightly in-sloping neck and a flaring to horizontal everted rim. Jars with this distinctive rim to neck morphology only occur in the lowest levels of the midden at structure N10-27, and in all but one case, have a red slip. (the remaining jar has an orange slip).

COARSE WARE

The frequency distribution of coarse ware vessel groups within the different stratigraphic contexts tell a similar story to that of the fine ware – the groups that are typical of the early end of the sequence decrease in frequency and are replaced by the increasing amounts of other vessel groups (Figure 7.25). For example, vessels with red-slipped rims or lips (Slipped Rim/Lip – Smoothed Body) and Red-Brown-Striated jars are more strongly associated with the deepest areas of the midden at structure N10-27, whereas Unslipped Utilitarian and Unslipped/Smoothed-Buff vessels are strongly associated with the upper levels and margins of this deposit, and clearly dominate the material from the midden at structure N10-9. This pattern appears to reflect a temporal trend in which the Unslipped Utilitarian and Unslipped/Smoothed Buff groups increase over time and replace the decreasing Slipped Rim/Lip – Smoothed Body and Red-Brown-Striated groups. As was the case with the fine ware vessel groups, this trend is most clearly seen in the relative frequencies of vessel groups within different areas of the midden at structure N10-27. Within the lowest level of the trench (LA717), which is where the deposit is deepest, vessels of the Slipped Rim/Lip – Smoothed Body and Red-Brown-Striated groups clearly predominate, comprising just over 90% of the recovered material. In contrast, Unslipped Utilitarian and Unslipped/Smoothed-Buff vessels are comparatively rare within this context. In the upper level of the trench (LA701), however, Unslipped Utilitarian vessels clearly predominate, whilst the other vessel groups comprise less than 15% of the material. This complete shift in the vessels groups that form the majority of coarse ware vessels within earlier vs. later stratigraphic contexts within the deposit also occurs in the squares that were dug in 25cm levels. The frequency distribution for the lowest 25cm of the squares is very similar to that of the lowest level of the trench. The middle and upper levels of the squares, however, give distributions that are more similar to the upper trench unit (LA701). Similarly, the frequency distribution of vessel groups at the margins of the deposit is practically identical to the upper level of the trench. Within the deposit at structure N10-9, Unslipped Utilitarian vessels clearly predominate and the Unslipped/Smoothed-Buff group is more strongly represented than within any stratigraphic context of the midden at structure N10-27. The Slipped Rim/Lip – Smoothed Body group is least well represented within the deposit at structure N10-9 than it is any other context and the Red-Brown-Striated group is entirely absent. The frequency distribution data reveals no particular temporal associations in relation to Washed/Smoothed jars, apart from the fact that they only occur in the midden at structure N10-27. No obvious temporal associations are

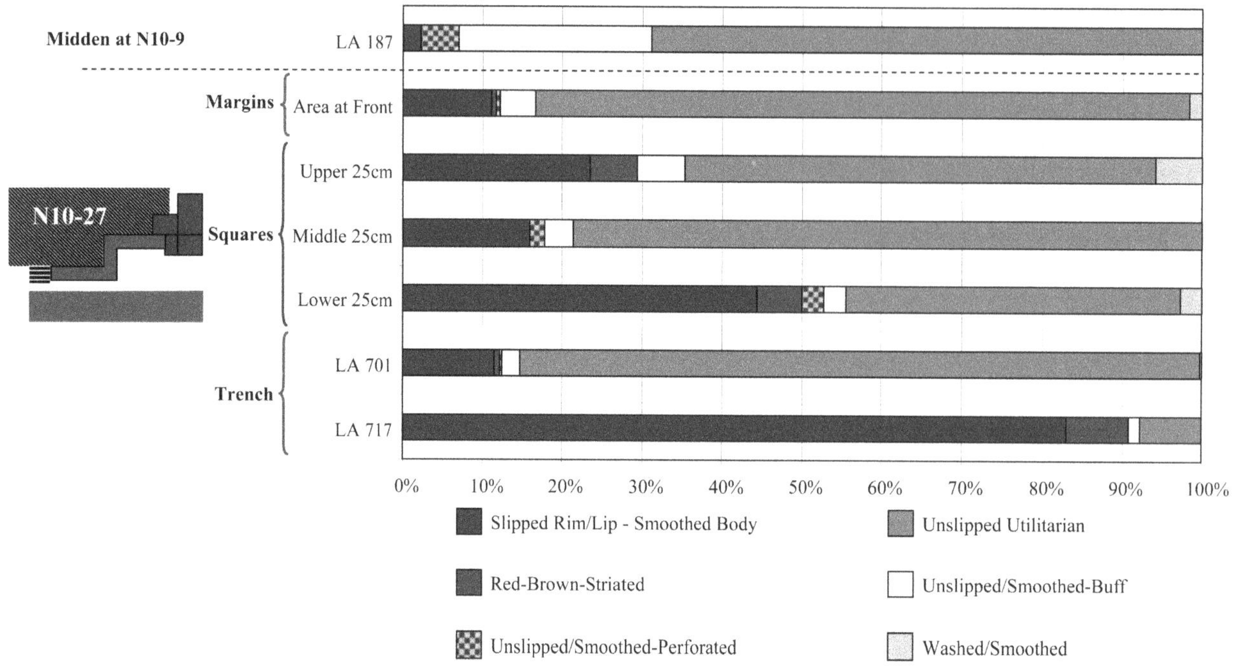

Figure 7.25: Frequency distribution of coarse ware groups within different stratigraphic contexts (see colour version on accompanying CD).

apparent in regard to the Unslipped/Smoothed-Perforated incense burners either, although they are slightly more frequent in the midden associated with structure N10-9.

As was discussed in the introductory section to the coarse ware, jars comprise over 90% of all of the vessels that occur within any particular group. Accordingly, jars form the dominant shape class among the coarse ware throughout the ceramic sequence. Among the jars forms that are typical of each of the groups, the Unslipped Utilitarian jars and the Washed Smoothed jars, stand out as the only instances in which the specific morphology of the rim to neck area is highly variable. The jars in these two groups are also comparatively crudely made. The jar forms comprising the other vessel groups, however, tend to be more internally consistent, in terms of their particular morphology, and many specific forms exhibit a high level of uniformity. Since the relative frequency of Unslipped Utilitarian jars increases over time, it would also seem that crudely made jars with a more variable rim to neck morphology become increasingly prevalent within the sequence. The strap handle jar is a form that occurs within two of the vessel groups (Unslipped/Smoothed-Buff and Unslipped with Appliqué). A single fragment of a grooved strap handle, which is a typical form within the Unslipped with Appliqué group, was recovered from the margins of the midden at structure N10-27, suggesting these jars may be more strongly associated with the later end of the sequence. Likewise, the strap handle jars of the Unslipped/Smoothed-Buff group also appear to be more strongly associated with the later end of the sequence. Handle fragments from these jars only occur in later stratigraphic contexts – i.e. within the upper levels and margins of the midden at structure N10-27 and in the deposit at structure N10-9. Accordingly, it would seem these jars emerge later on in the sequence when the Unslipped/Smoothed-Buff group increases in frequency.

Open coarse ware forms, which primarily include bowls, dishes and plates/lids, only occur in the Unslipped Utilitarian and the Unslippped/Smoothed-Buff groups. All of these forms are entirely absent from the lowest levels of the midden at structure N10-27, suggesting that they are more strongly associated with the later part of the sequence.

As with the open forms, comparatively rare vessel forms, such as pedestal-based censers, comals and the large basin, are more strongly associated with the later part of the sequence. The only rare forms that appear to be restricted to the early part of the sequence are the broad rimmed bowl and the in-sloping, flat-bottomed jar, both of which were recovered from the deepest area of the N10-27 midden (LA717).

From the point of view of vessel paste or fabric, the relative frequencies of the coarse ware vessel groups within different stratigraphic contexts indicate that the Unslipped Utilitarian group, which comprises vessels with sandy pastes, dramatically increases in frequency, whereas the frequency of two of the groups with calcite-tempered pastes (Slipped Rim/Lip – Smoothed Body and Red-Brown-Striated groups) dramatically decreases.

It would seem, therefore, that groups with sandy pastes increase in prevalence over time at the expense of the calcite-tempered groups (Slipped Rim/Lip – Smoothed Body and Red-Brown-Striated groups) that dominate the early part of the sequence. In addition, as these early calcite-tempered pastes decrease in frequency, the Unslipped/Smoothed Buff group, which is also dominated by calcite-tempered vessels, tends to increase. Since the jars comprising the different groups with calcite-tempered fabrics tend to exhibit a high level of morphological uniformity within particular form types, it would seem that the early 'standardized' jar forms are replaced by new standardized forms with a similar paste but a completely different morphology.

Continuity and Change in Vessel Forms and Decorative Treatments During the Terminal Classic to Early Postclassic Period

The patterns of continuity and change within the assemblage, in regard to vessel forms and decorative treatments, provide significant insight into trends in the style of ceramics that characterize the Terminal Classic to Early Postclassic period. These developments, which are summarized in Table 7.8, are of particular importance when considering a variety of issues that relate to patterns of ceramic production and consumption at the site, and specifically the ways in which these patterns might have changed over time. Based on these developments, the following observations can be made:

1) Within the fine ware component of the assemblage, there is a general loss of diversity in surface treatment through the sequence. The early part of the sequence is characterized by the presence of a number of different surface treatments, including Polychrome, Black on Red, Resist, Red Slipped and plain and decorated versions of Monochrome Black, which together comprise the majority of fine ware vessels. The prevalence of vessels with these surface treatments decreases over time, whilst vessels with Orange Slipped or Orange Slipped and Incised vessels become predominant. By the end of the sequence, fine ware vessels are almost exclusively orange-slipped with either plain surfaces or incised decoration.

2) Accompanying the loss of diversity in surface treatment, there appears to be a significant change in the overall approach to fine ware surface treatments and decoration. On a very general level, the technology of decoration shifts away from painting and the manipulation of slips to achieve desired colouring effects and towards a technology centred around the use of one colour and the post-slip, pre-firing alteration of the vessel surface. This trend is clearly seen in the apparent loss of painted fine ware such as Polychrome and Black on Red vessels and of the vessels with a resist surface treatment, and the substantial increase in Orange Slipped and Orange Slipped and Incised vessels. In the early part of the sequence, decorative treatments involving the alteration of a vessel's surface via gouging, stamping or groove-incising etc. is always done prior to slipping. The incised decoration that is typical at the end of the sequence, however, was always rendered after the vessel was slipped. Through this approach, the incised design is accentuated by the colour contrast created between the slipped surface and the underlying ceramic body exposed through the incisions. These later designs tend to be very intricate, occur almost exclusively on vessel exteriors and tend to cover most of the workable space on most vessel forms. The chile grinders are the only vessels with incising on the interior surface and these designs tend to be comparatively simple. Some vessels are further embellished through the addition of hand-modelled, appliqué heads and other body parts, and in the case of the footed vessels, through the addition of effigy feet. In the early part of the sequence, however, appliqué embellishments are entirely absent and complex designs and images are most often rendered on the interior surfaces of open forms such as dishes. In instances where it is the exterior surface that is decorated, as is the case with some bowls and vases, the scale of the design is comparatively small, and on monochrome vessels, it is somewhat obscured by the slip. In the early part of the sequence the dominant colours within the assemblage are red and black. Orange slips increase substantially over time, and by the end of the sequence, orange to reddish orange slips predominate.

3) Alongside the significant changes in surface treatment and decoration of fine ware vessels comes a general increase in the diversity of vessel forms that are seen. This general trend is clearly indicated by the introduction into the sequence of new forms, including drums, frying pan censers, composite vessels, whistles and pedestal-based jars. It could be argued that many of these later forms are more complex, morphologically, than those earlier in the sequence. The emergence of features such as hand-modelled appliqués, effigy handles and feet, segmented basal flanges and high pedestal bases exemplifies this trend towards increased complexity in vessel design. The intended function of most of these new forms also appears to fall outside the range of standard serving vessels or table wares, suggesting a general expansion in the kinds ceramic products produced during this time period. The introduction of the new vessel forms also appears to coincide with the decreased frequency and eventual loss of vases from the ceramic repertoire.

4) Among the coarse ware, there appears to be a dramatic shift in the approach to making storage jars. The early part of the sequence is dominated by calcite-tempered jars with red-slipped rims or lips (Slipped Rim/Lip – Smoothed Body group).

Table 7.8: Summary of temporal trends in vessel surface treatment and morphology for the Terminal Classic to Early Postclassic ceramic sequence.

	Early	➔ Late
Fine Ware		
Surface Treatment	- Polychrome Slipped and Painted - Black on Red Slipped and Painted - Monochrome Black - Monochrome Red Slipped - Red-Orange-Black Resist - Monochrome Red Slipped and Notched - 'interior-slipped' rounded bowls and dishes	- Monochrome Orange Slipped - Monochrome Orange Slipped ad Incised - Red to Orange Slipped-Notched and Incised - Monochrome Orange with Appliqué - Monochrome Gray (Plumbate) - Buff/Cream Slipped-Plain or Incised
Forms	- vase - flaring and outcurving bowls with flat bottoms - composite silhouette bowls with a flat or ring base - flaring and outcurving tripod bowls with flat bottoms and solid slab or nubbin feet - rounded dishes with a flat or ring base - composite silhouette dishes with a ring or low pedestal base - jars with a vertical to in-sloping neck and a flaring to horizontal everted rim	- drums - sieve - composite vessel - frying pan censer - whistle - rounded and outcurving bowls with rounded bases commonly with a circular concavity - chile grinders - tripod bowls with segmented basal flange - rounded and outcurving tripod bowls with hollow, oven to globular, conical or effigy feet - chalices with low and high pedestal bases - pedestal-based jars - 'neckless' jars - jars with strap handles
Coarse Ware		
Surface Treatment	- Slipped Rim/Lip-Smoothed Body - Red-Brown-Striated	- Unslipped Utilitarian - Unslipped/Smoothed-Buff - Unslipped with Appliqué
Forms	- well made jars with a 'standardized' morphology - broad rimmed bowl - in-sloping, flat bottomed jar	- well made jars with a 'standardized' morphology - crudely made jars with a variable rim to neck morphology - strap handle jars - bowls - dishes - plates/lids - pedestal-based censers - comals
Pastes	- calcite-tempered	- calcite-tempered - sandy pastes

5) The prevalence of unslipped jars with sandy fabrics (Unslipped Utilitarian group) substantially increases over time and these jars are predominant at the end of the sequence. The slipped jars with calcite-tempered fabrics are extremely well made. The slip that is used is of the same high quality as the fine ware slips and surfaces are well smoothed, suggesting that a considerable amount of time and effort was invested in finishing these vessels. Morphologically speaking, these jars are also fairly standardized in terms of their proportions

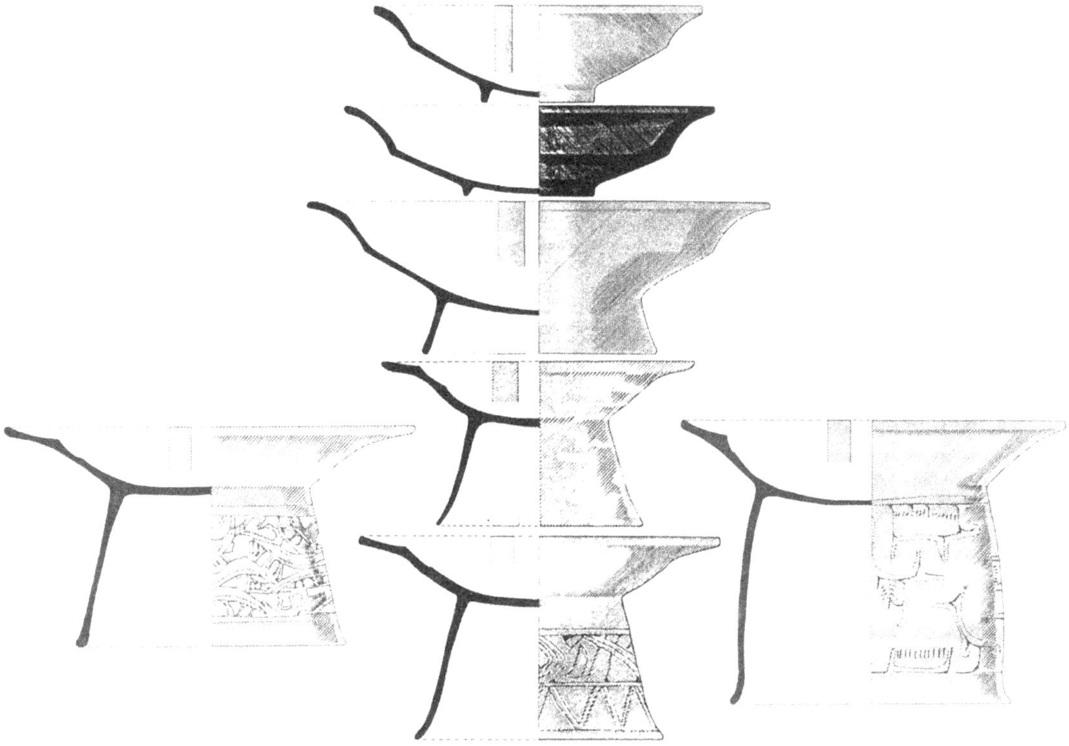

Figure 7.23: Composite silhouette dishes and chalices.

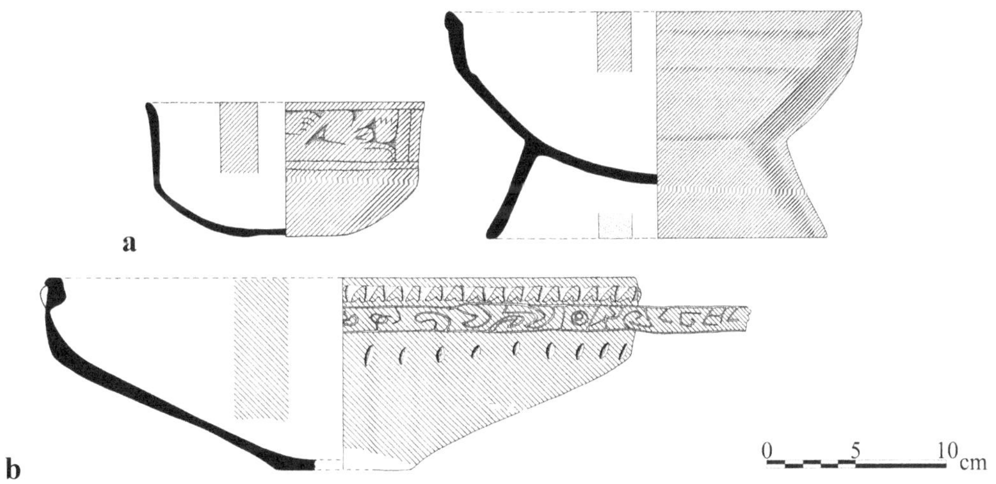

Figure 7.24: Fine ware vessels displaying a combination of early and late morphological and decorative characteristics: a) an early form with typical Early Postclassic incised decoration (LA68/6) (left), an early form with a pedestal base (LA122/2); b) an early form with a later style of decoration (LA124/1).

6) and basic form. In contrast, the jars with sandy fabrics are unslipped, poorly made and their surfaces only hastily smoothed over. Hence, a similar amount of effort does not seem to have been invested in finishing these jars. In addition, the morphology of the jars with sandy fabrics is highly variable, particularly in regard to the rim to neck area. This evidence would seem to suggest that during this time period there are marked changes in the manufacture and finishing of vessels that may well have served the same function.

7) As with the fine ware, with the increased prevalence of vessel groups that become predominant at the end of the sequence, comes the introduction of new vessel forms. These forms include bowls, dishes, plate/lids and a few comparatively rare forms such as pedestal-based censers and comals. The appearance of these

new forms within the coarse ware repertoire suggests an expansion in the range of coarse ware vessels being produced and/or consumed during this time period.

8) Against the backdrop of the obvious stylistic changes that occur during this time period, there is an element of continuity that cannot be ignored. For example, the presence of a large number of fine ware vessels, throughout the ceramic sequence, on which slip colour falls between red and orange seems to suggest a transitional phase between the colours that are typical of the early vs. the late end of the sequence. In addition, many of the fine ware groups contain conformal vessels, suggesting a gradual progression towards the corpus of vessel forms that dominates the end of the sequence. The composite silhouette bowls and the rounded dishes, for example, occur in four different vessel groups. Some of the vessel forms comprising the fine ware assemblage have morphologies that appear strongly related and seem to chart a sequence of change in specific kinds of vessels over time. The composite silhouette dishes and the chalices provide a vivid example of this kind of morphological change (Figure 7.23). Among the whole vessels there are also examples of vessels that appear to be intermediary within the sequence, in that they display a combination of early and late morphological and surface treatment attributes. Specific examples include LA68/6 and LA122/2 and LA124/1 (Figure 7.24). What emerges, therefore, is a very complex picture, in which obvious changes over time in vessel styles and certain approaches to making ceramics are offset against an undercurrent of continuity. The nature and meaning of these emergent patterns are explored in the next three chapters, which examine specific details of vessel manufacture during Terminal Classic to Early Postclassic Period.

CHAPTER 8
PETROGRAPHIC ANALYSIS OF THE TERMINAL CLASSIC TO EARLY POSTCLASSIC CERAMIC ASSEMBLAGE: COMPOSITIONAL AND TECHNOLOGICAL VARIABILITY AT THE MICROSCOPIC LEVEL

Although ceramic studies have been carried out at several sites in northern Belize with occupations spanning the Classic to Postclassic transition, detailed technological analyses of site assemblages using scientific techniques, have not been carried out until the present work at Lamanai. As a result, current explanatory models regarding ceramic change at sites situated in this area of the lowland region have been built upon the limited physical evidence that macroscopic examination of pottery brings to light. At this level of analysis it is difficult to accurately assess the mineralogy of ceramic fabrics, to identify their raw material ingredients and to link raw materials to potential sources. Without this information it is often difficult to discriminate locally produced and imported pottery, to characterize local pottery technology and to examine change over time in the interplay between stylistic and compositional variability. This information is of fundamental importance to achieving reliable reconstructions of patterns in the production and consumption of pottery items across time and space. Building upon the patterns and trends that were identified in the previous chapter with regard to continuity and change in stylistic characteristics of the Lamanai assemblage, this chapter presents the results of the microscopic analysis of compositional variation relating to provenance and paste technology. Compositional groups identified through thin section analysis are described and discussed and their relationships to the functional and stylistic categories considered in previous chapters are investigated. Diachronic patterns in local raw material choice and manipulation are examined and temporal trends in the kinds and origins of foreign ceramics entering the site are identified.

8.1. GENERAL COMPOSITION OF THE SAMPLE SET

A total of 646 vessels from the Terminal Classic to Early Postclassic assemblage were selected for petrographic analysis, representing 26% of the minimum of vessels recorded. This total includes all 152 of the whole and reconstructed vessels deriving from specialized deposits associated with architectural structures, 475 fragmentary vessels deriving from the temple middens associated with structures N10-27 and N10-9 and 18 different chalice base fragments recovered from the large smash-and-scatter offering covering the central stair of N10-9 (LA243) (Table 8.1). These chalice fragments were sampled in addition to the five reconstructed vessels recovered from this offering to examine compositional variation among stylistically similar vessels deposited at the same time as part of this comparatively large-scale offertory event.

As reported in Chapter 7, 96.1% (N=146) of the whole and reconstructed vessels are fine ware or serving/table ware and the remaining six are coarse ware vessels, including three large jars, a pedestal-based censer, a comal and a broad-rimmed deep bowl. Among the fragmentary vessels sampled from the midden associated with structure N10-27, 66.3% (N=232) are fine ware vessels and 33.7% (N=118) are coarse ware vessels, the large majority of which are large utilitarian storage jars (91.5%; N=108). Similarly, the majority of vessels sampled from the midden abutting N10-9 are serving vessels (71.4%; N=90), with storage jars representing the majority (63.9%; N=23) of the sampled coarse ware. Table 8.1 gives a breakdown of the vessels sampled for petrographic analysis from the two temple midden assemblages according to their surface treatment and form (see Table 6.7 for the breakdown for the whole and reconstructed vessel).

The disproportionately high representation of fine ware vessels within the sub-set of material sampled for petrographic analysis, considering that coarse ware and fine ware have near equal representation in the two temple middens, reflects the greater variability among the fine ware with regard to vessel surface treatment and morphology. Acknowledging this disparity between the compositions of the petrographic sample set vs. the actual assemblage, the observations reported in this chapter concerning diachronic patterns in the compositional characteristics of coarse ware vessels can be considered to be conservative.

8.2. FABRIC CLASSES, GROUPS AND SUBGROUPS

The petrographic analysis revealed seven main fabric classes, differentiated according to properties and characteristics of their groundmass and primarily by the composition and physical properties of their aplastic inclusions. These distinctions reflect basic differences in the types and/or combination of raw material ingredients that form the basis of these general paste categories. The classes, themselves, therefore, can be considered to reflect fundamentally different paste recipes in terms of the raw material ingredients involved. The main classes of fabrics recognised are:

Crystalline Calcite-Tempered
Dolomitic-Calcitic Marl-based
Grog-Mixed Carbonate
Quartz Sand
Sascab
Sascab-Quartz
Volcanic Glass

Table 8.1: A breakdown of the petrographic sample set.

CONTEXT	Fine Ware Vessels		Coarse Ware Vessels		Total	
	count	%	count	%	count	%
Whole Vessels – various contexts	146	96.1	6	3.9	152	100
Fragmentary Vessels Smash-and-Scatter Offering N10-9 (LA243)	18	100	0	0	18	100
N10-9 Midden	90	71.4	36	28.6	126	100
N10-27 Midden	232	66.3	118	33.7	350	100
Grand Total	486	75.2	160	24.8	646	100

Table 8.2: Breakdown of the vessels sampled for petrographic analysis from the two temple midden assemblages according to their surface treatment and form.

SURFACE TREATMENT		FORM	LA 187		N10-27 All Lots	
			Count	%	Count	%
Polychrome Slipped and Painted		vases	0	0.0	4	1.1
		bowls	0	0.0	11	3.1
		tripod bowls	0	0.0	2	0.6
		composite silhouette dishes	0	0.0	2	0.6
	(interior slipped)	dishes	0	0.0	20	5.7
	Total		0	0.0	39	11.1
Black on Red Slipped and Painted	Total	massive bowls	0	0.0	1	0.3
Red-Orange-Black Resist	Total	composite silhouette dishes	0	0.0	8	2.3
Monochrome Black (all decorative treatments)		vases	0	0.0	4	1.1
		bowls	0	0.0	5	1.4
	Total		0	0.0	9	2.6
Monochrome Red Slipped		bowls	0	0.0	6	1.7
		jars	1	0.8	8	2.3
		composite silhouette dishes	0	0.0	13	3.7
		Chalices	0	0.0	2	0.6
	(interior-slipped)	Rounded bowls	0	0.0	10	2.9
		Rounded dishes	0	0.0	3	0.9
	Total		1	0.8	42	12.0
Monochrome Orange Slipped		composite vessels	1	0.8	0	0.0
		Sieve	1	0.8	0	0.0
		Vases	0	0.0	2	0.6
		Bowls	14	11.1	21	6.0
		tripod bowls	4	3.2	7	2.0
		Jars	19	15.1	13	3.7
		Drums	1	0.8	5	1.4
		plate/lid	1	0.8	0	0.0
		composite silhouette dishes	0	0.0	7	2.0
		Chalices*	3	2.4	14	4.0
		Pedestal basis	0	0.0	4	1.1
	(interior slipped)	Rounded bowls	0	0.0	1	0.3
		Rounded dishes	1	0.8	9	2.6
	Total		45	35.7	83	23.7
Monochrome Red to Orange	(interior slipped)	Rounded bowls	0	0.0	3	0.9
		rounded dishes	0	0.0	0	0.0
	Total		0	0.0	3	0.9

Table 8.2: Breakdown of the vessels sampled for petrographic analysis from the two temple midden assemblages according to their surface treatment and form (continued).

SURFACE TREATMENT		FORM	LA 187		N10-27 All Lots	
			Count	%	Count	%
Monochrome Orange Slipped and Incised		bowls	15	11.9	19	5.4
		tripod bowls	11	8.7	1	0.3
		pedestal-based jars	4	3.2	1	0.3
		jars	3	2.4	1	0.3
		drums	0	0.0	2	0.6
		frying pans	3	2.4	2	0.6
		whistles	1	0.8	0	0.0
		chalices	6	4.8	6	1.7
	Total		43	34.1	32	9.1
Monochrome Orange Slipped and Groove-Incised		drum	0	0.0	1	0.3
		vases	0	0.0	2	0.6
	Total		0	0.0	3	0.9
Monochrome Orange to Red Slipped - Notched and Incised	Total	composite silhouette bowls	1	0.8	5	1.4
Total Monochrome Orange with Appliqué	Total	bowls	0	0.0	3	0.9
Total Buff/Cream Slipped - Plain or Incised	Total	bowls	0	0.0	2	0.6
Total Monochrome Grey (Plumbate or Slate ware)	Total	pedestal base	0	0.0	1	0.3
Total Monochrome Red Slipped and Notched	Total	large bowls	0	0.0	2	0.6
Total Fine Ware			**90**	**71.4**	**232**	**66.3**
Slipped Rim/Lip - Smoothed Body	Total	jars	1	0.8	29	8.3
Red-Brown-Striated	Total	jars	0	0.0	4	1.1
Unslipped/Smoothed-Perforated	Total	Incense burners	3	2.4	0	0.0
Unslipped/Smoothed-Buff		jars	16	12.7	8	2.3
		plate/lid	4	3.2	0	0.0
		bowls	0	0.0	2	0.6
		broad rimmed bowl	0	0.0	1	0.3
	Total		20	15.9	11	3.1
Unslipped Utilitarian		jars	6	4.8	66	18.9
		plates/lids	6	4.8	7	2.0
Unslipped Utilitarian	Total		12	9.5	73	20.9
Unslipped and Incised (appliqué impressed fillet)	Total	Jar	0	0.0	1	0.3
Total Washed/smoothed	Total	jars	0	0.0	1	0.3
Total Coarse Ware			**36**	**28.8**	**118**	**33.7**
GRAND TOTAL BY CONTEXT			126		350	

Notes: * Includes samples from dish and base portions

Four of these general fabric classes sub-divide into fabric groups (Crystalline Calcite-Tempered, Dolomitic-Calcitic Marl-based, Sascab, Sascab-Quartz, Volcanic Glass). Distinctions at the group level primarily relate to differences in the mineralogy and composition of the clay component and/or added constituents (temper), as well as the textural properties and features of the fabrics. These distinctions reflect geologically significant differences in the compositional characteristics of the raw materials involved and, in some cases, the treatment or processing of the raw materials. Accordingly, fabric groups can be regarded as reflecting apparent differences in the geological associations of the specific raw materials chosen for use, as well as paste preparation techniques. Three of the fabric classes (Quartz Sand, Grog-Mixed Carbonate, and Sascab) contain only a single fabric group. The Quartz Sand and Grog-Mixed Carbonate classes, both of which are represented by a large number of vessels, are characterized by a high level of internal variability in comparison to the other fabric groups. The continuous and overlapping nature of the internal variation within these fabric classes, with respect to a range of compositional and textural characteristics, did not permit reproducible sub-divisions or 'groupings' below the class level. In contrast, the Sascab class, being represented by just two samples, is comparatively internally consistent. The discrimination of these fabrics at the class level underscores their compositional dissimilarity to other classes of fabrics in terms of the mineralogy of the raw material ingredients chosen for use.

Several of the fabric groups contain one or, in two cases, two subgroups. These subgroups are similar in most respects to the main group of fabrics to which they relate, except for one or two differences in the compositional characteristics or features of the groundmass. In most instances, these differences relate primarily to the added constituents of the fabrics, distinguished by the presence, in one or a few samples, of an additional tempering material such as grog. The subgroups highlight geologically or technologically significant variation within the fabric group, whilst underscoring the equally important characteristics that unite the group and distinguish it from other groups and classes.

The main fabric classes are summarised here, whilst detailed descriptions of their associated fabric groups, organized alphabetically by class, are presented in Appendix X.

Petrographic Characteristics of the Main Fabric Classes and Their Associated Fabric Groups

CRYSTALLINE CALCITE-TEMPERED

The Crystalline Calcite-Tempered class represents 27.2% (N=176) of the sample set, comprising 10 fabric groups which can be characterized as calcareous clays tempered with freshly crushed crystalline calcite (Table 8.2 and Figures 8.1 and 8.2). The groups are distinguished primarily by reference to the mineralogy of their fine and coarse fractions, as well as by their textural properties. Calcite C is particularly distinctive, as these fabrics also contain added micrite (or sascab temper) deriving from a parent material that is clearly independent of the accompanying crystalline calcite. These fabrics thus contain two tempering materials. Four of the calcite-tempered fabric groups (Calcite A, B, C and H) are consistent geologically with raw material resources available in the immediate vicinity of the site. In the case of Calcite H, the presence in some fabrics of possible organic temper and calcite mosaics containing chalcedony is suggestive of differences in paste recipes that may reflect the work of different potters and, possibly, differences in the origin of these ceramics. This group, therefore, likely contains both locally produced and non-local vessels. Another distinctive feature of the local calcite-tempered fabric groups is that they each contain a petrographic subgroup distinguished by the presence of grog temper.

Of the six remaining fabric groups within the Crystalline Calcite-Tempered class at least five clearly derive from non-local raw materials (Calcite E, F, G, I and J). A non-local provenance for Calcite D is less certain since the calcite temper has features which suggest a non-local provenance, whilst the clay component is generally consistent with clays available in the immediate vicinity of the site. Since the ceramics in group D overlap stylistically with those that occur in the calcite-tempered groups for which a local provenance is fairly certain, a non-local provenance for Calcite D cannot simply be assumed. The different calcite temper in Calcite D fabrics might also reflect the work of different potters or the use of slightly different raw materials over time.

The geological characteristics of all calcite-tempered groups designated non-local, with the exception of Calcite I, indicate connections to inland areas of northern Belize, where calcareous soils directly overlay Tertiary limestone and, possibly, to soils and rock formations located further to the northeast and extending into the Yucatan area of Mexico. Calcite I and J, which are each represented by only one sample, are compositionally unique within the sample set as a whole. Calcite J is distinguished by the large amount of Fe-oxides, in the form of nodules, staining and segregations, in addition to the presence of silicified limestone. These inclusions suggest a connection to the Orange Walk Group formation prevalent in areas north of the New River Lagoon, which characteristically contains a lithified carapace. The main distinguishing feature of Calcite I is the presence of distinctive amorphous concentration features (ACFs) in the clay component. Since similar ACFs were not observed in any of the other fabrics analysed, the compositional uniqueness of this vessel may indicate a provenance outside the general area to which other crystalline calcite-tempered ceramics are geologically connected.

DOLOMITIC-CALCITIC MARL-BASED

Representing 4.8% (N=31) of the sample set, the Dolomitic-Calcitic Marl-based class comprises four fabric groups characterized by a fine-textured, well-sorted groundmass dominated by dolomitic to calcitic grains (Table 8.4, Figure 8.3). The frequent inclusions and the textures of these fabrics links them to sandy-textured marls associated with weathered limestone, dolomitic limestone and dolomite. Their texture suggests a basic untempered clay, perhaps highly processed to achieve the uniform, fine texture. The four groups within this class are distinguished by their mineralogy, and by the form of the calcite, partially dolomitized, and dolomite grains, particularly their crystallinity.

For example, the fabrics in group A are dominated by rhombic cryptocrystalline grains of dolomite, whereas the fabrics in Group B are dominated by rhombic, cryptocrystalline calcite. Groups C and D comprise dolomitic fabrics, containing predominantly equant grains of calcite and dolomite, as well as partially dolomitized grains, and crystalline grains predominate. Group C fabrics are further distinguished by a greater quantity of

Table 8.3: Summary of mineralogy, distinguishing features, paste technology and provenance of the fabric groups comprising the Crystalline Calcite-Tempered class.

Fabric Group	Inclusions (in order of abundance)	Distinguishing Features	Paste Technology	Associated Provenance
A	calcite, qtz, cc-fc calcite mosaics R-VR =- micrite, pqtz, chal, chert	- angular, rhombic to irregular-shaped fragments of calcite - rare to very rare micrite and calcite mosaics	a calcareous clay, containing few siliceous inclusions tempered with colourless, coarsely crystalline calcite	*local* – similarities to clays that form directly below the ground's surface at the site
Petrographic subgroup – grog tempered variant				
B	calcite, fc-cc calcite mosaics, qtz, micrite VR = pqtz, chal, chert	-bimodal (calcite predominate lower mode) - grainy appearance (discrete calcite grains dominate the matrix) - common mosaics of finely to coarsely crystalline calcite Very few to rare micrite lumps	a calcareous clay containing discrete calcite grains, and lesser quantities of other minerals, tempered with and finely to coarsely crystalline calcite	*local* - similarities to clays at the site that are associated with weathering limestone
Petrographic subgroup – grog tempered variant				
C	calcite, micrite, fc-cc calcite mosaics, qtz VR = pqtz, chal, chert.	- grainy appearance (discrete calcite grains dominate the matrix) - common micrite inclusions, calcite mosaics	a calcareous clay containing discrete calcite grains, and lesser quantities of other minerals, tempered with sascab and finely to medium crystalline limestone	*local* - similarities to clays at the site that are associated with weathering limestone
Petrographic subgroup – grog tempered variant **Petrographic subgroup i** – distinguished due to the highly micritic clay matrix				
D	calcite, mc-cc calcite mosaics, micrite, qtz VR = qtz with microlites etc., pqtz , chal, chert, shell, oolites	- shell fragments - quartz containing microlites, vacuoles and fragments of volcanic glass, and calcite mosaics containing these inclusions	a calcareous clay containing shell fragments and rounded siliceous inclusions tempered with a white, sugary textured, medium to coarsely crystalline calcite (containing distinctive quartz)	*non-lcoal?* – similarities to samples local Yalbac clays analysed, but rare shell fragments are distinctive. The temper is non-local

Table 8.3: Summary of mineralogy, distinguishing features, paste technology and provenance of the fabric groups comprising the Crystalline Calcite-Tempered class (continued).

Fabric Group	Inclusions (in order of abundance)	Distinguishing Features	Paste Technology	**Associated Provenance**
E	calcite, mc&fc calcite mosaics, micrite, qtz VR – qtz and pqtz with microlites etc, finely crystallized pqtz, oolites	- mono- and polycrystalline quartz containing microlites, volcanic glass frags. and vacuoles - co-occurrence of angular and rounded, equant calcite grains	A calcareous clay containing carbonate sand (rounded inclusions of calcite and micrite) and quartz tempered with a white, sugary-textured medium crystalline and finely crystalline calcite	*non-local* - presence of carbonate sand in the clay component suggests a connection to areas adjacent to the north east coast of northern Belize, extending into the Yucatan (stylistic links well-known to Yucatecan wares)
F	calcite, mc calcite mosaics, micrite, qtz R-VR= finely crystallized pqtz with microlites etc.,chert, chalcedony	- well-sorted - contain several different types of siliceous inclusions -finely crystallized polycrystalline quartz containing microlites, volcanic glass frags and chalcedonic quartz co-occurrence of rounded and angular calcite inclusions	A calcareous clay containing carbonate sand (rounded inclusions of calcite and micrite) and siliceous inclusions, tempered with a white, sugary-textured medium crystalline calcite	*non-local* - presence of carbonate sand in the clay component suggests a connection to areas adjacent to the north east coast of northern Belize, extending into the Yucatan
G	calcite, , vcc-cc calcite mosaics, qtz VR pqtz, chal, mudstone (?)	- angular, rhombic to irregular-shaped inclusions of crystalline calcite - few to common ferromanganiferrous nodules (Acfs) - calcite mosaics containing chalcedony	a calcareous clay containing few to common ferromanganiferrous nodules and a small quantity of quartz, tempered with colourless, coarsely crystalline to very coarsely crystalline calcite containing chalcedony	*non-local* – composition, mineralogy and textural properties of clay component suggest a connection to deep clay beds overlying horizons of weathering limestone west of the New River
H	calcite, qtz, mc-cc mosaics, micrite R-VR= pqtz, chal, clino, shell	- bimodal (calcite predominates upper mode) - large, angular to irregular-shaped inclusions of crystalline calcite - greater quantity and comparatively large siliceous inclusions - presence of carbon residues in some samples Variation in the frequency, presence/absence of inclusion types (particularly siliceous)and the roundness of siliceous inclusions	a calcareous clay containing a significant quantity of siliceous rock and mineral inclusions tempered with colourless, medium to coarsely crystalline calcite. Some samples may also have organic temper. In sample 504 calcite mosaics contain chalcedony	*local and non-local*- general similarities to clays that form directly below the ground's surface at the site fabrics. Samples containing a greater quantity of chalcedony and polycrystalline quartz may derive from quartzy-calcareous soils situated between the eastern coast and the New River.
Petrographic subgroup – grog tempered variant				

Table 8.3: Summary of mineralogy, distinguishing features, paste technology and provenance of the fabric groups comprising the Crystalline Calcite-Tempered class (continued).

Fabric Group	Inclusions (in order of abundance)	Distinguishing Features	Paste Technology	Associated Provenance
I	calcite, qtz R-VR=pqtz, micrite, chert, feldspar	- common distinctive ferromanganiferrous nodules - fine feldspar inclusions in the clay component	a calcareous clay containing common, and often large-sized, ferromanganiferrous nodules and a small quantity of quartz, chert, feldspar and perhaps calcite, tempered with crystalline calcite	*non-local* - dissimilarity to fabric groups that are generally consistent with the geology of northern Belize suggest that this group's provenance lay outside of this general region. The mineralogy of the fabric suggests a connection to areas underlain by limestone.
J	calcite, micrite containing calcite and ferrous nodules etc., qtz, mc-cc cacite mosaics, silicified limestone/mudstone, limestone R=pqtz	- frequent lumps of micrite that are rusty-brown in colour and contain grains of cryatalline calcite and ferrous nodules and segregations - presence of fragments of silicified limestone and limestone fragments comprising micrite, crystalline calcite and silicified limestone/mudstone - common ferrous nodules and segregations in the clay component	a calcareous clay containing quartz, tempered with limestone	*non-local* - the presence of silicified limestone/mudstone suggests a connection to the Orange Walk Group formation prevalent in areas north of the New River Lagoon, which characteristically contain a lithified carapace

Key for Tables 8.3 – 8.6
R = rare; VR = very rare; qtz = quartz; pqtz = polycrystalline qartz; cc = coarsely-crystalline;
fc = finely-xrystalline; mc = microcrystalline; chal = chalcedony; clino = clinopyroxene;
dol. = dolomitic; ortho = orthoclase; plagio = plagioclase; vg = volcanic glass; frags = fragments

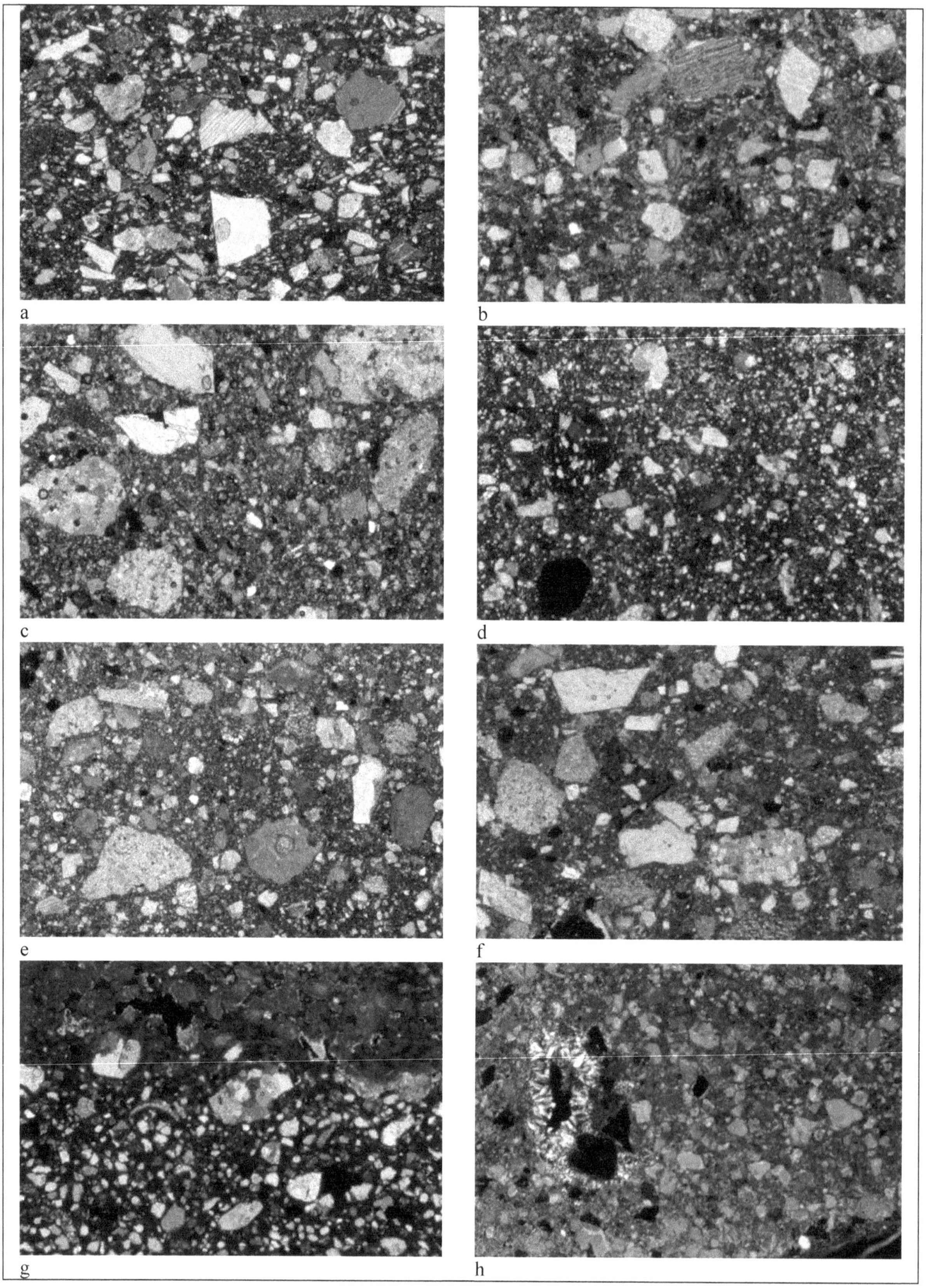

Figure 8.1: Crystalline Calcite-Tempered fabric groups (x25): a) Calcite A; b) Calcite A subgroup – grog-tempered variant; c) Calcite B; d) Calcite B subgroup – grog-tempered variant; e) Calcite C; f) Calcite C subgroup – grog tempered variant; g) Calcite D; h) Calcite E (crystalline calcite inclusions are stained pink). Field of view = 3mm. (see colour version on accompanying CD)

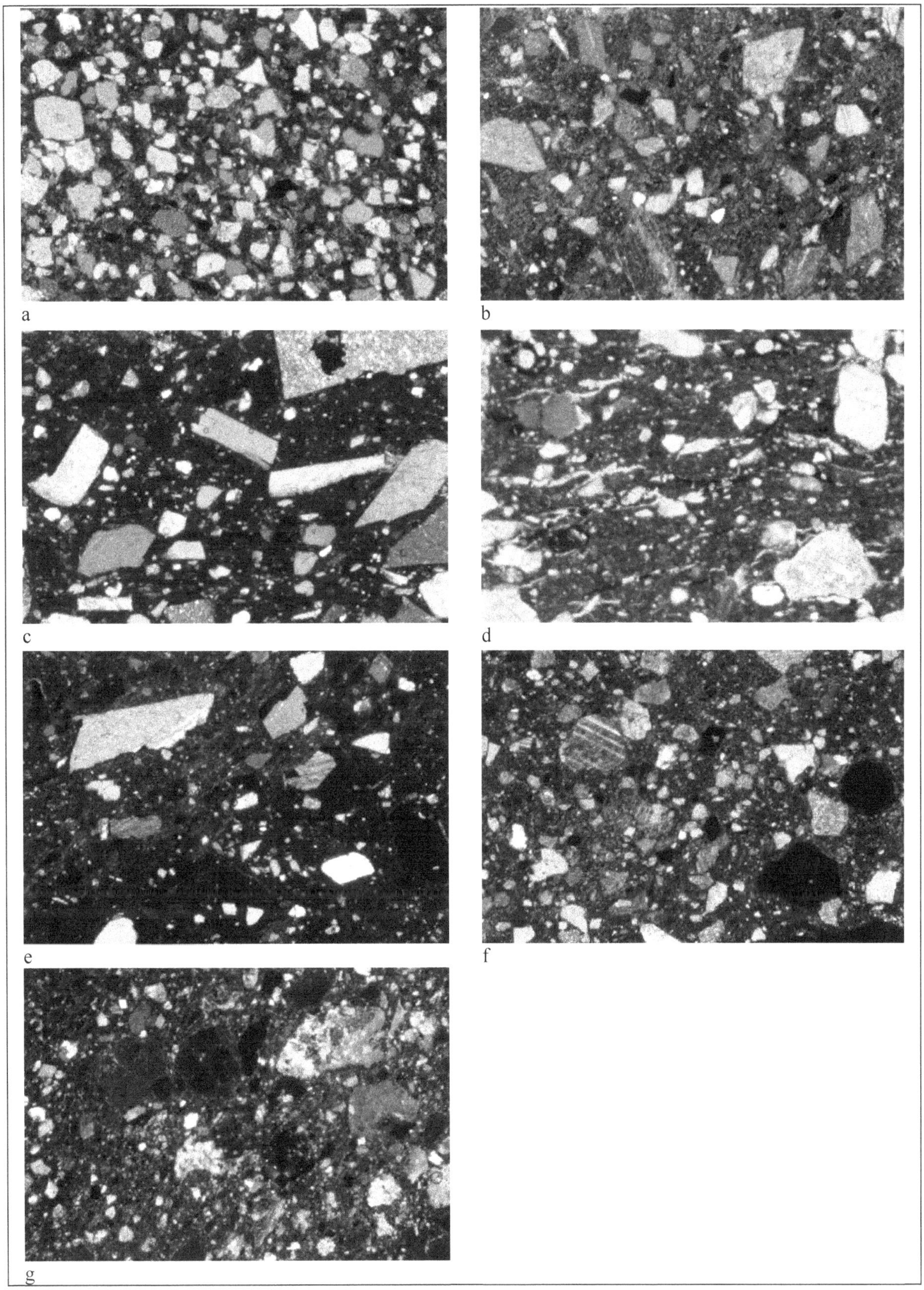

Figure 8.2: Crystalline Calcite-Tempered fabric groups (x25): a) Calcite F; b) Calcite G; c) Calcite H; d) Calcite H containing possible organic temper; e) Calcite H subgroup – grog-tempered variant; f) Calcite I; g) Calcite J. Field of view = 3mm. (see colour version on accompanying CD)

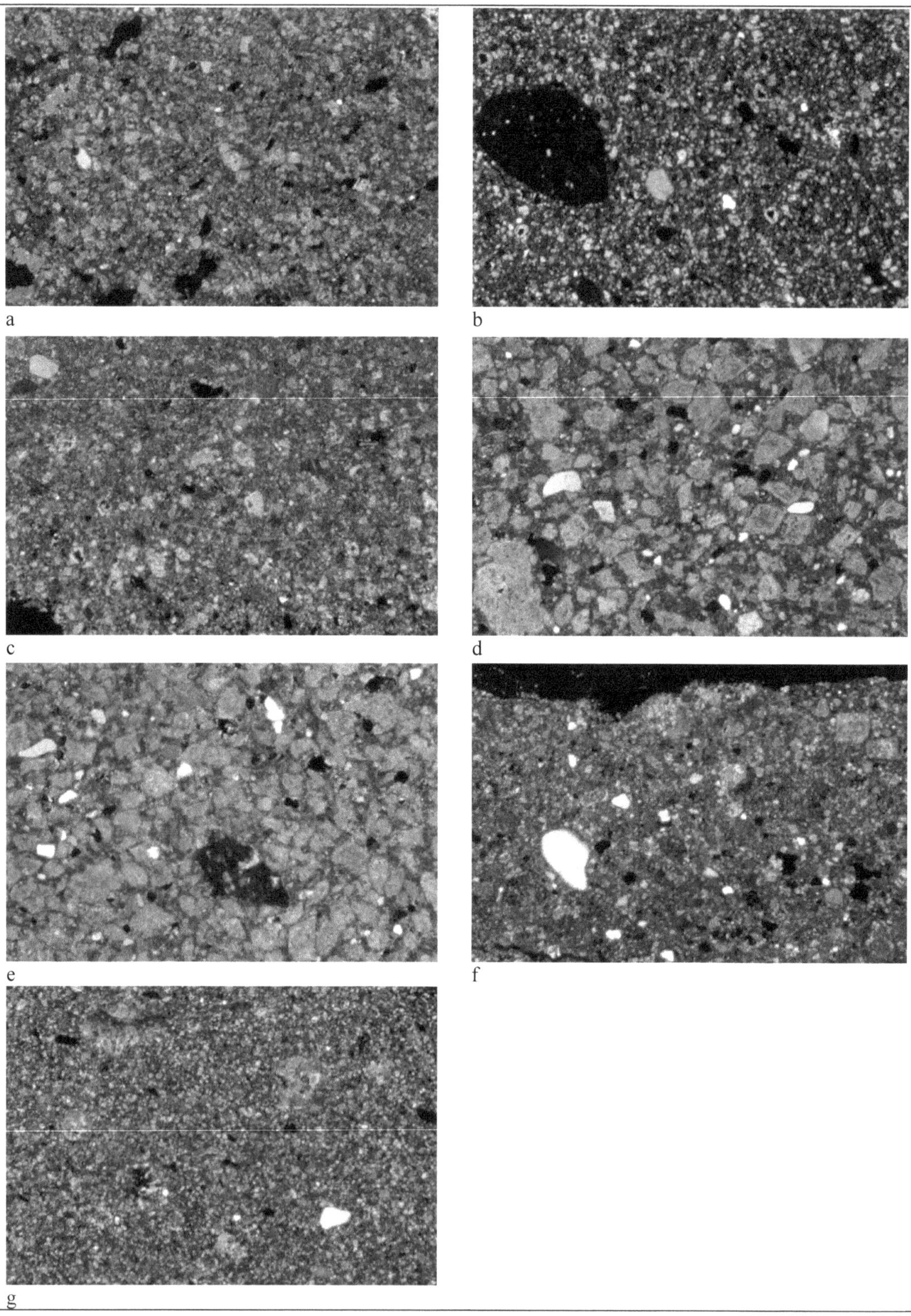

Figure 8.3: Dolomitic/Calcite Marl-based fabric groups (x25): a) Marl A (crystalline calcite inclusions are stained pink); b) redced Marl A fabric (crystalline calcite inclusions are stained pink); c) Marl A subgroup showing more frequent crystalline-calcite inclusions (stained pink) and shell fragments; d) Marl B; e) Marl B subgroup - grog-tempered variant; f) Marl C (crystalline calcite inclusions are stained pink); g) Marl D (crystalline calcite inclusions are stained pink). Field of view = 3mm. (see colour version on accompanying CD)

Table 8.4: Summary of mineralogy, distinguishing features, paste technology and provenance of the fabric groups comprising the Dolomitic-Calcitic Marl-based class.

Fabric Group	Inclusions (in order of abundance)	Distinguishing Features	Paste Technology	**Associated Provenance**
A	dolomite and dolomitic grains and mosaics, qtz, chert, chal, finely crystallized pqtz, calcite, VR=- feldspar dol. limestone	- well sorted, unimodal, fine-textured groundmass - cryptocrystalline, rhombic gains of dolomite dominate the groundmass -siliceous inclusions are comparatively rare	sandy-textured marl deriving from weathered dolomite, likely untempered and perhaps highly processed to achieve the uniform, fine texture	*non-local* – associated with clays that form in association with horizons of weathering dolomite. Likely connected to dolomite present in east and northeast northern Belize (Cayo Group?). Fabrics are similar to the numerically dominant fine ware fabric at Late Classic Altun Ha.
Petrographic subgroup – distinguished due to greater quantity of crystalline calcite present (common)				
B	calcite, qtz, micrite R-VR = cc calcite mosaics feldspar(ortho),pqtz, chal, chert, amphibole, micrite	-well sorted, unimodal - cryptocrystalline, rhombic gains of calcite dominate the groundmass -siliceous inclusions are comparatively rare and small-sized	sandy-textured highly calcareous marl, likely untempered and perhaps highly processed to achieve the uniform, fine texture	*non-local* – associated with coastal deposits containing calcareous sands mainly consisting of cryptocrystalline grains composed dominantly of micrite in northeastern northern Belize, extending into southern Yucatan
Petrographic subgroup – grog tempered variant				
C	calcite, dolomite, dolomitic grains, qtz, micrite, mc mosaics of the above VR =shell, anhydrite	- well sorted, unimodal - equant to irregular-shaped of crystalline calcite and dolomite dominate the groundmass - rounded quartz inclusions - comparatively greater quantity of quartz 152 contains grog	sandy-textured dolomtic marl containing quartz sand, likely untempered and perhaps highly processed to achieve the uniform, fine texture	*non-local* – associated with coastal deposits adjacent to the northeast coast of northern Belize, extending into southern Yucatan and adjacent areas, where sandy marls overly dolomitic limestone (Cayo Group?)
D	calcite, dolomite, dolomitic grains, qtz, micrite, cc & fc mosaics of the above VR = qtz & pqtz containing vacuoles, pqtz, chert, chal feldspar	- well sorted, unimodal - equant to irregular-shaped of crystalline calcite and dolomite dominate the groundmass - both coarsely and finely crystalline aggregates are present - presence of distinctive quartz and polycrystalline quartz containing vacuoles and Fe-oxide residues	sandy-textured dolomtic marl containing a small quantity of siliceous rock and mineral inclusions, likely untempered and perhaps highly processed to achieve the uniform, fine texture	*non-local* – associated with clays that form in association with horizons of weathering dolomitic limestone in coastal and adjacent areas of northern Belize - possibly connected with Cayo Group formation (?).

quartz and other siliceous rock, and Group D, by more frequent polycrystalline quartz and dolomitic mosaics. These differences clearly relate to the geology and provenance of the raw material used in each case. Two of the groups contain fabrics tempered with grog, and in the case of group B these fabrics are distinguished as a petrographic subgroup.

The presence of dolomite and dolomitic inclusions in three of the marl-based fabric groups (A, C and D) clearly differentiates these fabrics geologically from local raw material resources, indicating these ceramics were not produced locally. Group B fabrics also appear to be non-local, based on the physical properties of the distinctive micritic inclusions they contain. The geology of the dolomitic marl-based fabric groups indicates a connection to the northeast coast of northern Belize and southern Yucatan and adjacent areas, including the lower reaches of the New River, where sandy marls overlie dolomite and dolomitic limestone (most likely of the Tertiary Cayo Group). It is noteworthy that Group A fabrics are striking similar to the numerically dominant fabric type associated with fine wares at Late Classic Altun Ha (Howie et al. 2004; Howie-Langs 1999), suggesting that the similar fabrics at Lamanai might derive from this particular area of northern Belize. The compositional properties of group B associate these fabrics with coastal deposits containing calcareous sands, which mainly consist of cryptocrystalline grains composed dominantly of micrite.

GROG-MIXED CARBONATE

The Grog-Mixed Carbonate class represents 35.4% (N=229) of the sample set and constitutes the most technologically complex of the fabric classes, comprising an intricate mixture of multiple raw material ingredients. These fabrics are characterised by common inclusions of grog, or crushed pottery, occurring together with varying amounts of micrite, crystalline calcite (mosaics, discrete grains and spar fragments) and quartz, and can generally be described as constituting calcareous clay tempered with grog, varying amounts of sascab, crystalline calcite and perhaps, sand (Table 8.5 and Figures 8.4 and 8.5). There is considerable variation across this class with regard to the compositional characteristics and physical properties of the groundmass.

The high level of compositional variation appears to relate to natural differences in the clay component (deriving from differences in the geological contexts within which the clays were formed), as well as to the specific combination and proportions of the raw materials put into the paste. Clays with slightly different, but mineralogically related, compositions appear to be represented. The difference among them may relate most directly to the geological and/or environmental context from which they derive. No significant dissimilarity can be demonstrated, on geological grounds, to raw material resources that occur on the west side of the lagoon (in the immediate vicinity of the site), which suggests that these ceramics were produced locally. Nonetheless, the area of Pleistocene alluvium situated directly east of the site can be discounted as the source of the associated clays due to their obvious geological dissimilarity. The distinctive compositional characteristics of the grog inclusions in subgroup B may reflect a non-local provenance for this particular vessel.

Additional variation across this class with respect to the physical properties of the groundmass relate to differential firing, as indicated by major differences in the optical activity of the clay matrix as well as the presence/absence of dark-coloured cores that vary in both colour and thickness. The continuous nature of the variation within this class with respect to a range of compositional and physical attributes made it impossible to subdivide these fabrics below the class level, into reproducible groups. In addition, the presence of a thick black core in many samples (more than 60%) obscured other characteristics and properties of the groundmass. The main characteristics that unite these fabrics are the presence and relative frequency of the grog inclusions, their general rock and mineral content and the presence of multiple tempers. These fabrics also exhibit comparatively fine textures in comparison to other local fabric groups, rarely containing inclusions larger than coarse sand. Consequently, their complex paste recipe might have been accompanied by a more careful preparation of raw materials.

QUARTZ SAND

The Quartz-Sand class comprises 14.6% (N=94) of the sample set. These fabrics are characterized by predominant siliceous inclusions, occurring together with igneous-related accessory minerals (Table 8.5 and Figure 8.6), as well as by a rarity of carbonate inclusions. These fabrics are mineralogically and compositionally consistent with clayey soils that form in association with deep deposits of Pleistocene alluvium throughout northern Belize. Igneous-related accessory minerals such as olivine, clinopyroxene and feldspar emphasize a connection to the Maya Mountains as the original source of the alluvium. Nonetheless, some of the fabrics are virtually identical to the clay samples taken from the area of Pleistocene alluvium (Boom Subsuite soils) situated directly east of Lamanai, across the New River Lagoon. The striking similarity between some of the archaeological specimens and the samples of local clays taken from both inland and creek-side localities within this particular alluvial deposit suggests that at least some of the fabrics within this class derive from local raw materials.

There is considerable variation among the Quartz-Sand fabrics with regard to the textural and physical properties of the groundmass, as well as their rock and mineral content. This variation is suggestive of differences in the provenance of the raw materials. For example, fabrics containing greater amounts of chert and chalcedony probably derive from clays formed in areas of Pleistocene alluvium situated east of Lamanai, between the coast and the Northern Lagoon drainage system, since these rock types are characteristically more prevalent in the soils of this particular area of northern Belize. In addition, fabrics containing less frequent, better sorted and more rounded

Table 8.5: Summary of mineralogy, distinguishing features, paste technology and provenance of Grog-Mixed Carbonate, Quartz Sand, Sascab and Sascab-Quartz classes.

Fabric Group	Inclusions (in order of abundance)	Distinguishing Features	Paste Technology	Associated Provenance
Grog-Mixed Carbonate Class	micrte, qtz, calcite, finely crystalline calcite mosaics, pqtz, chal, chert, limestone frags. VR=feldspar, amphibole, shell	- common grog inclusions - co-occurrence of grog lumps of micrite and fragments and mosaics of crystalline calcite - comparatively fine-textured - high internal variation - distinctive firing horizons and commonly optically inactive to slightly active	a calcareous clay tempered with grog and varying amounts of sascab and crystalline calcite. The comparatively fine texture of the groundmass indicates a careful or more rigorous processing of raw materials.	*local* – connections to Yalbac clays, both those that form directly below the ground's surface and those associated with horizons of weathering limestone, as well as Filipe Subsuite clayey soils associated with wash deposits of Pleistocene alluvium situated on the north side of the site
Subgroup A – variant with a highly micritic clay and containing comparatively less and larger grog inclusions. Petrographic similarities to Sascab-Quartz A **Subgroup B** – variant distinguished by the distinctive composition of the grog it contains. The dominant type of grog is composed on fine grains of calcite or dolomite. This grog is petrographically similar to fabrics of Dolomitic-Calcitic Marl-based A, which may suggest a non-local provenance for this vessel.				
Quartz Sand Class	qtz, pqtz, chert, feldspar (alkali and plagio), VR= sandstone, micrite, olivine, clino, shell, amphibole, gypsum, calcite	- predominant siliceous inclusions occurring together with igneous-related accessory minerals - very rare, and predominantly absent carbonate inclusions high internal variation	Sandy low to non-calcareous clay, possibly tempered with sand in some cases.	*local and non-local* – multiple different provenances. Consistent with clayey soils that form in association with deep deposits of Pleistocene alluvium that occur throughout northern Belize. Some fabrics are virtually identical to the clay samples taken from the area of Pleistocene alluvium (Boom Subsuite soils) situated directly east of Lamanai, across the New River Lagoon, suggesting local manufature. Fabrics containing greater amounts of chert and chalcedony likely derive from flinty clays associated with alluvial deposits situated between the coast and the Northern Lagoon drainage system. Fabrics containing comparatively fewer inclusions that are comparatively well-sorted and rounded, shell and micrite mostly likely derive from clays associated with waterways.
Subgroup – crystalline calcite-tempered variant				

Table 8.5: Summary of mineralogy, distinguishing features, paste technology and provenance of Grog-Mixed Carbonate, Quartz Sand, Sascab and Sascab-Quartz classes (continued).

Fabric Group	Inclusions (in order of abundance)	Distinguishing Features	Paste Technology	Associated Provenance
Sascab Class	micrite, calcite, mosaics of finely crystalline calcite VR, qtz, shell, limestone (comprising micrite and fc calcite aggregates)	- bimodal (micrite lumps comprising upper mode) - inclusion types other than micrite of crystalline (and fine grains) - highly micritic matrix	highly micritic clay with rare vary fine inclusions of quartz tempered with sascab	*non-local* – predominance of fine inclusions of micrite and crystalline calcite in the clay component suggests a connection to clays that form in association with horizons of weathering limestone. Since the clay contains very little quartz, and the sascab temper contains none, an association with Cretaceous limestone is suggested. Cretaceous limestone outcrops at three principal locations north of the Maya Mountains: below an escarpment situated west of the southern end of the Rio Bravo, below the Booth's River escarpment, and along the western edge of the New River Lagoon
Sascab-Quartz Class				
Sascab-Quartz A	qtz, micrite, calcite, pqtz R-VR=pqtz, chal, feldspar (alkali and plagio), amphibole, clino, shell, oolites	- dominant to common quartz inclusions - frequent to few lumps of micrite - very rare amphibole, clinopyroxene and feldspar	a sandy calcareous clay, tempered with sascab	*local* - similarities to clay associated with wash deposits of Pleistocene alluvium situated at the north side of the site adjacent to Barber Creek(Filipe Subsuite). The sascab temper is compositionally and visually identical to samples taken from the outcrop southwest of the central precinct. This sascab is associated with cretaceous limestone
Subgroup – grog tempered variant characterized by a comparatively fine-textured groundmass				
Sascab-Quartz B	qtz, micrite, fc calcite mosaics, chal, chert, feldspar (ortho) R-VR= pqtz, biotite and sandstone	clay matrix is brick-red in colour presence of biotite and sandstone fragments micrite lumps contain finely crystalline calcite aggregates, quartz and biotite	a sandy calcareous clay containing biotite tempered with sascab comprising micrite, finely crystalline calcite, quartz and biotite inclusions	*Non-local*- The clay component is consistent with Xaibe Subsuite soils that overly Orange Walk Group limestone in the Corazal District, between the Rio Hondo and Corozal Town. The presence of the quartz inclusions, their quantity and roundness suggests an association with alluvial deposits. The presence of quartz inclusions in the micrite lumps suggests a connection to sascab associated with the Tower Hill or Orange Walk Group limestones characteristic of areas to the north of the New River Lagoon.

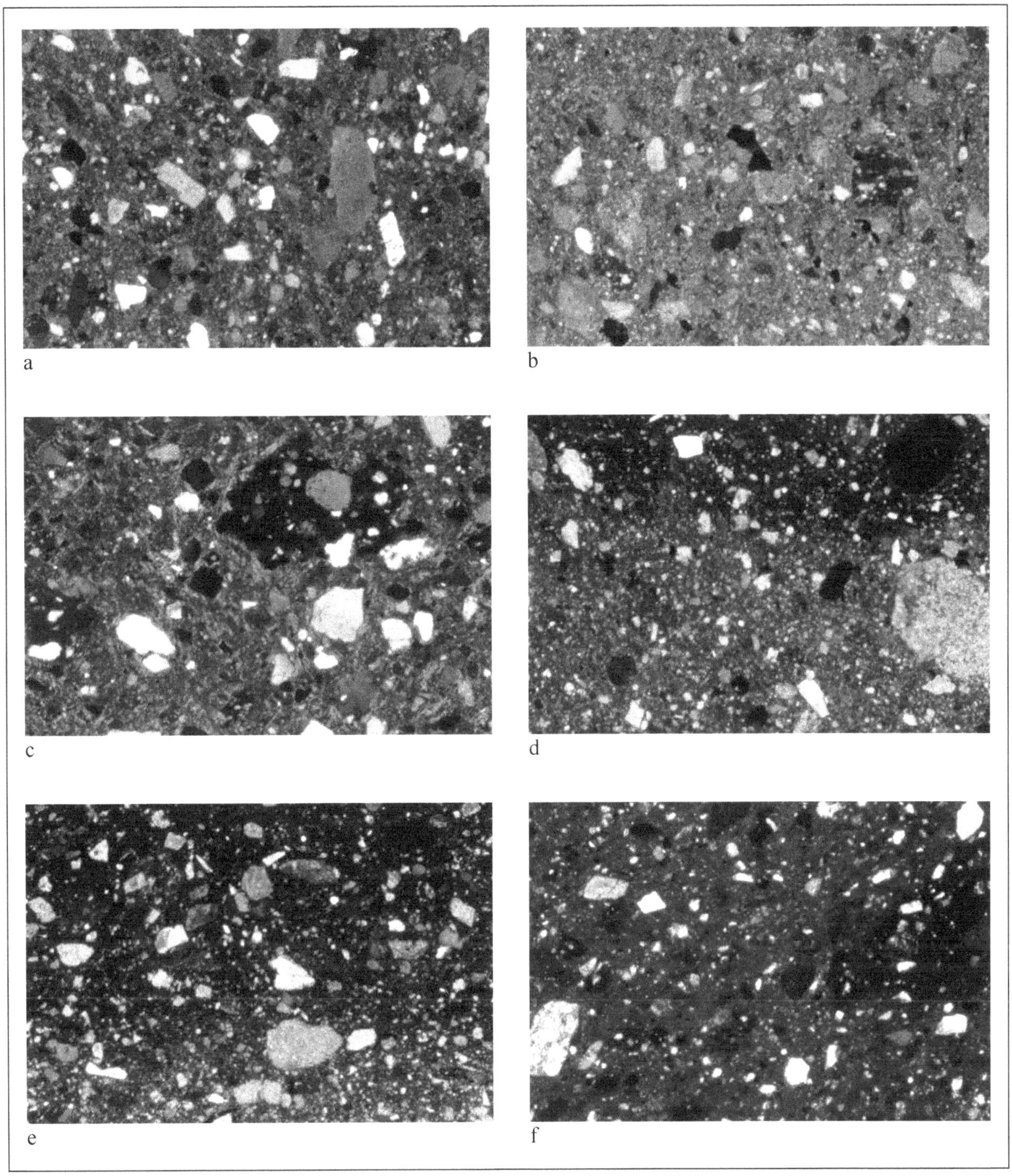

Figure 8.4: Grog-Mixed Carbonate fabrics (x25): a) fabric containing comparatively more quartz; b) fabric containing less quartz and more calcite and micrite; c) fabric containing rare micrite; d) fabric with a matrix dominated by fine calcite grains and containing finely crystalline mosaics; e) fabric containing a dark-coloured core that obscures compositional features and properties of the groundmass; f) comparatively high fired fabric with an optically inactive clay matrix. Field of view = 3mm. (see colour version on accompanying CD)

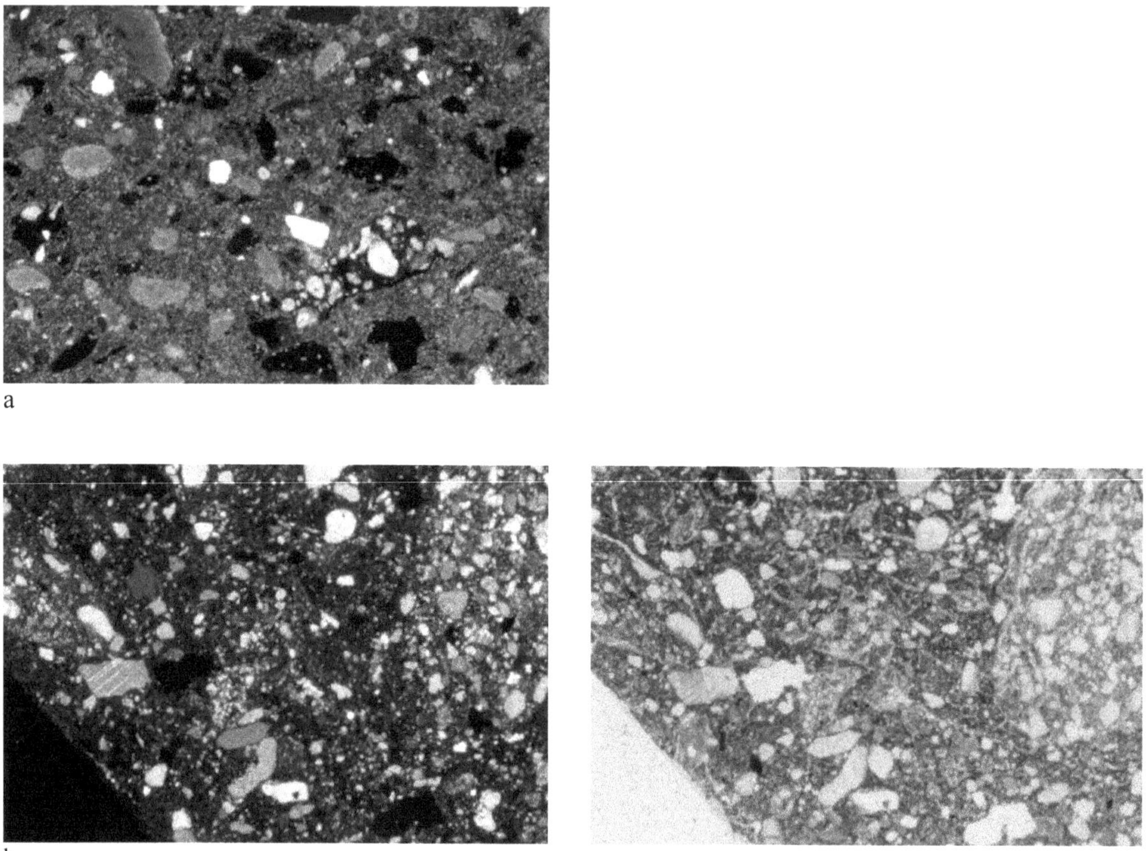

Figure 8.5: Grog-Mixed Carbonate fabrics (x25): a) Subgroup A showing comparatively large inclusions of grog and with a highly micritic clay component; b) Subgroup B showing distinctive grog likely deriving from a marl-based fabric c) same fabric as b) in PPL (grog inclusions are more easily distinguished). Field of view = 3mm. (see colour version on accompanying CD)

inclusions mostly likely derive from clays associated with waterways (creeks, streams and rivers). The presence of shell fragments and lumps of micrite in some samples also suggests a connection to river or creek deposits. A bimodality of grain size in some fabrics suggests the addition of a little sand as temper, while crystalline calcite temper is present in the sample distinguished as a petrographic subgroup. Based on these apparent differences among the fabrics, and their geological and technological significance, it is postulated that the Quartz-Sand class comprises both local and non-local fabrics, with some differences also in the approach to paste manufacture. The characteristic that unites the group is that all of the fabrics derive from the same general type of clay. The continuous nature of the variation among these fabrics does not permit more discrete petrographic distinctions below the class level and hence, more discrete provenance distinctions.

SASCAB

This class of fabrics, which is represented by only two samples (0.3% of the sample set), can be characterized as a highly micrtic clay tempered with sascab (Table 8.5 and Figure 8.7). The main distinguishing characteristic of these fabrics is the rarity of quartz and any other inclusion type, except for micrite lumps and fine inclusions of crystalline calcite. The clay component is dominated by fine inclusions of micrite and crystalline calcite, suggesting a connection to clays that form in association with horizons of weathering limestone. The fact that the clay contains very little quartz, and the sascab temper contains none, is suggestive of an association with Cretaceous limestone, as opposed to the Teritary formations, since sascab deposits associated with the latter characteristically contain a significant quantity of sand. Cretaceous limestone formations outcrop at only three principal locations north of the Maya Mountains: below an escarpment situated west of the southern end of the Rio Bravo, below the Booth's River escarpment, and along the western edge of the New River Lagoon. Nonetheless, samples of local clay associated with Cretaceous outcrops generally contain more quartz than is present in these archaeological fabrics. Considering the distinctive nature of the Sascab fabrics within the sample set as a whole, it is most likely that these two vessels are of non-local origin.

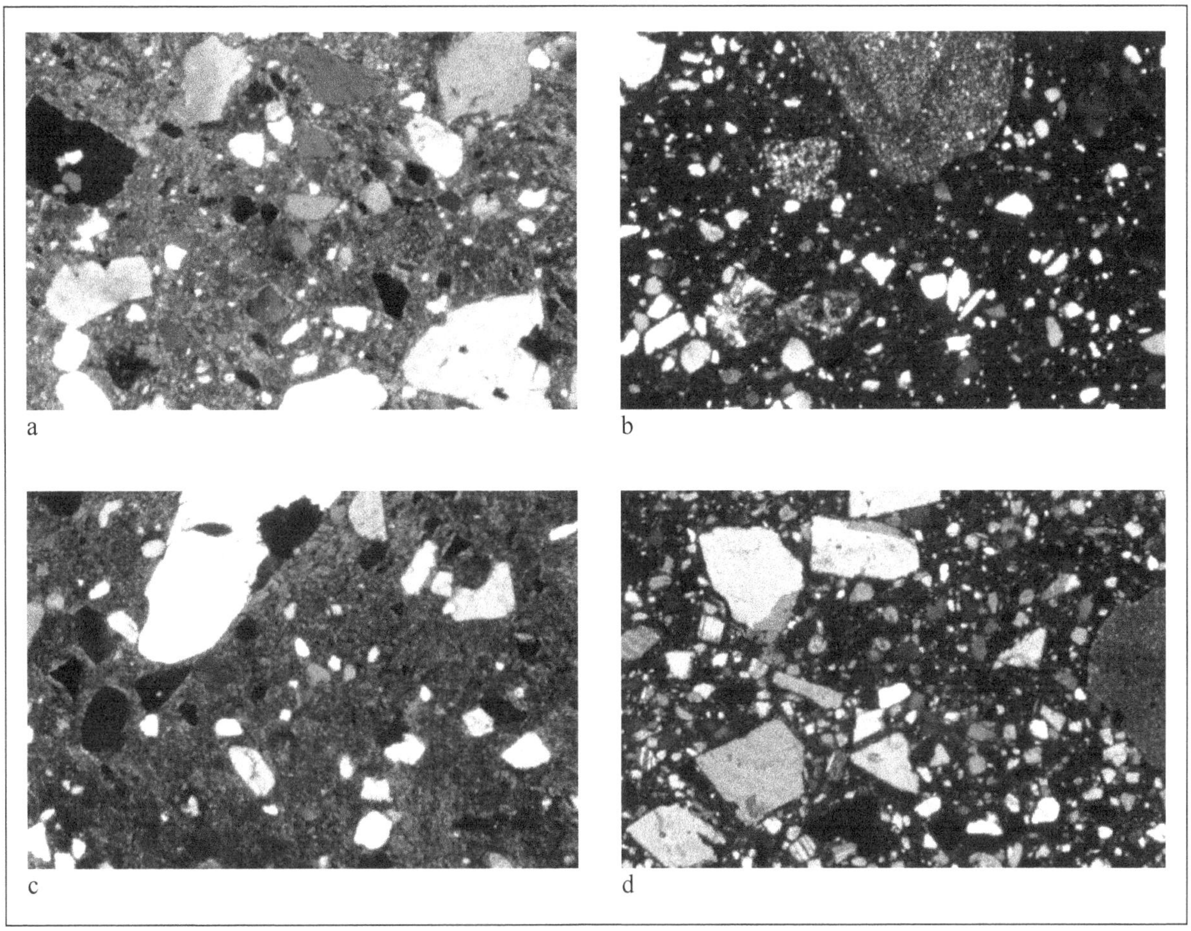

Figure 8.6: Quartz Sand fabrics (x25): a) fabric identical to local clays (see Plate VII.27); b) fabric containing large amounts of chert and chalcedony; c) fabric containing comparatively few, well-sorted inclusions; d) Quart Sand subgroup – crystalline calcite-tempered variant. Field of view = 3mm. (see colour version on accompanying CD)

SASCAB-QUARTZ

The Sascab-Quartz class represents 15.2% (N=98) of the sample set and comprises two fabric groups, distinguished based on significant differences in their mineralogy (Table 8.5 and Figure 8.7). Sascab-Quartz fabrics comprise a sandy calcareous clay tempered with sascab that has been comminuted into micrite lumps due to crushing of the parent rock. Sascab A, the larger of the two groups, is geologically consistent with raw material resources available in the local area. The clay component of these fabrics is mineralogically and compositionally consistent with the Filipe Subsuite soils that occur in association with wash deposits containing Pleistocene alluvium situated directly north of the site, adjacent to Barber Creek (King et al. 1992, see also Chapter 5). Samples of local clay deriving from the western bank of Barber Creek bear striking similarities to these fabrics, except that inclusions of micrite are rare in the clay samples. The sascab temper in tmhese fabrics is identical in appearance and composition to samples of local sascab associated with Cretaceous limestone. The temper and the geological samples both contain ghosts of gastropods and foraminifera and quartz inclusions are generally absent. Also noteworthy is that Sascab-Quartz A contains a subgroup of fabrics, distinguished by their finer texture and presence of grog temper.

The distinctive mineralogy of the Sascab-Quartz B fabrics indicates that they derive from non-local raw material resources. The presence of biotite in both clay and temper, along with the presence of other igneous-related accessory minerals in the clay, suggests a connection to micaceous soils typical of areas of central Peten bordering the Rio Hondo. The ultimate source of the igneous-related minerals is most likely the Sierra Madres of Chiapas, Mexico. The connection to geological formations in the vicinity of the Rio Hondo is supported by the predominance of rounded quartz inclusions in the clay, which suggest an alluvial deposit, and also by the presence of quartz in the sascab temper. The presence of quartz in the sascab suggests a connection to the Tower Hill or Orange Walk Group formations typical in areas to the north of the New River Lagoon and adjacent to Chetumal Bay, which characteristically contain quartz.

VOLCANIC GLASS

The Volcanic Glass class represents just 2.5% (N=16) of the sample set. The fabrics within this class are distinguished mineralogically by the presence of volcanic

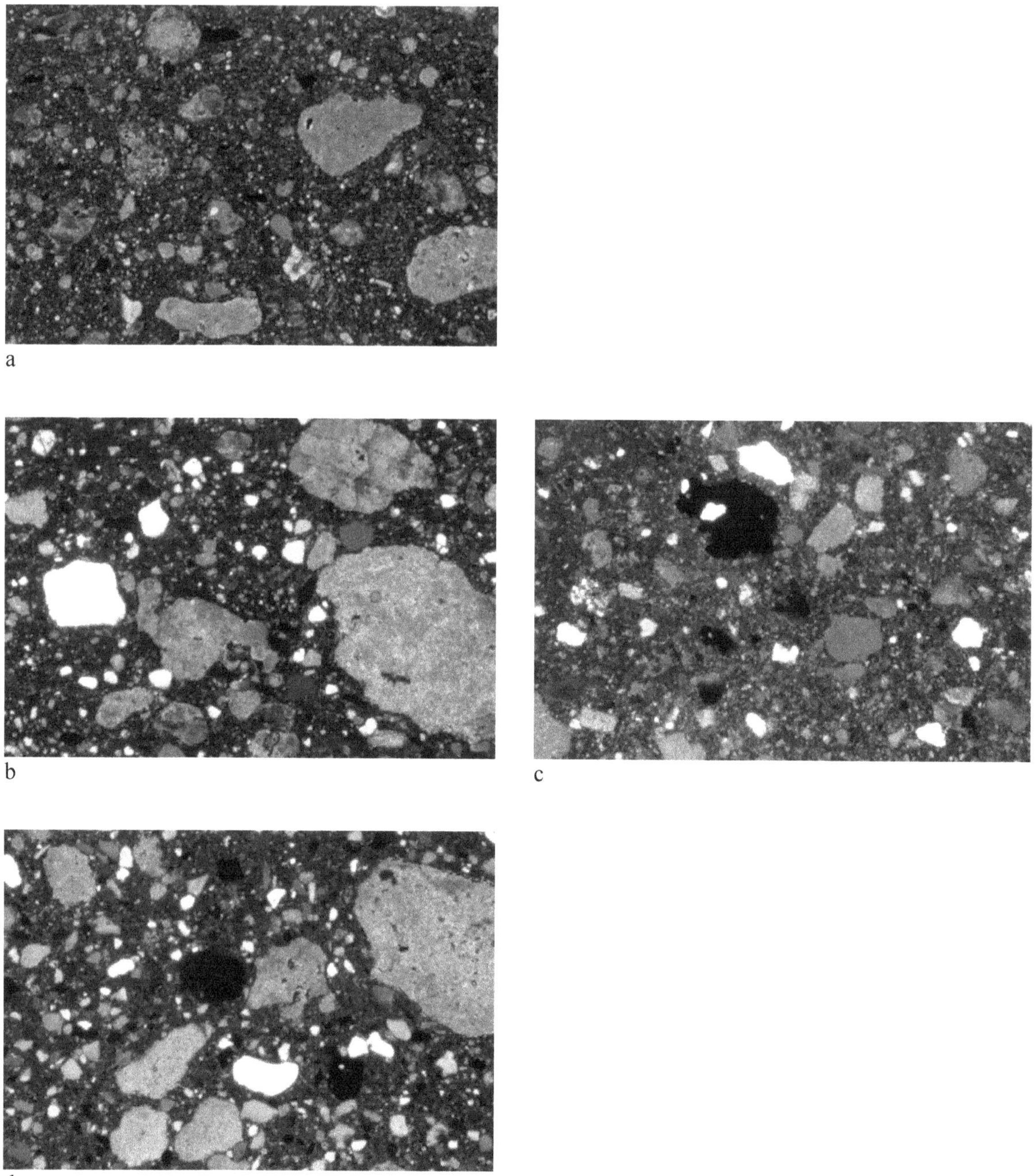

Figure 8.7: Sascab and Sascab-Quartz fabrics (x25): a) Sascab fabric; b) Sascab-Quartz A; c) Sascab-Quartz A subgroup – grog-tempered variant; d) Sascab-Quartz B. Field of view = 3mm. (see colour version on accompanying CD)

glass fragments and tuff, deriving from unconsolidated and lithified forms of volcanic ash. The presence of volcanic ash in lowland ceramics is conventionally interpreted as an intentional tempering practice (e.g., Jones 1986; Simmons and Brem 1979; Shepard 1964). It is generally held that volcanic ash, presumed to originate from deposits widespread in the Guatemala Highlands, was widely traded throughout lowland region specifically for pottery production, especially during the Late Classic period (e.g. Simmons and Brem 1979; see Arnold 1985:59 for an opposing view). The widespread production of volcanic ash-tempered ceramics is reflected in this class at Lamanai, which comprises five fabric groups with different geological characteristics, indicating fundamental differences in their raw materials, and thus different provenances.

Four of the groups within the Volcanic Glass class are represented by a single sample, whilst Volcanic Glass A comprises several samples, including a petrographic subgroup distinguished by slight mineralogical differences in the clay component (Table 8.5 and Figure 8.8). The distinctive mineralogical and compositional characteristics of the Volcanic Glass groups clearly

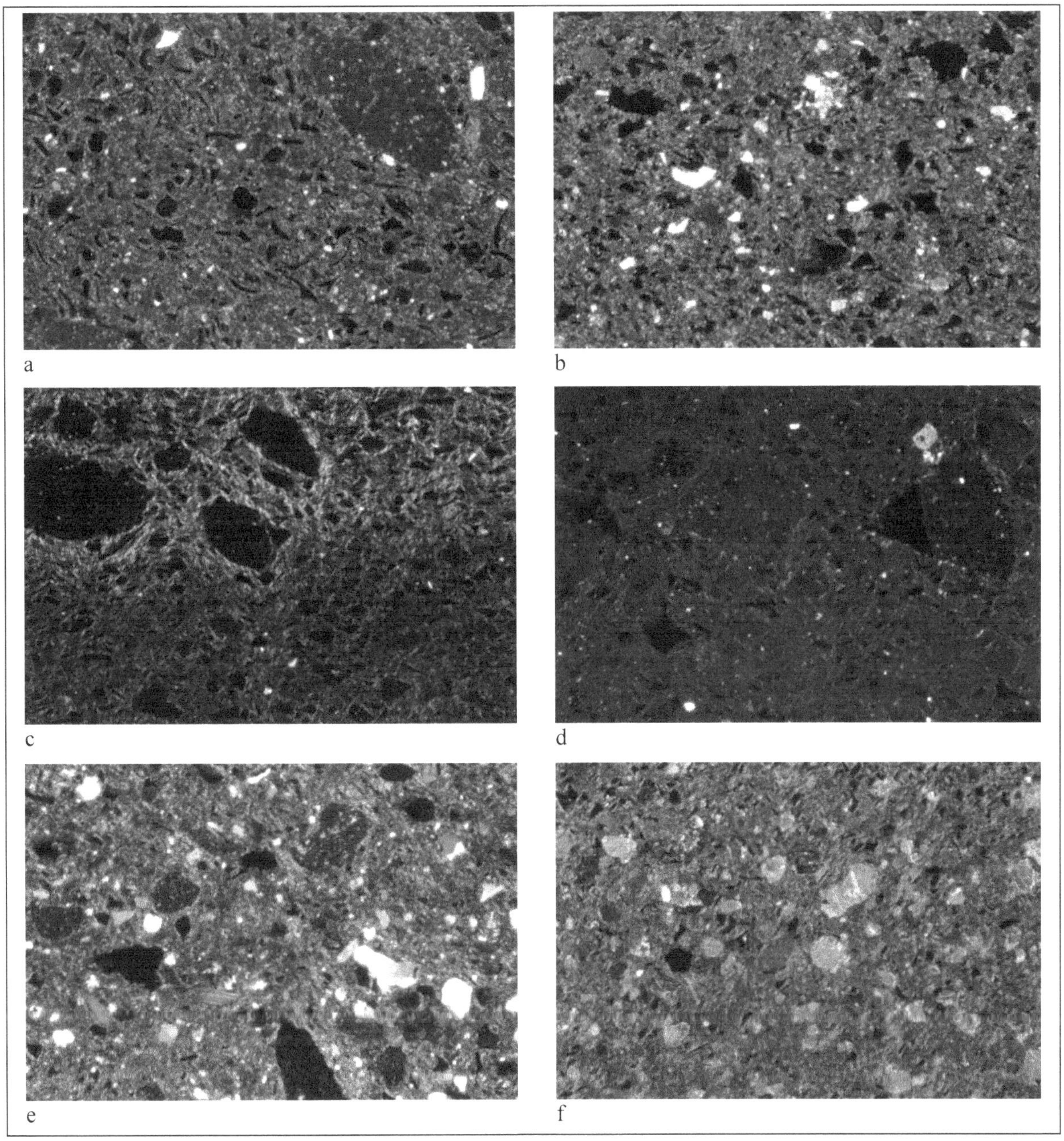

Figure 8.8: Volcanic Glass fabrics (x25): a) Volcanic Glass A; b) Volcanic Glass A subgroup showing more frequent rock and mineral inclusions; c) Volcanic Glass B; d) Volcanic Glass C; e) Volcanic Glass D; f) Volcanic Glass E (crystalline calcite inclusions are stained pink). Field of view = 3mm. (see colour version on accompanying CD)

Table8.6: Summary of mineralogy, distinguishing features, paste technology and provenance of the fabric groups comprising the Volcanic Glass class.

Fabric Group	Inclusions (in order of abundance)	Distinguishing Features	Paste Technology	Associated Provenance
A	vg & tuff frags, qtz, micrite, biotite, feldspar VR = calcite	- sickle and lunate shaped frags of volcanic glass and equant to elongated tuff frags - micritic matrix containing lumps of micrite	micritic calcareous clay tempered with volcanic ash	*non-local* – from outside of northern Belize/Yucatan region but associated with area underlain by limestone, possibly Central Peten?
Petrographic subgroup – distinguished due to presence of polycrystalline quartz ferromanganiferrous nodules, more frequent calcite and rare biotite				
B	vg & tuff frags, qtz, feldspar, R-VR = mica, amphibole, basalt	- sickle and lunate shaped frags of volcanic glass and equant to elongated tuff frags - slightly bimodal appearance due to presence of large tuff fragments - very fine mineral inclusions - presence of muscovite, amphibole and basalt	calcareous clay, likely heavily processed to remove rock and mineral inclusions, tempered with volcanic ash	*non-local* – unknown but from outside of northern Belize/Yucatan region
C	vg & tuff frags, qtz, feldspar, biotite, micrite VR = chal	- sickle and lunate shaped frags of volcanic glass and equant to elongated tuff frags - olive green clay matrix (XP) - bimodal appearance due to presence of large, angular tuff fragments - presence of chalcedony	calcareous clay containing chalcedony and quartz, likely heavily processed to remove larger rock and mineral inclusions, tempered with volcanic ash (crushed tuff?)	*non-local* – unknown but from outside of northern Belize/Yucatan region. Stylistic links to Yucatan area.
D	vg frags, qtz, igneous and volcanoclastic rock frags, feldspar (alkali and plagio), biotite, pqtz R = tuff frags, muscovite	- sickle and lunate shaped frags of volcanic glass - slightly bimodal appearance due to large rock and quartz inclusions - presence of igneous and volcanoclastic rock fragments - orangish- to reddish-brown clay matrix	non calcareous clay containing igneous and volcanoclastic rock fragments, tempered with volcanic ash	*non-local* – connected geologically to the Bladen Volcanic Member of the Santa Rosa Group (metasedimentary rocks), situated on the southern margin of the Maya Mountains due to the presence of igneous and volcanoclastic rock fragments and frequency of quartz inclusions

Table 8.6: Summary of mineralogy, distinguishing features, paste technology and provenance of the fabric groups comprising the Volcanic Glass class (continued).

Inclusions (in order of abundance)	Distinguishing Features	Paste Technology	Associated Provenance
calcite, vg frags, tuff frags R-VR = chal, qtz, biotite	- sickle and lunate shaped frags of volcanic glass and equant to elongated tuff frags - dominant, rounded, equant and irregular shaped inclusions of crystalline calcite - presence of chalcedony and quartz	calcareous clay containing rounded calcite inclusions (carbonate sand), chalcedony and quartz, tempered with volcanic ash	*non-local* – the clay component, containing carbonate sand, chalcedony and quartz, suggests a geological connection to coastal areas of northeastern northern Belize, extending into the Yucatan

reflect differences in their provenance. For example, the geological characteristics of the clay component of Volcanic Glass E, suggest a connection to coastal areas of northeastern northern Belize and the Yucatan. In contrast, Volcanic Glass D can be linked to volcanoclastic rocks from the Bladen Volcanic Member of the Santa Rosa Group (metasedimentary rocks), situated on the southern margin of the Maya Mountains. The provenance of the other three groups is less certain, due to the absence of diagnostic mineral and rock inclusions, particularly in the case of Volcanic Glass B and C. Nonetheless, these fabrics clearly differ from each other, and their geology is inconsistent with clays local to Lamanai, indicating these ceramics were not manufactured locally. In addition, the vessel representing Volcanic Glass C, conforms stylistically to 'Slate ware', suggesting that it may derive from the Yucatan.

The Relative Frequencies of Different Fabric Types and Their Provenance and Temporal Associations

Class Frequencies - The sample set of vessels analysed petrographically divides into seven main fabric classes, representing 24 distinct fabric groups, which includes the three fabric classes that are represented by a single group. Figure 8.9 presents the relative frequencies of the main fabric classes based on the number of vessels they contain and as percentages of the total number of samples analysed. As can be observed, the numerically dominant fabric type within the sample set is the Grog-Mixed Carbonate class, representing 35% of the samples analysed. Crystalline Calcite-Tempered fabrics are the next most abundant, representing 27.2% of the sample set, followed by the Sascab-Quartz and the Quartz Sand classes, which represent 15.2% and 14.6%, respectively. Dolomitic-Calcitic Marl-based and Volcanic Glass fabrics are rare by comparison, representing just 4.8% and 2.5%, respectively, of the samples analysed. The Sascab fabrics, represented by just two samples, constitute the least common fabric type.

Provenance Associations – Local vs. Non-local -
Table 8.7 presents a summary of the fabric groups and their provenance ascription (local vs. non-local), including their relative frequencies within the sample set. The fabric groups ascribed as local comprise 68.7% of the vessels analysed. Two fabric groups contain ceramics that derive from non-local as well as locally available raw materials. These groups include group H of the Crystalline Calcite-Tempered class and the Quartz Sand class. Together, these fabric groups comprise 17% of the sample set. Although, based on the current body of evidence and geological information, the exact number of samples deriving from local raw materials cannot be determined for these groups, the fact that they contain ceramics consistent with local raw materials, indicates that the actual proportion of local vs. non-local ceramics within the sample set, is far greater than the 68.7% minimum determined based on the local fabric groups alone. There are 16 fabric groups suggested to be non-local. These groups include five of the Crystalline

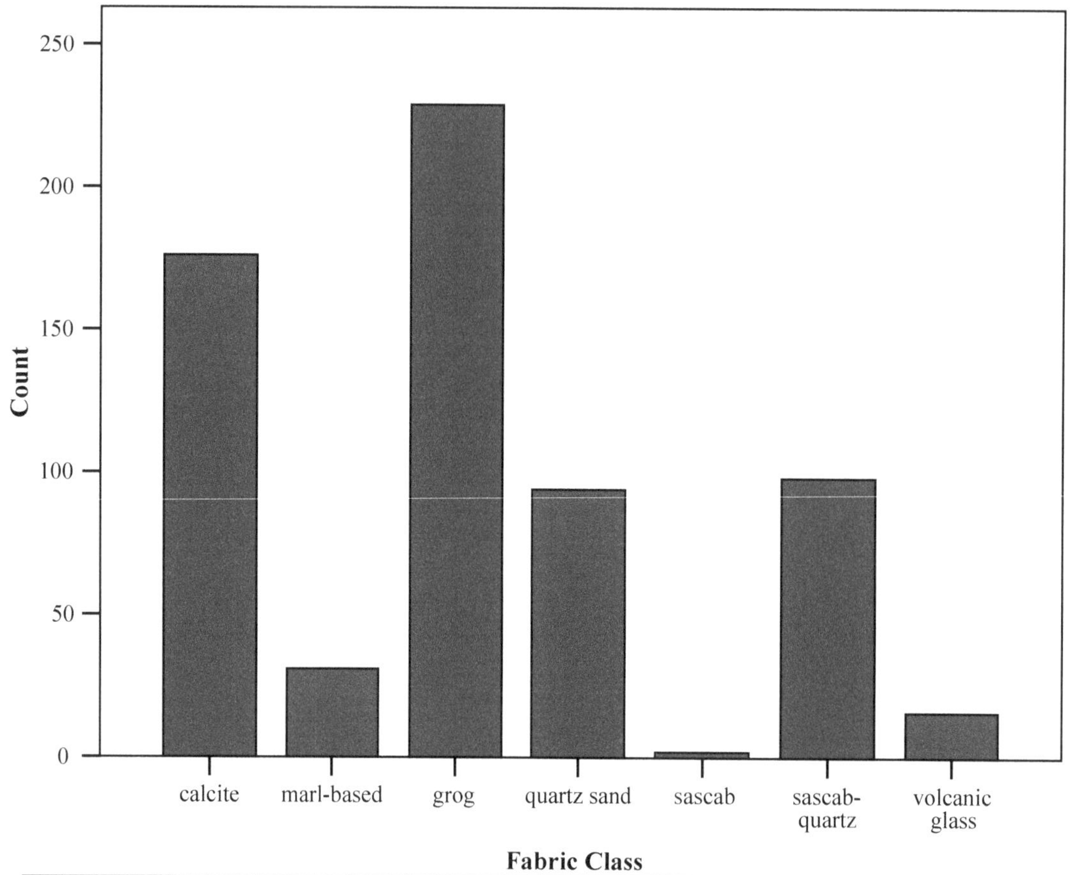

Ware	Crystalline Calcite-Tempered	Grog-Mixed Carbonate	Dolomitic-Calcitic Marl-Based	Quartz Sand	Sascab	Sascab-Quartz	Volcanic Glass	Total
Fine	17.0	35.4	4.8		0.3	15.2	2.5	75.2
Coarse	10.2			14.6				24.8
Total	27.2	35.4	4.8	14.6	0.3	15.2	2.5	100.0

Figure 8.9: Fabric class frequencies based on counts and percent of total number of analyzed samples.

Calcite-Tempered fabric groups (E, F, G, I and J), Sascab-Quartz B, the Sascab class and all of the fabric groups within the Dolomitic-Calcitic Marl-based and the Volcanic Glass classes. The non-local fabric groups, together, comprise only 11.1% of the sample set. Provenance could not be determined with confidence for group D of the Crystalline Calcite-Tempered Class, as well as the sample distinguished as subgroup B of the Grog-Mixed Carbonate Class. These compositionally ambiguous fabrics comprise just 3.1% of the sample set. If the relative frequencies of local vs. non-local ceramics within the sample set can be assumed to accurately reflect assemblage-wide patterns, then an important conclusion that can be drawn is that the large majority of pottery comprising the Lamanai assemblage was of local manufacture.

Temporal Associations and Trends - The frequency distribution of the main fabric classes by stratigraphic context within the two temple middens is presented in Table 8.8. The basic pattern that emerges is that the Crystalline Calcite-Tempered class, the Volcanic Glass class and the Sascab-Quartz class significantly decrease over time, whilst the Grog-Mixed Carbonate class and Quartz Sand class increase. Also significant is that the frequency of the Dolomitic-Calcitic Marl-based class, albeit represented by comparatively few samples, remains quite stable. The changes in the frequencies of different fabric classes reflect the fact that the Crystalline Calcite-Tempered, Volcanic Glass and Sascab-Quartz classes are strongly associated with early vessel styles (Terminal Classic), whereas the Grog-Mixed Carbonate class is more strongly associated with Early Postclassic styles (Table 8.8). In comparison, the Dolomitic-Calcitic Marl-based class contains both early and late styles within the sequence. Also apparent is that there is little change in fabric diversity over time. That is, although certain fabric classes clearly dominate the early and later ends of the sequence and class frequencies change, a similar range of different fabric types (classes) occurs throughout the sequence. This continuity in the diversity of fabrics stands in direct contrast to the loss of stylistic diversity with regard to surface treatment that occurs.

Table 8.7: Frequencies of fabric classes and associated groups and their general provenance association.

General Provenance	Fabric Class/Group/Subgroup	Number of Samples	% Total	% within Provenance
Local	Crystalline Calcite Tempered A (includes subgroup)	34	5.3	7.7
	Crystalline Calcite Tempered B (includes subgroup)	32	5.0	7.2
	Crystalline Calcite Tempered C (includes 2 subgroups)	55	8.5	12.4
	Grog-Mixed Carbonate (includes subgroup A)	228	35.3	51.4
	Sascab-Quartz A (includes subgroup)	95	14.7	21.4
Total		444	68.7	100.0
Groups containing local and non-local vessels	Calcite H (includes subgroup)	16	2.5	14.5
	Quartz Sand (includes subgroup)	94	14.6	85.5
Total		110	17.0	100.0
Non-local	Crystalline Calcite Tempered E	3	0.5	4.2
	Crystalline Calcite Tempered F	8	1.2	11.1
	Crystalline Calcite Tempered G	7	1.1	9.7
	Crystalline Calcite Tempered I	1	0.2	1.4
	Crystalline Calcite Tempered J	1	0.2	1.4
	Dolomitic-Calcitic Marl-Based A (includes subgroup)	17	2.6	23.6
	Dolomitic-Calcitic Marl-Based B (includes subgroup)	7	1.1	9.7
	Dolomitic-Calcitic Marl-Based C	2	0.3	2.8
	Dolomitic-Calcitic Marl-Based D	5	0.8	6.9
	Sascab	2	0.3	2.8
	Sascab-Quartz B	3	0.5	4.2
	Volcanic Glass A (includes subgroup)	12	1.9	16.7
	Volcanic Glass B	1	0.2	1.4
	Volcanic Glass C	1	0.2	1.4
	Volcanic Glass D	1	0.2	1.4
	Volcanic Glass E	1	0.2	1.4
Total		72	11.1	100.0
Possibly Non-local	Crystalline Calcite Tempered D	19	2.9	95.0
	Grog-Mixed Carbonate subgroup B	1	0.2	5.0
Total		20	3.1	100.0
Grand Total		646	100.0	

Table 8.8: A breakdown of fine ware stylistic groups within the different fabric classes (Note: % = % within fabric class).

Stylistic Group	Crystalline Calcite-Tempered		Grog-Mixed Carbonate		Dolomitic-Calcitic Marl-Based		Sascab		Sascab-Quartz		Volcanic Glass		Total	
	#	%	#	%	#	%	#	%	#	%	#	%	#	%
Polychrome Slipped and Painted	4	3.6	1	0.4	4	12.9	2	100.0	29	29.6	8	50.0	48	9.9
Black on Red Slipped and Painted	0	0.0	0	0.0	0	0.0	0	0.0	7	7.1	0	0.0	7	1.4
Red-Orange-Black Resist	9	8.2	0	0.0	1	3.2	0	0.0	0	0.0	0	0.0	10	2.1
Monochrome Black	8	7.3	0	0.0	17	54.8	0	0.0	0	0.0	0	0.0	25	5.1
Monochrome Red Slipped	48	43.6	4	1.7	2	6.5	0	0.0	15	15.3	1	6.3	70	14.4
Monochrome Red Slipped and Groove-Incised	2	1.8	0	0.0	0	0.0	0	0.0	0	0.0	0	0.0	2	0.4
Monochrome Red Slipped and Notched	3	2.7	0	0.0	0	0.0	0	0.0	0	0.0	0	0.0	3	0.6
Monochrome Red to Orange Slipped (interior slipped)	0	0.0	0	0.0	0	0.0	0	0.0	2	2.0	0	0.0	2	0.4
Monochrome Orange Slipped and Groove-Incised	0	0.0	1	0.4	0	0.0	0	0.0	0	0.0	3	18.8	4	0.8
Buff/Cream Slipped	2	1.8	0	0.0	0	0.0	0	0.0	1	1.0	1	6.3	4	0.8
Monochrome Brown	2	1.8	0	0.0	0	0.0	0	0.0	0	0.0	0	0.0	2	0.4
Monochrome Red Slipped and Incised	1	0.9	0	0.0	0	0.0	0	0.0	0	0.0	0	0.0	1	0.2
Monochrome Red Slipped and Model-Carved	1	0.9	0	0.0	0	0.0	0	0.0	0	0.0	0	0.0	1	0.2
Monochrome Grey (Plumbate)	1	0.9	0	0.0	0	0.0	0	0.0	0	0.0	0	0.0	1	0.2
Monochrome Orange Slipped	27	24.5	107	46.7	6	19.4	0	0.0	37	37.8	2	12.5	179	36.8
Monochrome Orange Slipped and Incised	2	1.8	107	46.7	1	3.2	0	0.0	7	7.1	1	6.3	118	24.3
Monchrome Orange to Red Slipped - Notched and Incised	0	0.0	6	2.6	0	0.0	0	0.0	0	0.0	0	0.0	6	1.2
Monochrome Orange with Applique	0	0.0	3	1.3	0	0.0	0	0.0	0	0.0	0	0.0	3	0.6
Total	110	100.0	229	100.0	31	100.0	2	100.0	98	100.0	16	100.0	486	100.0

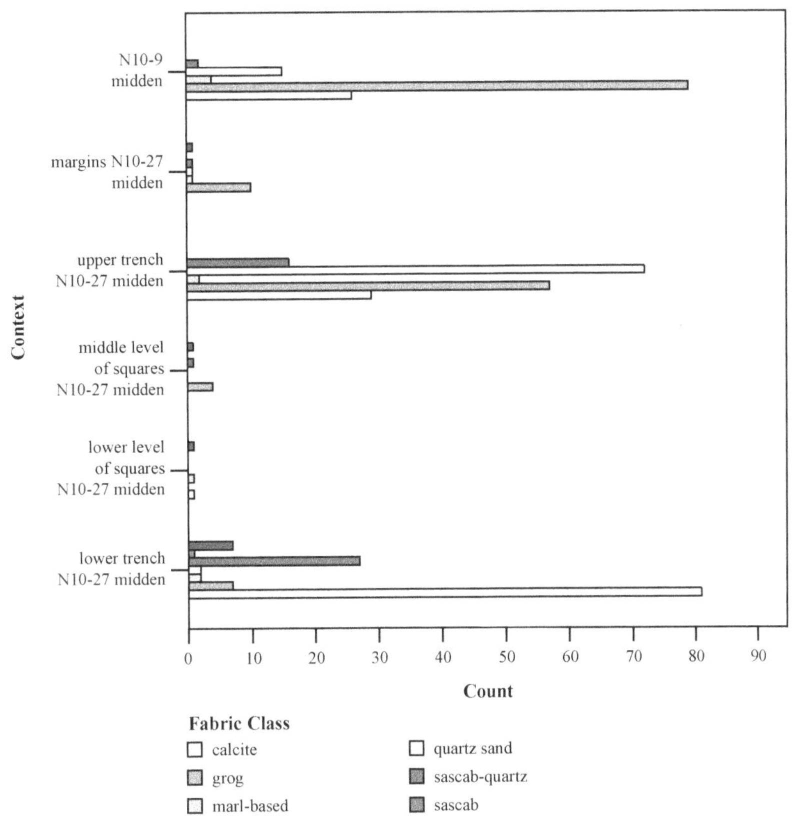

Figure 8.10: Temporal association of fabric classes based on their relative frequencies within different stratigraphic contexts. (see colour version on accompanying CD)

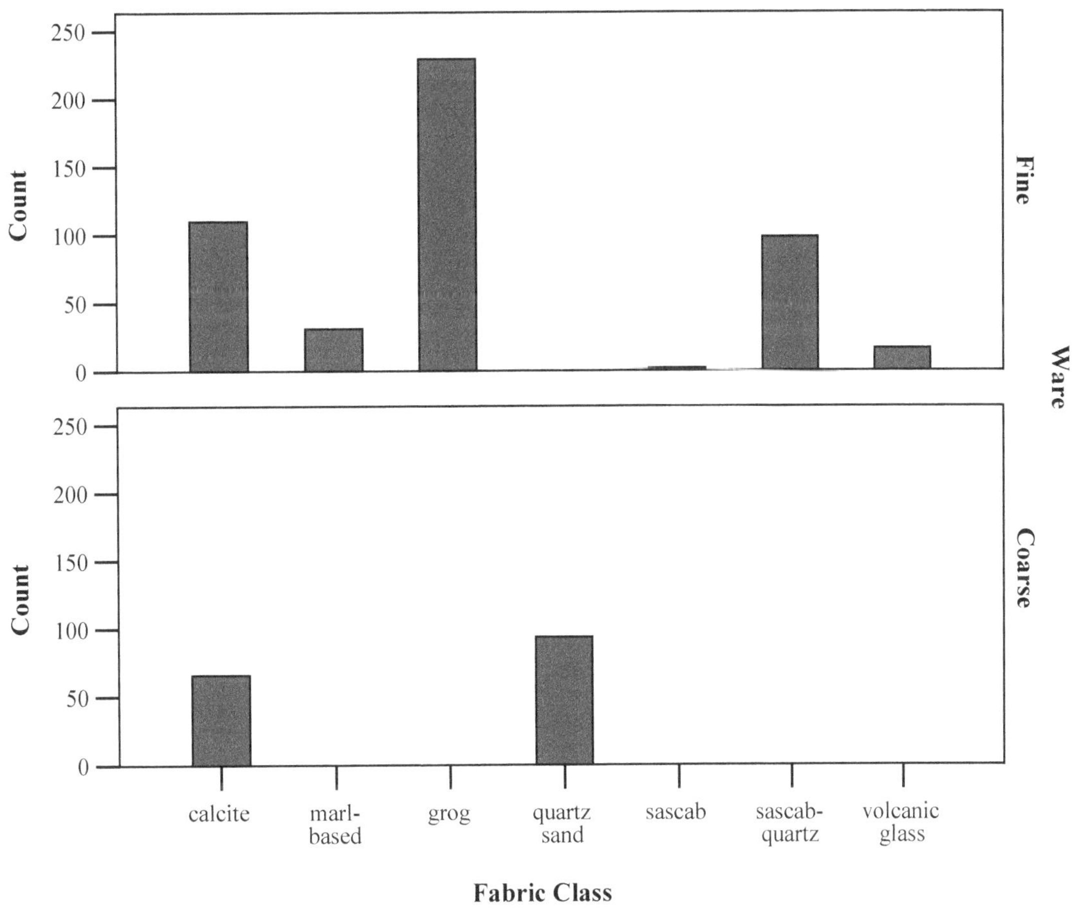

Figure 8.11: The frequency distribution of fine ware and coarse ware according to fabric class.

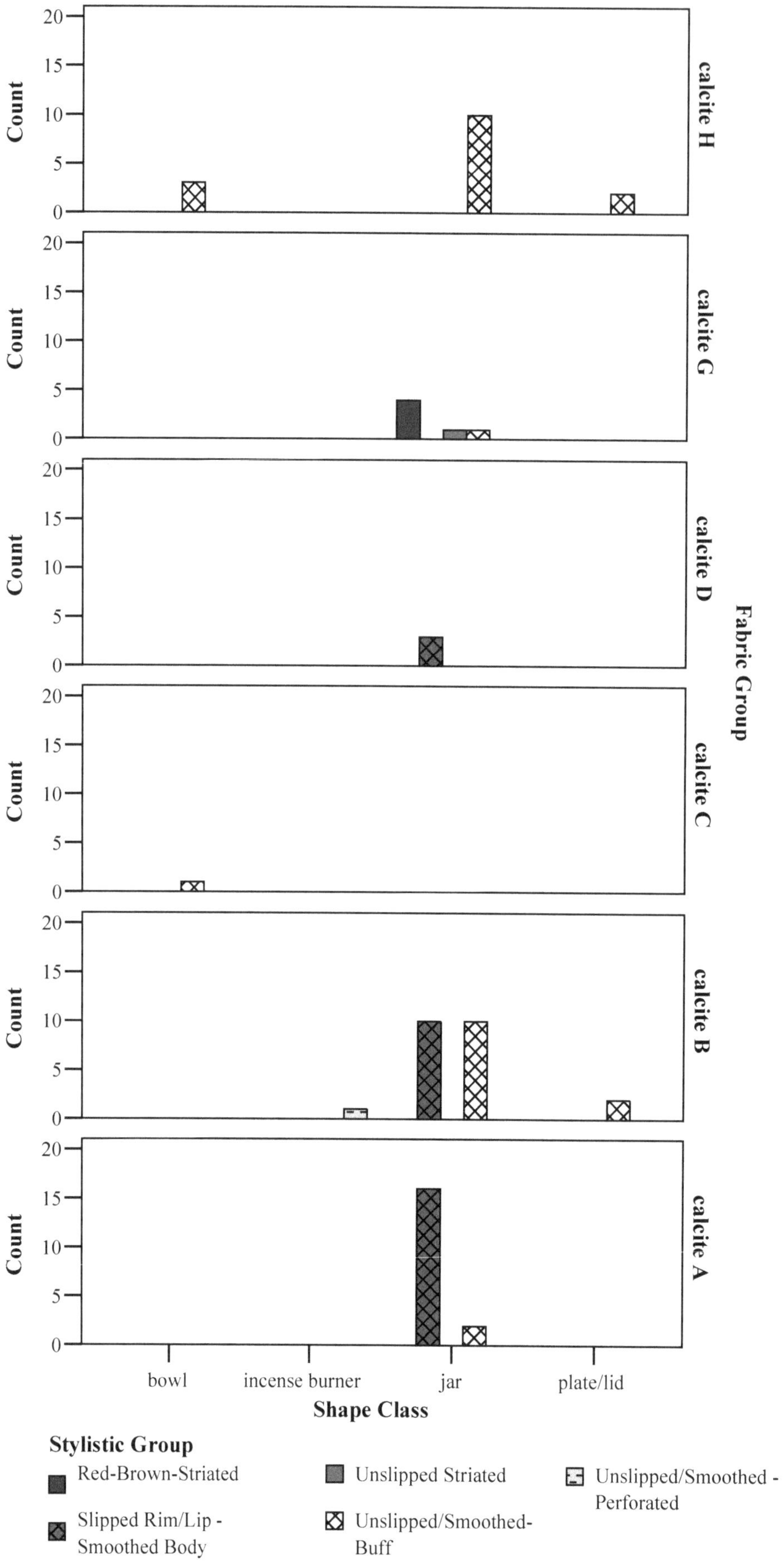

Figure 8.12: Frequency distribution of stylistic groups according to form within Crystalline Calcite-Tempered fabric groups. (see colour version on accompanying CD)

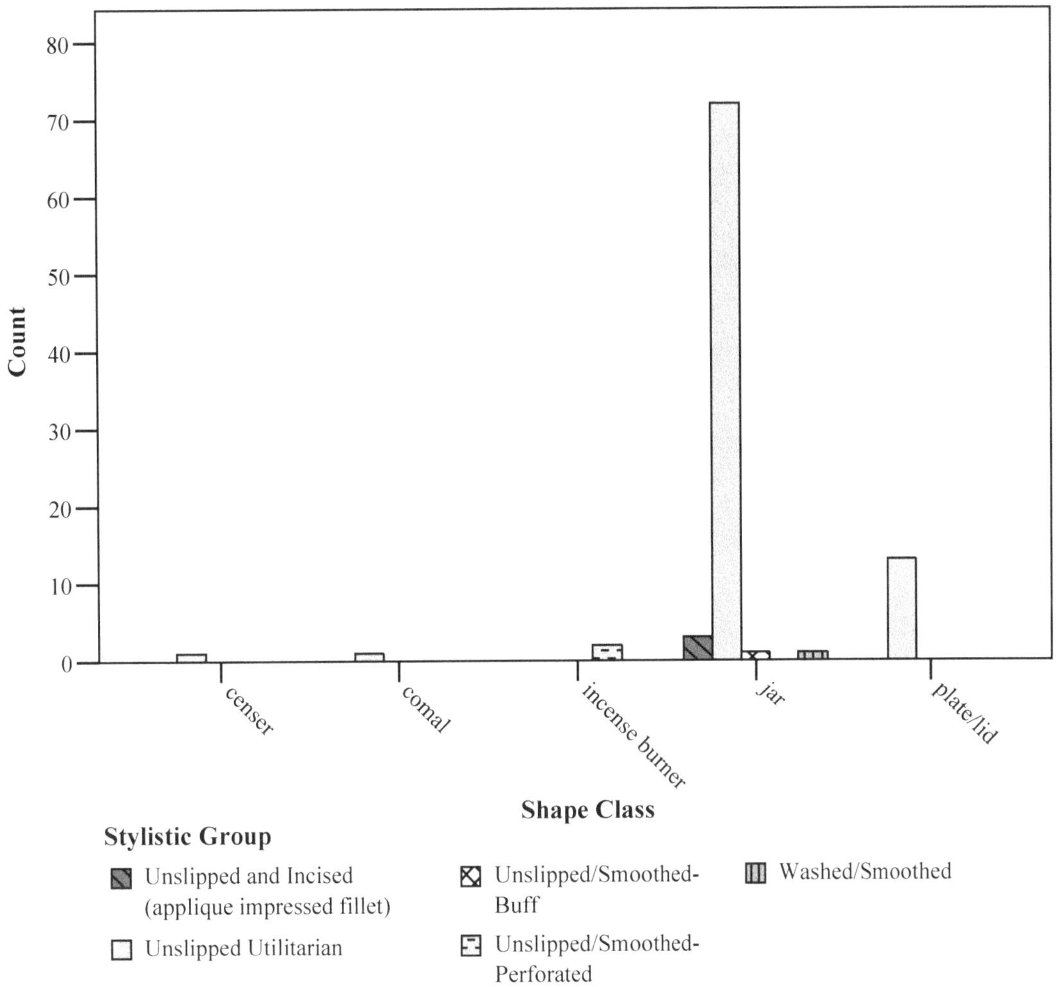

Figure 8.13: Frequency distribution of stylistic groups according to form within the Quartz Sand class. (see colour version on accompanying CD)

8.3. FUNCTIONAL AND STYLISTIC ASSOCIATIONS OF THE FABRIC CLASSES

When the main fabric categories are considered in light of the functional associations and the surface treatment characteristics of the vessels they contain, several significant patterns emerge. These patterns can be considered to reflect behaviourally meaningful correlations between the main fabric types (classes) and, in some cases, fabric groups, and the functional and stylistic categories discussed in previous chapters.

Figure 8.11 presents a frequency distribution of the main fabric classes for fine ware and coarse ware. Fine ware includes serving/table wares (a range of service and drinking vessels), musical instruments (drums and whistles) and special purpose vessels presumably used in ritual (sieves, frying pan censers, incense burners, 'torches' and 'stands'), whilst coarse ware primarily includes vessels with broadly utilitarian functions relating to food preparation and storage, but in this assemblage also includes incense burning paraphernalia (incense burners and a censer). As can be observed, all of the fabric classes except the Quartz Sand class contain fine ware vessels. In contrast, coarse ware vessels are confined to just two classes: the Crystalline Calcite-Tempered class and the Quartz Sand class, with incense burning vessels occurring in both. This pattern indicates a strong correlation between vessel composition at the class level, and vessel function. It is also evident that the fine ware, or vessels used in service and ritual, exhibits greater compositional diversity than vessels serving broadly utilitarian functions. Another significant pattern is that all of the fabric classes except the Quartz Sand class contain service and drinking vessels, while in all cases except three (a drum and the only 'torch' and 'stand' within the assemblage), musical instruments and special purpose vessels are restricted to the Grog-Mixed Carbonate class.

When considering the relationship between compositional and stylistic variability within these general functional categories, several additional observations can be made.

Fine ware - All of the fabric classes, except the Sascab Class, which comprises two polychrome bowls, contain multiple stylistic groups (Table 8.8). In all cases, however, one or two stylistic groups comprise the majority of vessel within a particular class. For examples, close to 70% of the vessels within the Crystalline Calcite-

Tempered Class are monochrome red or orange slipped. In comparison, monochrome orange slipped vessels with or without incised decoration comprise over 90% of the vessels in the Grog-Mixed Carbonate class. More than 50% of the ceramics in the Dolomitic-Calcitic Marl-based class are black slipped, whereas polychrome and orange slipped vessels comprise over 60% of both the Sascab-Quartz and the Volcanic Glass classes. In the case of the Volcanic glass class, however, the orange slipped vessels tend to have groove incising. It is notable that while all of the fabric classes, except the Sascab class, contain orange slipped vessels, they comprise over 95% of the Grog-Mixed Carbonate class. Accordingly, there appears to be some correlation between vessel composition at the class level and the stylistic categories presented in Chapter 7. The nature of the relationship between vessel composition and style among the fine ware is explored further in subsequent sections.

Coarse Ware - Although restricted to two fabric classes, coarse ware vessels occur in six of the ten fabric groups comprising the Crystalline Calcite-Tempered class. These groups include: the local groups A, B, C and H, Group G, which is non-local and Group D, for which provenance is ambiguous. As can be seen in Figure 8.12, jars are the predominant or exclusive form in all of the calcite-tempered fabric groups except Calcite C, with different jar styles predominant in the local and non-local groups. In comparison, the Quartz Sand class is dominated by one stylistic group, in which different form types exhibit high morphological variability (Figure 8.13). In addition, this class also contains a greater range of jar styles than the Calcite-Tempered groups. The greater stylistic variation among jars with Quartz Sand Fabrics is in keeping with the fact that this class contains vessels made using many different clays deriving from both local and non-local sources.

8.4. Local Approaches to Paste-Making: Temporal Trends and Stylistic Correlations

Petrographic analysis of the Lamanai assemblage has revealed two fabric classes and five fabric groups that are geologically compatible with raw material resources that occur in the local area (Crystalline Calcite-Tempered groups A, B, C and H, group A of the Sascab-Quartz class, the Grog-Mixed Carbonate class and the Quartz Sand class). The petrographic differences among the local groups relate to the selection of raw material ingredients – i.e. different choices in paste-making. We might refer to the four different classes of local fabrics as 'traditions', as they occur at the site over extended periods of time. Fabrics strikingly similar to the crystalline calcite-tempered fabric groups and Sascab-Qaurtz A fabrics occur as early as the Late Preclassic period (Powis and Howie n.d.; Powis et al. 2002; Powis 2002). Similarly, recent macroscopic analysis conducted by the author of Middle and Late Postclassic assemblages at the site has indicated that Grog Mixed Carbonate and Quartz Sand fabrics continue into these later time periods.

The temporal association of the local fabric groups is shown in Figure 8.14. As can be seen, the calcite-tempered groups A, B and C and Sascab-Quartz A dominate the earliest stratigraphic context (the lower trench lot from N10-27) and their proportional representation decreases significantly in the most recent stratigraphic context (the upper trench lot and margins of N10-27 and in the midden at N10-9). In contrast, the Grog-Mixed Carbonate and the Quartz Sand classes and group H of the Crystalline Calcite-Tempered class are poorly represented in the earliest stratigraphic contexts and clearly dominate temporally later contexts. Accordingly, the former groups can be characterized as 'early' fabric types within the sequence that decrease in prevalence over time, whilst the latter classes and group can be considered as fabric types that become predominant by the end of the sequence, seemingly at the expense of the earlier fabric types.

The Local Crystalline Calcite-Tempered Tradition

The Crystalline Calcite-Tempered fabric groups derive from a basic paste recipe that involves calcareous clay mixed with freshly crushed crystalline calcite. All of the groups appear to derive from clay resources that occur at the site and in the surrounding inland area characterized by Yalbac soils, and they are all tempered with forms of crystalline calcite that generally occur as veins or nodules in the underlying limestone bedrock and associated horizons of weathering limestone. Groups A and H appear to derive from local clays that form directly below the ground's surface (at the site and in the surrounding area), and the calcite temper used in both derives from the colourless, coarsely-crystalline form of calcite that occurs locally. The essential difference between these two groups is the greater quantity of quartz in the fabrics comprising Group H, as well as their comparatively coarse-textured groundmass. In addition, three of the fabrics within group H contain carbon residues that exhibit a consistent morphology and are associated with channel voids. These residues may represent an organic temper that was incorporated into the paste in addition to calcite temper.

A main characteristic that distinguishes Calcite B and C from Calcite A and H is the predominance of fine grains of crystalline calcite in the clay component of B and C fabrics, which gives them a distinctly 'grainy' appearance. The high aplastic content of the groundmass links them to local clays that form in association with horizons of weathering limestone. These clayey horizons directly underlay the clays associated with Calcite A and H. Consequently, it is quite likely that the primary factor responsible for the compositional differences observed among the groups with respect to the clay component is the specific geological context in which these clays were formed or, alternatively, the relative depth below the ground's surface from which they were procured. A further characteristic that distinguishes groups B and C is the physical properties of the crystalline calcite used for temper. In these cases, the temper derives from the finely crystalline form of the mineral, which has a distinctive

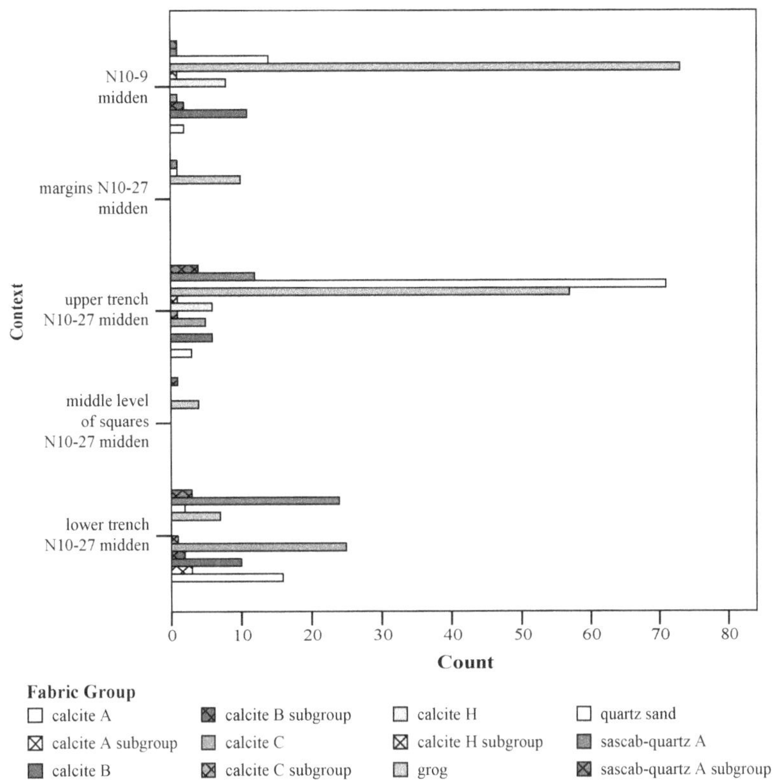

Figure 8.14: Frequency distribution of local fabric groups and classes within different stratigraphic contexts. (see colour version on accompanying CD)

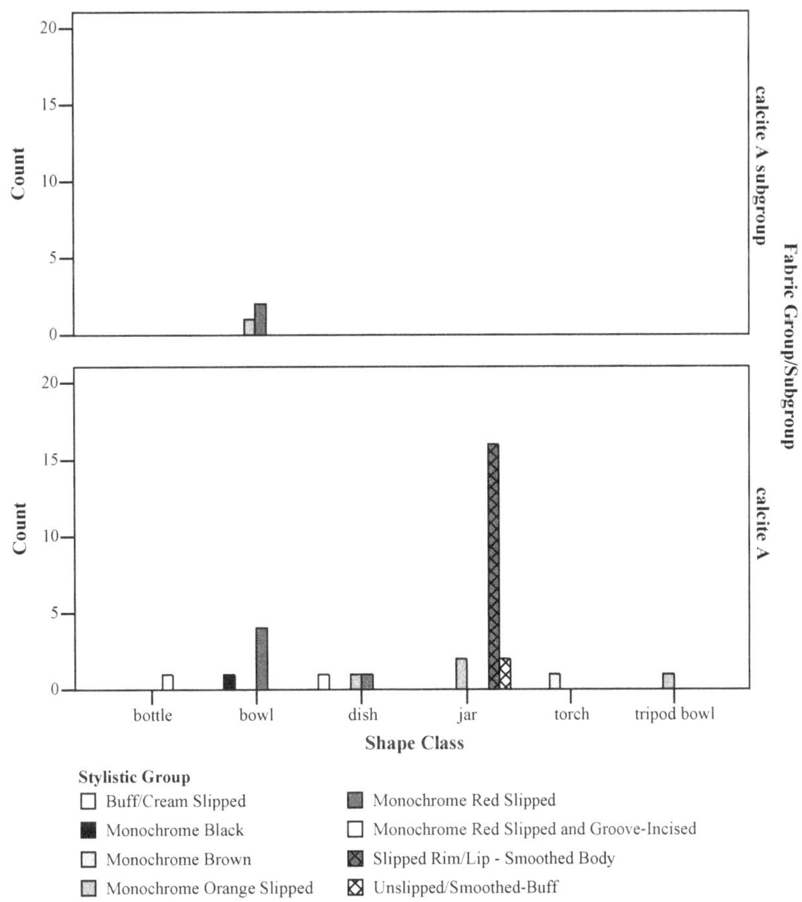

Figure 8.15: Vessel styles within group A of the Crystalline Calcite-Tempered class summarized according to surface treatment and form. (see colour version on accompanying CD)

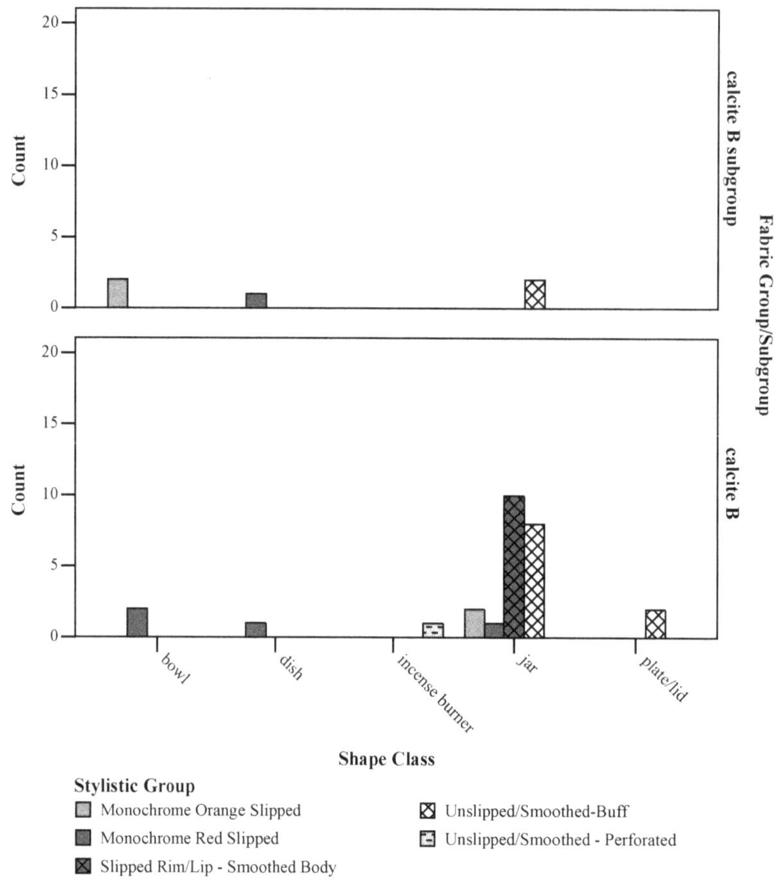

Figure 8.16: Vessel styles within group B of the Crystalline Calcite-Tempered class summarized according to surface treatment and form. (see colour version on accompanying CD)

sugary texture and is white in colour. The main characteristic that distinguishes Calcite C from Calcite B is that Calcite C fabrics contain sascab temper, in addition to crystalline calcite. In fact, these two fabric groups are so similar in all respects that C could possibly be described as a 'sascab-tempered version' of Calcite B.

Groups A, B and C are internally homogeneous, except for the samples distinguished within their petrographic subgroups due to the presence of grog temper. In comparison, group H exhibits significant internal variation, which in some cases appears to reflect provenance differences related to clay sources. This group also contains a grog-tempered variant. In each of the groups, the grog-tempered variant contains the same vessel styles as the main group (see accompanying figures below).

The close compositional relationship between the local calcite-tempered fabric groups is reflected by an overlap in the vessel styles they contain. The predominant types within groups A and B are well made coarse ware jars of the Slipped Rim/Lip-Smoothed Body and Unslipped/Smoothed Buff stylistic groups and rounded fine ware bowls with a red to orange slip that extends to just below the rim on the exterior surface ('interior-slipped bowls') (Figures 8.15 and 8.16). Composite silhouette dishes, which constitute the predominant vessel type in Calcite C, are also present in these two groups (Figure 8.17).

The slipped jars and bowls, which might be considered to reflect a 'service set', in the sense that the red and cream exteriors of these vessels express a particular and distinctive stylistic convention, represent early vessel styles within the sequence, whereas the unslipped jars clearly reflect later jar styles. These unslipped jars are dominant in Calcite H. Given the similarities between groups A, B and H, this overlap in the jar styles seems to document a stylistic shift in a local tradition of calcite-tempered storage jars. The petrographic differences among these groups clearly relate to natural variation in clays due to differences in their geological context, and therefore, might reflect the exploitation of different clay resources over time, within a circumscribed geological zone.

The close petrographic and stylistic relationship between groups A and B requires further discussion. These groups contain the same range of Terminal Classic vessel styles and yet they are clearly distinguished, petrographically, by their raw materials. The basic question that emerges is whether these groups can be considered to reflect the activities of contemporaneous potters that use slightly different raw materials or changes over time in raw materials related to source or resource exploitation. It

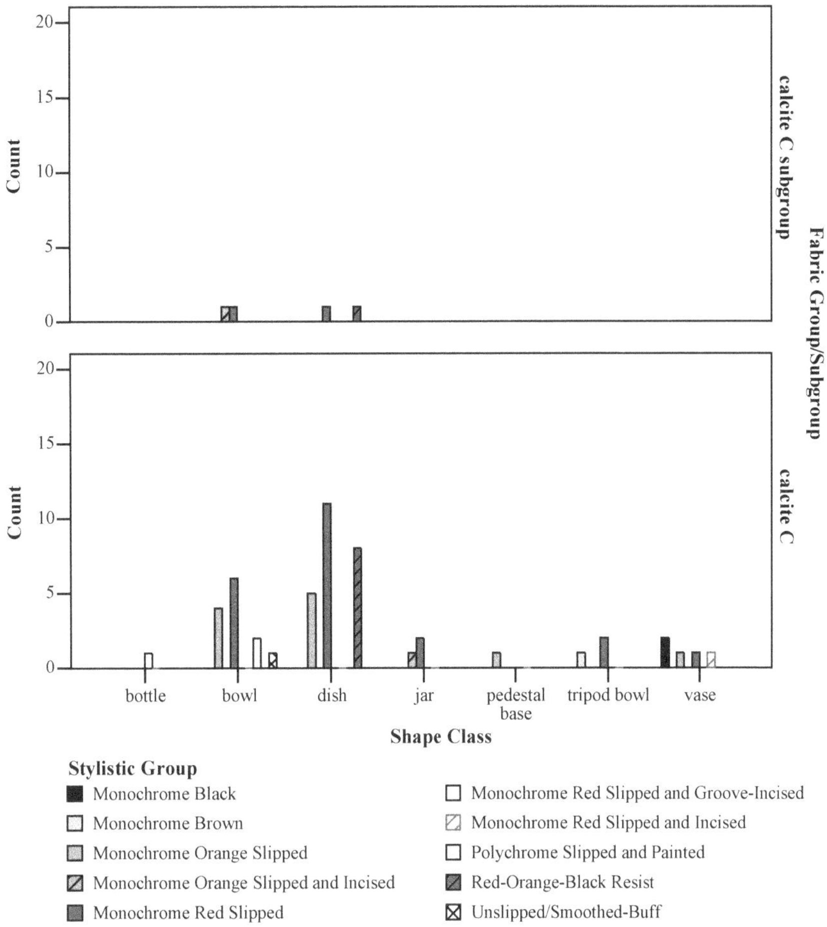

Figure 8.17: Vessel styles within group C of the Crystalline Calcite-Tempered class summarized according to surface treatment and form. (see colour version on accompanying CD)

could be argued that either of these explanations is equally plausible and, supportable. And certainly, a more fine-grained chronological control over these fabric groups than the current data permit, would be required to justify any distinction made in this regard. The relationship between these two groups is interesting nevertheless, because of the insight that might be gained into potters' understanding and use of the raw material resources available them in the local area. For example, did observable differences in the physical properties of mineralogically equivalent raw materials have any significance or meaning to local potters? Did they differentiate, or make a qualitative distinction, in any way, between the colourless coarse-textured form of calcite and the white sugary-textured form as a raw material to produce temper? Furthermore, did the natural colour and texture of different clays have any significance to potters if they fired to the same colour and could both be used successfully? The petrographic evidence suggests that these differences *were* recognized by local potters, and potentially informed their choice of raw materials, since there is no indication that the different clays and calcite tempers were used interchangeably or combined in the same paste. Rather, groups A and B represent two distinct combinations of raw materials, or paste recipes, based on different, but geologically related, clays and temper source materials.

The presence of sascab temper in Calcite C fabrics clearly distinguishes this group from groups A, B and H, indicating a slightly different approach to paste-making within the general calcite-tempered tradition. The compositional distinctiveness of this fabric group is mirrored in the types of vessels it contains (Figure 8.17). Storage jars and other coarse ware forms are entirely absent from this group and composite silhouette dishes and bowls with different surface treatments predominate. The dishes have either monochrome slipped surfaces, ranging in colour from deep red to orange or feature a bichrome decorative effect produced through a resist technique. The bowls have either red or orange slips or polychrome painted decoration. Considering the internal petrographic consistency of this fabric group, the occurrence of conformal vessels with different surface treatments is quite significant. Chronological variation might be a contributing factor, but the contemporaneous use of different decorative treatments can not be discounted, as conformal vessels with different decorative treatments occur together in primary contexts such as burials at Lamanai.

Despite the apparent differences in the predominant vessel types between Calcite C and the other calcite-tempered groups, a close stylistic relationship to groups

A and B is indicated by the stylistic overlap between these three groups. For example, Calcite A and B also contain composite silhouette dishes, albeit very few, and all three groups contain rounded 'interior slipped' bowls – the stylistic counterpart of the storage jars with the red slipped rims. These bowls are visually indistinguishable in terms of their morphological and surface treatment attributes, suggesting interaction, if not a close relationship, among the potters who produced them.

The Local Sandy-Sascab-Tempered Tradition - Sascab Quartz A

Sascab-Quartz A fabrics are clearly based on a completely different set of local raw materials than the calcite-tempered groups and, therefore, can be considered to reflect a different, yet contemporaneous, paste-making tradition. The paste recipe involved can be described as sandy calcareous clay mixed with a freshly crushed, solidified form of sascab. As with the calcite-tempered fabric groups, this group contains a petrographic subgroup, which is distinguished by the presence of grog temper and a comparatively fine-textured groundmass.

The petrographic distinctiveness of the fabric group is reflected in the vessel styles it contains (Figure 8.18). The large majority of vessels within the main group are morphologically identical rounded dishes with different surface treatments, including polychrome painted decoration on an orange or buff slipped background, black painted decoration on a red slipped background and monochrome red to orange and buff slipped. On all of these dishes the slip extends to just below the rim area on the exterior surface ('interior slipped'). Dishes with polychrome decoration are numerically dominant, followed by monochrome slipped dishes. These dishes can be considered to represent a functional equivalent of the calcite-tempered composite silhouette dishes and, therefore, they might be interpreted as reflecting a different and contemporaneous local tradition of dish-making. As was suggested for Calcite C, the occurrence of conformal vessels with different surface treatments can be attributed either to chronological variation or the contemporaneous use of different decorative treatments by potters working within the Sascab-Quartz A paste-making tradition. Also significant is that four composite silhouette dishes and a chalice (pedestal-based dish) are present in the main group of Sascab-Quartz A fabrics. Two of the composite silhouette dishes have polychrome decoration that is identical, stylistically, to that which occurs on the rounded dishes, and interestingly, these two vessels are the only examples of polychrome decorated composite silhouette dishes that occur within the assemblage. These dishes could be interpreted as an instance of potters imitating the products of other local producers, but they might also be considered as evidence of direct social interaction among local potters working within different manufacturing traditions. The presence of 'interior slipped' bowls that are stylistically identical to the more usual calcite-tempered vessels is another example of stylistic overlap that is suggestive of these sorts of behaviour.

The occurrence of the chalice within the main group of Sascab-Quartz A fabrics is significant as it is the only instance, within the corpus of ceramics analysed petrographically, of a chalice that does not contain grog temper. Within the grog-tempered subgroup of Sascab-Quartz A, all of the dishes are chalices, and the remaining vessels within the subgroup conform to Early Postclassic stylistic conventions, in both surface treatment and morphology. The presence of Early Postclassic vessel styles within both the subgroup and the main group of Sascab-Quartz A indicates an important connection, on both stylistic and compositional grounds, to the Grog-Mixed Carbonate class, which nearly exclusively comprises Early Postclassic vessel styles. The co-occurrence of Early Postclassic style vessels, particularly chalices, with Sascab-Quartz A (subgroup) and Grog-Mixed Carbonate fabrics, suggests that these fabric groups overlap temporally.

Chronological variation is undoubtedly at least partly responsible for the sharp contrast in vessel styles that occur in the main group of Sascab A fabrics and the petrographic subgroup. Given the compositional similarity of the fabrics comprising the main group and the subgroup and the stylistic evidence for time depth among the vessels that occur, Sascab-Quartz A could be interpreted as documenting changes over time in paste preparation techniques and in the style of vessels produced within this local pot-making tradition. The petrographic variation within the group, which relates to the presence of grog temper and textural differences in the groundmass (the subgroup is comparatively fine-textured), reflects the adoption of an additional tempering material, as well as a different, if not more careful, processing of the raw material ingredients of the paste.

The Grog-Mixed Carbonate Tradition

The Grog-Mixed Carbonate class reflects a tradition of fine ware manufacture that emerges during the Terminal Classic period. The vessels within this class are almost exclusively orange-slipped, commonly with incision and other secondary decorative treatments (Figure 8.19). All of the new vessels forms that are introduced into the sequence, such as drums, chilli grinders and frying pan censers, pedestal-based jars and tripod bowls with a segmented basal flange, are represented within this class, and several of these new forms are restricted to it (e.g. drums, composite vessels, whistles, frying pan censers and sieves). Accordingly, this class is strongly associated with Early Postclassic vessel styles. The presence of three monochrome composite silhouette dishes and a polychrome composite silhouette bowl provides evidence of a stylistic overlap with the local fabric groups in which 'early' or Terminal Classic vessel styles predominate.

Petrographically this class of fabrics is characterized by significant internal heterogeneity, which has been interpreted as related to natural differences in the clay component, reflecting the use of compositionally different local clays, as well as variation in the specific

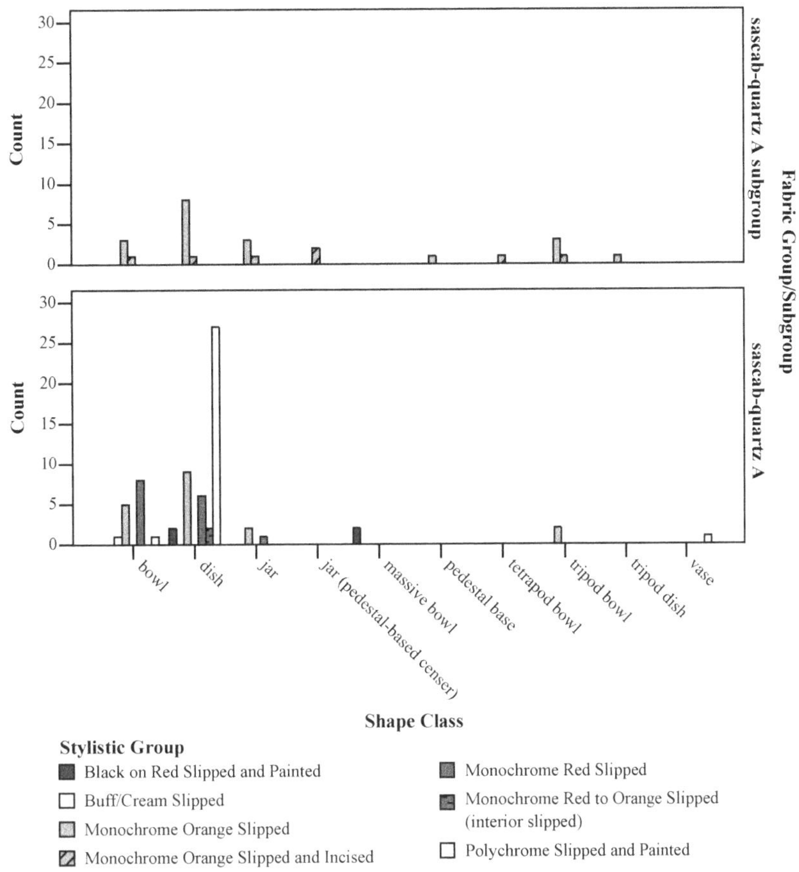

Figure 8.18: Vessel styles within Sascab-Quartz A summarized according to surface treatment and form. (see colour version on accompanying CD)

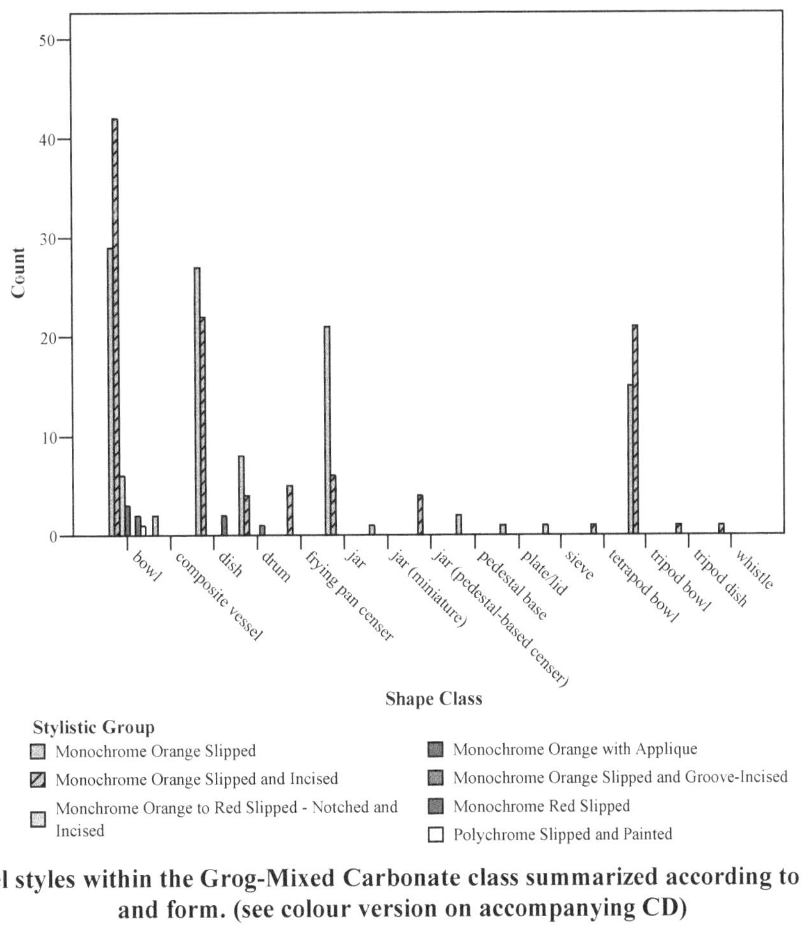

Figure 8.19: Vessel styles within the Grog-Mixed Carbonate class summarized according to surface treatment and form. (see colour version on accompanying CD)

combination of raw materials and their relative proportions. Although not unprecedented in the other local fabric groups, the use of multiple tempering materials, including freshly crushed pot sherds, sascab and both the white and colourless forms of crystalline calcite, is particularly distinctive of this approach to paste-making. These fabrics are also comparatively fine-textured, suggesting a more careful or rigorous processing of raw materials. Connections to Yalbac clays, both those that form directly below the ground's surface and those associated with horizons of weathering limestone, as well as Filipe Subsuite clayey soils, can be demonstrated petrographically. The connection to Filipe soils is particularly apparent with regard to the samples comprising subgroup A, which constitute an internally consistent subgroup characterized by a highly micritic groundmass containing a significant quantity of rounded siliceous inclusions or 'quartzy' sand. These fabrics also contain comparatively few large grog inclusions.

Another distinctive characteristic of this fabric class is the large percentage (>70%) of samples containing black cores. Unlike the firing horizons characteristic of other local fabric groups, the colour of core rarely grades into the lighter colour of the margins (i.e. they are sharply defined). The preponderance of black cores and the distinctive character of these firing horizons may suggest an important difference in firing methods. The high frequency of fabrics with an optically inactive or only slightly optically active clay matrix, which is often accompanied by a 'channely' microstructure due to bloating pores and 'shrink voids' in association with the grog inclusions, lends further support to this interpretation.

The high level of internal variation within this fabric class might be considered to reflect the interchangeable use of different clays and forms of carbonate temper by potters working within this general paste-making tradition. The sheer volume of Early Postclassic style vessels at the site suggests that the variation is at least partly chronological. It is likely, however, that some compositional variation among these fabrics relates to the paste preparation methods of different potters. This is suggested by the fact that different and potentially geographically separated clay resources are represented in this fabric class. It seems unlikely that a single group of potters would simply choose to exploit multiple clays and source locations at any one given time.

The Quartz-Sand Tradition

The Quartz-Sand class represents a tradition of coarse ware manufacture that appears in the Lamanai sequence during the Terminal Classic period. The mineralogy of these fabrics indicate that they derive from clays formed in association with deep deposits of Pleistocene alluvium, and a connection to the deposit bordering the east side of the New River Lagoon is demonstrated through the striking similarity between many of the archaeological samples and the clay samples taken from this area. The Quartz Sand class contains the only local fabrics that derive from raw material resources situated on the opposite side of the lagoon. Hence, the appearance of this fabric type at the site can be considered to reflect the introduction of a significantly different approach to coarse ware production, based on previously unexploited raw material resources.

The Significance of Grog-Tempered Variants

A final issue concerns the potential significance of the petrographic subgroups within the Crystalline Calcite-Tempered and Sascab Quartz A fabric groups, which were distinguished primarily due to the presence of grog temper. The comparative rarity of fabrics containing grog in the early part of the sequence and the predominance of the Grog-Mixed Carbonate class by its end is indicative of a strong temporal trend toward the prevalence of grog-tempering as a practice. In addition, the presence of grog temper as a compositional attribute is strongly associated with vessel forms and surface treatments that become the standard by the Early Postclassic period. This close relationship is clearly reflected in the surface treatment and morphology of the vessels comprising the subgroup of Sascab A in comparison to the main group. Nonetheless, grog temper is also present in a small number of vessels reflective of early styles within the sequence. Within the calcite-tempered fabric groups, these grog-tempered vessels are virtually identical in terms of surface treatment and morphology to their non-grog-tempered counterparts. Similarly, the composite silhouette dishes and bowl that occur within the Grog-Mixed Carbonate are also visually indistinguishable from the more usual, calcite-tempered versions of these vessels. The occurrence of these grog-tempered variants of early vessel styles provides evidence that the use of grog as a tempering material did not necessarily coincide with an alteration in the stylistic characteristics of the vessels being produced. The implication is that this change in local paste-making practices, as indicated by the increasing prevalence of grog-tempering throughout the sequence, occurred independently of the incorporation of new vessel forms and surface treatments into the local ceramic repertoire.

8.5. Temporal Trends in the Influx of Non-Local Ceramics to the Site

Thin section analysis has produced several, compositionally distinct fabric groups that are geologically incompatible with raw material resources available in the local area. These include: five groups within the Crystalline Calcite-Tempered class, Sascab-Quartz B, the Sascab class, the five groups comprising the Volcanic Glass class, and the four groups comprising the Dolomitic-Calcitic Marl-Based class. Since each of these groups is petrographically distinctive in its own right, it is suggested that they derive from different points of origin. Some of the groups represent different and geographically separated manufacturing areas, as indicated by their geology (e.g. Volcanic Glass D and E), whist other appear to reflect different manufacturing

Table 8.9: Vessel style within the non-local fabric groups and classes.

Fabric Group/Class		bowl	dish	drum	jar	pedestal base	stand	tripod bowl	tripod dish	vase	Total
Crystalline Calcite-Tempered											
Group E	Buff/Cream Slipped	1									1
	Monochrome Grey (Plumbate)					1					1
	Monochrome Red Slipped and Notched	1									1
	Total	2				1					3
Group F	Monochrome Black	3								2	5
	Monochrome Orange Slipped				1						1
	Monochrome Red Slipped	1									1
	Polychrome Slipped and Painted									1	1
	Total	4			1					3	8
Group G	Monochrome Orange Slipped				1						1
	Red-Brown-Striated				4						4
	Unslipped Striated				1						1
	Unslipped/Smoothed-Buff				1						1
	Total				7						7
Group I	Monochrome Red Slipped and Model-Carved	1									1
Group J	Monochrome Red Slipped				1						1
Dolomitic-Calcitic Marl-based											
Group A	Monochrome Black	11								6	17
Group B	Monochrome Orange Slipped	1			2				1		4
	Monochrome Orange Slipped and Incised						1				1
	Monochrome Red Slipped				2						2
	Total	1			4		1		1		7
Group C	Monochrome Orange Slipped			1				1			2
Group D	Polychrome Slipped and Painted	4									4
	Red-Orange-Black Resist		1								1
	Total	4	1								5

Table 8.9: Vessel style within the non-local fabric groups and classes (continued).

Fabric Group/Class	bowl	dish	drum	jar	pedestal base	stand	tripod bowl	tripod dish	vase	Total
Black on Red Slipped and Painted		3								3
Polychrome Slipped and Painted	2									2
Volcanic Glass										
Monochrome Orange Slipped	1								1	2
Monochrome Orange Slipped and Groove-Incised									3	3
Monochrome Orange Slipped and Incised	1									1
Polychrome Slipped and Painted	2						2		2	6
Total	4						2		6	12
Monochrome Red Slipped							1			1
Buff/Cream Slipped	1									1
Polychrome Slipped and Painted	1									1
Polychrome Slipped and Painted									1	1

conventions or the use of different raw material resources that occur within the same broad geological zone (e.g. the groups with geological links to coastal areas of northern Belize and southern Yucatan). The Quartz Sand class and group H of the Crystalline Calcite-Tempered class also contain vessels deriving from non-local raw materials.

In many cases, the compositional distinctiveness of the non-local fabric groups, as well as the petrographic differences among them, are mirrored in the stylistic characteristics of the vessels they contain (Table 8.9). For example, Sascab-Quartz B exclusively comprises red-slipped rounded dishes with black painted decoration, and the two vessels comprising the Sascab class are similarly decorated polychrome composite silhouette bowls. Similarly, the model-carved vessel representing group I of the Crystalline Calcite-Tempered class is stylistically unique within the assemblage as a whole. Also evident is that Monochrome orange vessels with groove-incising are confined to Volcanic Glass A, Red-Brown Striated storage jars are restricted to group G of the Crystalline Calcite-Tempered class and Monochrome Black vessels only occur in Dolomitic-Calcitic Marl-based A and Calcite E.

The provenance associations of the different non-local fabric groups are presented in Table 8.10. As can be observed, non-local ceramics within the Lamanai assemblage derive from several different areas within the lowland region and from different localities within northern Belize and the Yucatan. In addition, at least one vessel, represented by Volcanic Glass D, has geological connections to the Maya Mountains area. Due to the absence of diagnostic rock and mineral fragments in Volcanic Glass A, it is not possible to assign a provenance based on the geology of the associated raw materials. In addition, this group contains a range of different vessel styles, and so stylistic criteria provide no further insight. Nonetheless, the clay component of these fabrics is calcareous, suggesting a connection to an area underlain by limestone as the source of the clay, and igneous-related inclusions, which would suggest a connection to the Maya Mountains or Sierra Madres, are absent. In addition, these fabrics are clearly dissimilar to the fabric groups that are geologically compatible with clay resources located in northern Belize and southern Yucatan. It is well recognized that volcanic ash tempering is characteristic of Late and Terminal Classic fine wares at sites in Central Peten and much of this area is underlain by limestone. Accordingly, by process of elimination, it is possible, but not demonstrable geologically (based on the present body of evidence and information), that the vessels comprising Volcanic glass A originate from Central Peten.

When the functional and temporal associations of the non-local fabric groups are considered, three significant patterns emerge with regard to local consumption of foreign ceramics. These patterns can be considered to reflect changes over time in the kinds of foreign ceramic products entering the site and from which areas of the lowland region they originated. Firstly, the large majority of non-local vessels are drinking vessels and jars used for storing (and serving) liquids and other food stuffs (Figure 8.20). When considering the provenance associations of the different non-local fabric groups, it is also evident that fine ware vessels derive from a range of geographically widespread localities, whereas coarse ware jars derive mainly from northern Belize. Secondly, the frequency of foreign fine ware, including drinking vessels, decreases over time, whilst the frequency of

Table 8.10: Non-local fabric groups and their provenance associations.

Fabric Group/Class	Provenance Association based on Geological Characteristics
Crystalline Calcite-Tempered	
Group E	Areas adjacent to north east coast of northern Belize and in Yucatan, with stylistic links to Yucatan
Group F	Areas adjacent to north east coast of northern Belize and in Yucatan
Group G	Inland areas of northern Belize, likely surrounding or west of the New River
Group I	Unknown, but likely outside of northern Belize
Group J	Inland areas of northern Belize north of the New River Lagoon
Dolomitic-Calcitic Marl-based	
Group A	Coastal areas of northeastern eastern northern Belize - similarity to numerically dominant fabrics at Late Classic Altun Ha
Group B	Coastal areas of northeastern northern Belize, extending into southern Yucatan
Group C	Coastal areas of northeastern northern Belize, extending into southern Yucatan
Group D	Coastal areas of northeastern northern Belize, extending into southern Yucatan
Sascab-Quartz B	Corozal District likely in proximity to the Rio Hondo
Sascab	Inland areas of northern Belize
Volcanic Glass	
Group A	Outside of northern Belize/Yucatan, possibly Central Peten
Group B	Unknown but clearly non-local
Group C	Stylistic links to Yucatan
Group D	Maya Mountains with specific connections to Bladen Volcanic Member of the southern fringe
Group E	
Quartz Sand	Inland areas of northern Belize, specific links to areas east of the Northern Lagoon drainage system
Crystalline Calcite-Tempered H	Inland areas of northern Belize, specific links to areas east of the Northern Lagoon drainage system

coarse ware jars appears to increase (Figure 8.21). Even though these figures include the two groups containing both local and non-local coarse ware jars, an increase in non-local storage jars is clearly indicated. Thirdly, ceramics deriving from inland areas of northern Belize, coastal areas of north eastern northern Belize, extending into southern Yucatan, and with stylistic connections to the Yucatan area continue throughout the sequence (Figure 8.22). In contrast, vessels deriving from the Maya Mountains area and which potentially derive from Central Peten only occur in the earliest stratigraphic contexts. Although the sample sizes are admittedly small for some stratigraphic contexts, these general patterns within the sequence can be considered to be accurate, given the sampling strategy that was employed in the selection of vessels for thin section analysis. A major implication of these findings is that a reorientation of economic and social contacts towards the northern lowlands is not demonstrated, as has been suggested for sites in northern Belize (e.g. Masson and Mock 2004). Rather, the evidence indicates that these contacts were ongoing by the beginning of the sequence and simply continued into Early Postclassic period. This and other aspects of community level patterns in the consumption of foreign ceramics will be explored and elaborated upon in greater detail in Chapter 11.

8.6. Summary

The petrographic study of the Lamanai assemblage has revealed several significant patterns relating to local pot-making activities as well as the consumption of local and foreign pottery items. These patterns are summarized as follows:

1) Thin section analysis produced seven different fabric classes, representing general paste recipes characterized by the types of raw material ingredients involved, and 24 fabric groups that reflect geological differences related to raw material sources.

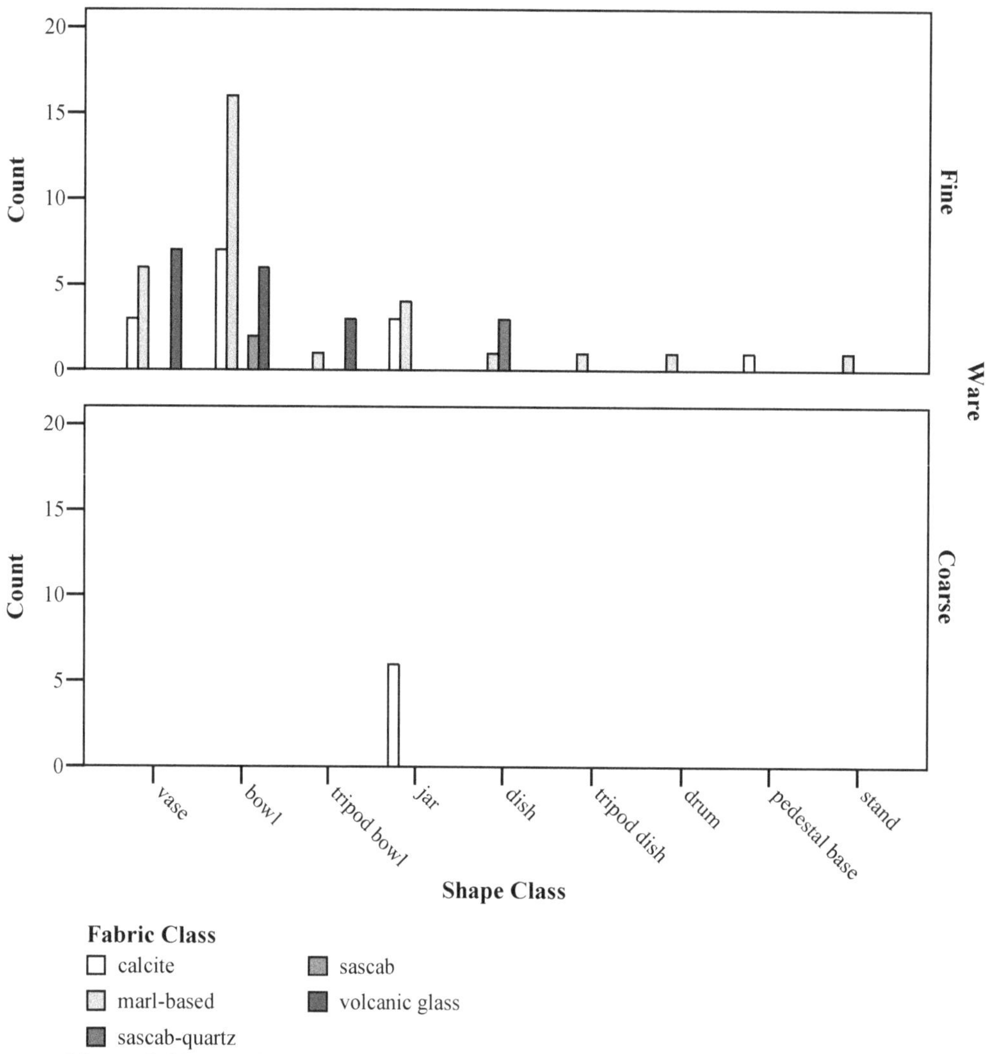

Figure 8.20: The frequency of vessel forms within the non-local fabric groups. (see colour version on accompanying CD)

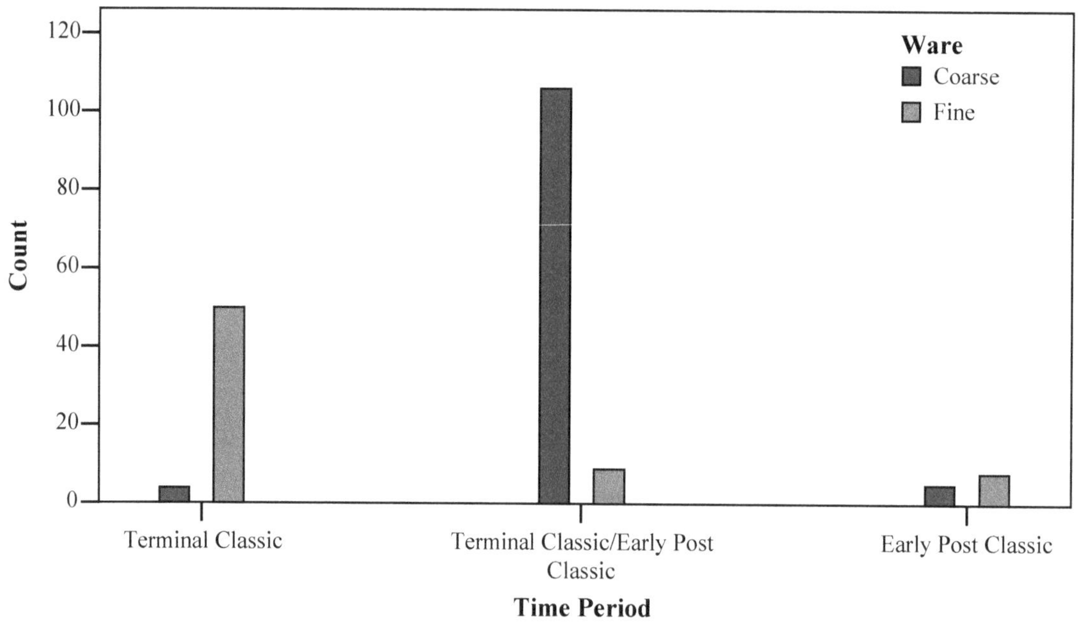

Figure 8.21: Temporal trends in the frequency of non-local fine ware and coarse ware vessels according to stylistic and contextual criteria (counts reflect totals for non-local groups and groups containing both local and non-local ceramics). (see colour version on accompanying CD)

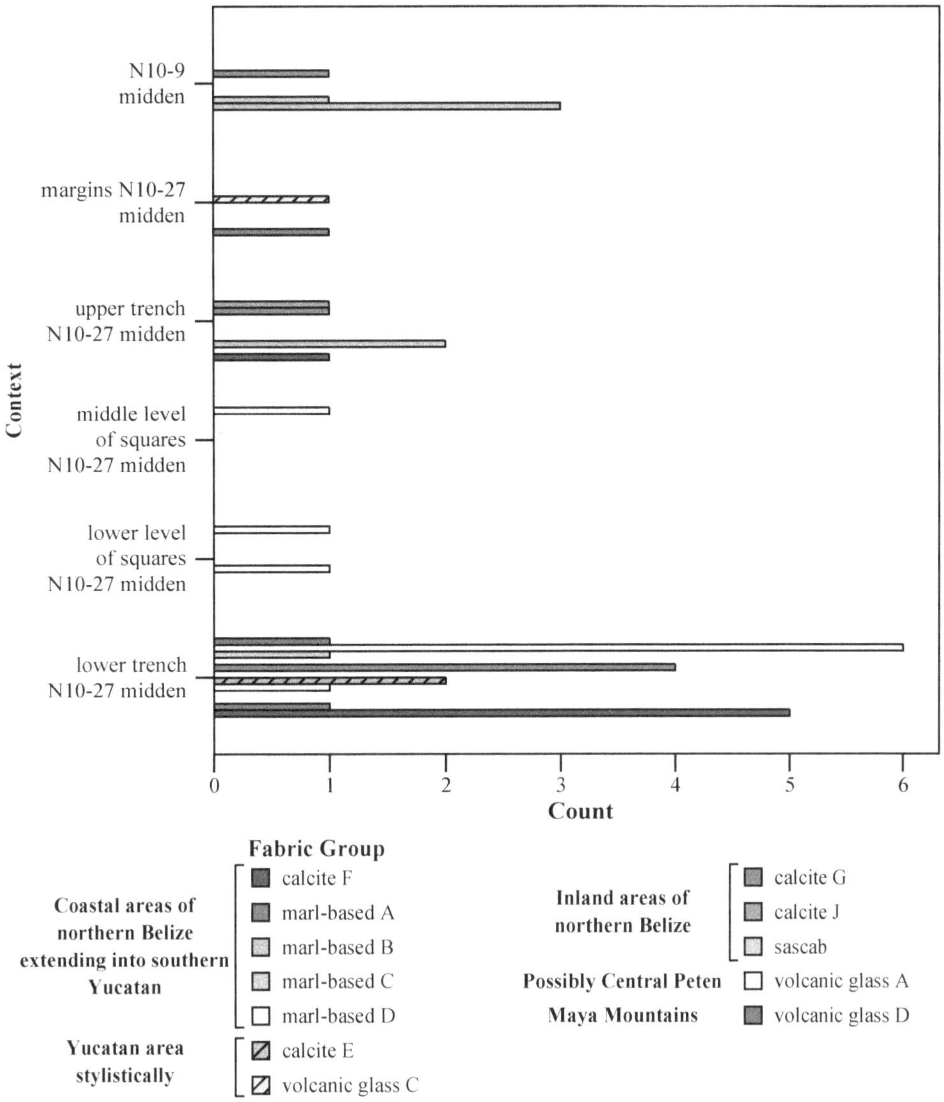

Figure 8.22: Stratigraphic associations of non-local groups summarized by provenance. (see colour version on accompanying CD)

2) The large majority of the ceramics within the archaeological contexts analyzed were manufactured locally.

3) Certain fabric classes clearly dominate the early and later ends of the sequence and class frequencies change over time. Nonetheless, a similar range of different fabric types, defined on the class level, occurs throughout the sequence, indicating continuity in the diversity of fabrics that occur over time.

4) There is a strong correlation between vessel composition at the class level and vessel function, indicating that vessels used in service and ritual tend to have different compositions than those serving broadly utilitarian functions. In addition, fine ware exhibits greater compositional diversity than coarse ware.

5) Four different local traditions of paste-making are represented. Within these traditions, Group C of the Crystalline Calcite-Tempered class and Sascab Quartz A reflect different contemporaneous approaches to fine ware manufacture that give way to the significantly different approach represented by the Grog-Mixed Carbonate Class. Groups A and B of the Crystalline Calcite Tempered tradition also reflect early approaches to vessel manufacture within the sequence, but both coarse ware and fine ware were made using these pastes. Group H of the Crystalline Calcite-Tempered tradition appears to reflect a developmental outgrowth of these

earlier approaches to calcite-tempered coarse ware manufacture. Occurring alongside the crystalline calcite-tempered tradition of coarse ware manufacture is the Quartz Sand class, which represents a different local tradition of coarse ware manufacture that is introduced into the sequence during the Terminal Classic period.

6) The multiple, internally consistent local approaches to fine ware paste-making develop into a single approach characterized by high internal variation. This latter approach, is comparatively complex, involving multiple tempering materials and a more careful or rigorous processing of the raw material ingredients.

7) Changes within the sequence with regard to local paste-making practices, particularly the use of grog temper, appear to have occurred independently of the changes in the stylistic characteristics of the vessels being produced.

8) Vessels that were not produced locally derive from several different areas within the lowland region. Ceramics that originate from the Maya Mountains and possibly Central Peten only occur with the earliest stratigraphic levels, whilst ceramics deriving from different localities in northern Belize and Yucatan occur throughout the sequence.

9) The large majority of foreign ceramics are drinking vessels and storage jars. The frequency of fine wares, including drinking vessels, decreases over time, whilst that of storage jars increases. In addition, fine ware vessels derive from geographically widespread localities within the lowland region, whilst storage jars derive mainly from northern Belize.

These patterns have several implications for current understanding of the nature of ceramic change and during the Classic to Postclassic transition, as well as the kinds of factors influencing distributional patterns and stylistic trends on both the local and regional levels. Their potential archaeological and behavioural significance, when considered in light of the broader setting of community activities within which the manufacture, use and deposition of ceramic items took place, will be explored in Chapter 11, which examines the human and material dimensions of continuity and change in the local ceramic record.

CHAPTER 9
CHEMICAL VARIABILITY WITHIN THE LAMANAI ASSEMBLAGE

Chemical analysis by neutron activation offers an independent means of investigating and characterizing compositional variability, generating complementary information to that generated through thin section petrography. Comparison of ceramic groups distinguished on the basis of chemical, petrographic and visual criteria provides important insight into the relationship between mineralogical, geochemical and stylistic aspects of ceramic variation. This information is fundamental to interpretation of ceramic groupings with regard to characterizing and discriminating their technological and provenance characteristics (Day et al. 1999).

Neutron activation analyses undertaken by Arnold and colleagues of ethnographic pottery and constituent raw materials from contemporary communities in Guatemala (Arnold et al. 1991) and the Yucatan Peninsula of Mexico (Arnold et al. 1999) have shown that the pottery produced by different communities exploiting similar raw material resources within a delimited 'resource area' can be distinguished chemically. Their Mexican case study is particularly significant as it demonstrates that different production centres located within close geographic proximity may produce distinct chemical profiles, even when they use they same clay. In this case, the chemical distinctiveness of the products of each community derives from the fact that they use tempering materials containing clay minerals, which they procure from their own local resource area (Arnold et al. 1999:81-82). Chemical analysis thus revealed the corporate behaviour of communities of potters.

In light of these findings, Arnold et al. suggest that it should be possible to recognize source-related chemical subgroups in the analysis of archaeological pottery, even when the pottery derives from multiple sources within a broadly geologically homogeneous environment. They recommend that such a study might begin by inspecting the fine-structuring of chemical groups produced through cluster or principal components analysis, searching for subgroups within the main groups identified.

The potential to differentiate pottery with similar mineralogical compositions through chemical analysis is of particular relevance to the present study since the potential problems posed by a geologically homogeneous source area reside on two different scales. Firstly, ceramics identified as 'locally manufactured' through thin section analysis, derive from different combinations of local clays and temper materials but the associated deposits are proximate and, in some cases, potentially adjoining. Secondly, the large majority of the pottery identified as 'non-local' is mineralogically similar to the local ceramics, deriving from the same broad source area underlain by limestone. An issue that is central to the present study, therefore, is whether it is possible to discriminate mineralogically similar pottery deriving from a single site on these two scales of source-related differentiation. More specifically, is it possible to chemically distinguish the products of local potters working within different paste-making traditions, as well as the products of different non-local production centres, given their similar mineralogy and the fact that similar approaches to paste-making or 'paste recipes' are sometimes employed? As Day et al. (1999) have shown, it is sometimes impossible to attribute chemical variability to either technology or provenance, especially when great similarity exists between raw materials found in different locations and approaches to paste preparation are akin (also see Hein et al. 2004a).

With these fundamental issues in mind, the chemical compositional analysis aimed to address the following questions:

How do the compositional groups identified through chemical analysis correspond with those produced by thin-section analysis?

Does chemical analysis discriminate the local and non-local fabric groups identified in the previous chapter?

Does chemical analysis also discriminate the different local paste-making traditions suggested in the previous chapter?

How does the chemistry of local clays compare to that of the pottery identified as 'locally produced' based on petrographic criteria?

Is it possible to link the local fabric groups identified in the previous chapter to extant clay resources?

We will return to these questions at the end of the chapter after the results of the statistical analysis of the chemical data have been discussed. The analytical strategy recommended by Arnold et al. (1999) has been employed to discriminate groups and subgroups potentially reflecting chemical differences related to source. It should be noted, however, that calcium was included in both the cluster and principal component analysis. Although the authors recommend that calcium and strontium should be discounted due their diluting/enrichment effect, its removal during analyses of the Lamanai data set did not alter group membership or the chemical relationships between the groups. Due to the high overall variability of the present data set, the groups derived from cluster analysis were evaluated on the basis of their internal variability. As a measure of this internal variability, the standard deviation of the mean value was used.

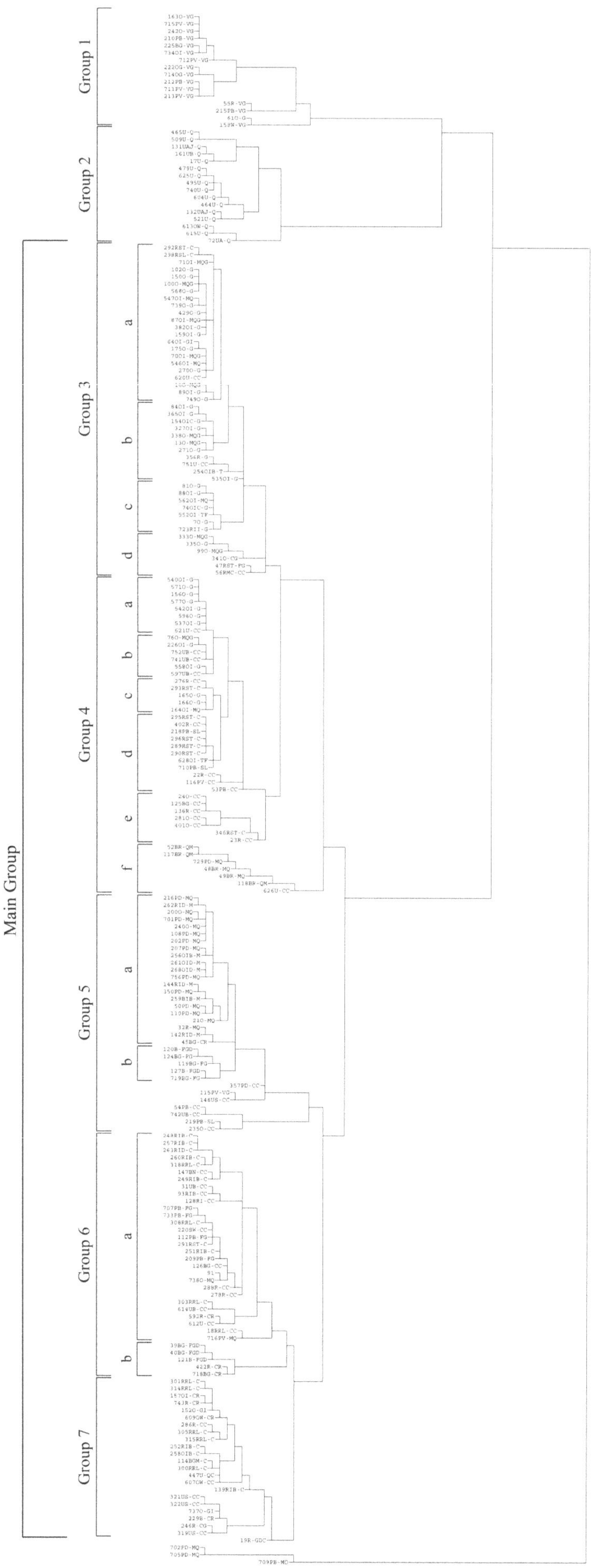

Figure 9.1: Dendrogram of the Lamanai data set showing the three principal clusters, and the chemical groups and subgroups. (see larger scale version on accompanying CD)

9.1. BASIC STRUCTURE OF THE DATA SET

As can be seen in the dendrogram presented in Figure 9.1, cluster analysis of the ceramic samples produced a main cluster containing the large majority of the samples analysed and two smaller clusters (Groups 1 and 2). The main cluster is separable into at least five groups (Groups 3-7), and with the exception of Group 7, each of these groups is separable into two to several chemical 'sub-clusters' or subgroups.

The average element values and standard deviations (actual and in %) for the seven chemical groups are given in Table 9.1. It can be observed that for each group the standard deviations of the concentrations of all of the elements are quite high, generally well above 15%. In fact, Group 5 is the only group for which multiple elements have standard deviations below 15% (e.g. Sm, Lu, Eu, Fe, Th and Sc). Accordingly, overall variability within the data set is high and the emergent chemical groups are characterized by high internal variability.

The removal of chemical outliers, containing exceedingly high or low concentrations (not necessarily in one direction) for three or more elements, from the ends of the group clusters did not significantly reduce the internal variation within individual groups, except in the case of Group 1. The removal of four outliers from the end of this group reduced the standard deviations of the majority of elements to below 15%, and to below 10% for Cr, Hf, Fe, and Eu (Table 9.1). These outliers can be considered as chemically related to the remaining samples comprising Group 1, but their chemical profiles are significantly different in important respects. For example, samples 55 and 158 contain comparatively high concentrations of La, Ce, Th, Sc and Fe, whereas sample 215 contains comparatively low concentrations of La and Ce, but high concentrations of Sc and Fe (see Appendix VI). Sample 610 is separated from the main group due to its low Na content. The different concentrations for the rare earth elements, in particular, (i.e. La, Ce and Th) most likely reflect natural chemical differences in the clay component of these pastes (see Glascock et al. 2004 and Bishop et al. 1982 for a discussion).

With the exception of Groups 1 and 5, the within-group variation in element concentrations greatly exceeds the acceptable range of below 10-15 %, which is the conventional 'benchmark' that is considered to denote a coherent chemical group or profile. In most cases, however, inspection of the fine-structuring of each group resolves them into subgroups that exhibit an acceptable range of variation for most elements. The exceptions are Groups 2 and 7, which are not separable into relatively chemically homogenous subgroups.

The identified subgroups within Groups 3, 4, 5 and 6 of the main cluster can be considered to represent chemically discrete groupings within these larger groups, since differences in the concentrations of particular elements are apparent even when standard deviations across the subgroups are taken into account. As will be discussed below (Section 9.4), in certain cases these chemical distinctions can be related to differences in 'paste recipes' among locally *and* non-locally produced ceramics, when stylistic and petrographic information is considered. In other cases, however, such connections are less than clear.

The stylistic and petrographic characteristics of the vessels comprising each group and their component subgroups, when present, are summarized in Tables 9.2 to 9.8. In each case, chemical outliers are marked with an asterisk. Since their removal from the respective groups or subgroups does not significantly decrease within-group variation in element concentrations, they have been included in the group summaries. These outliers were removed for the principal components analysis, however, to achieve a better separation of the groups and subgroups in the component plots.

9.2. CHEMICAL CHARACTERISTICS OF THE IDENTIFIED GROUPS AND THEIR PETROGRAPHIC AND STYLISTIC CORRELATES

Group 1 - Group 1 is the smallest of the groups identified, containing 12 samples, not including the four outliers. The samples within this group contain comparatively high concentrations of Na, Cs, and Rb and comparatively low Ca (Table 9.1). This group exclusively comprises non-local polychrome and plain and groove-incised orange slipped bowls and vases with Volcanic Glass A fabrics (Table 9.2). Perhaps not surprisingly, three of the four outliers within Group 1 also derive from non-local ash-tempered vessels, corresponding to the Volcanic Glass groups B, C and D. As was discussed in the previous chapter, these fabric groups are represented by only one sample each and are clearly differentiated from Group A fabrics on mineralogical grounds. Hence, the compositional differences observed among the volcanic ash-tempered fabrics on the microscopic level are also reflected in their chemistry. The high concentrations of the alkali metals Na, Rb and Cs in Group 1 derive from the fact that these pastes contain an appreciable quantity of alkali feldspar, which is naturally present in the ash or tuff used for temper.

Group 2 - Group 2, comprising 16 samples, is also a comparatively small chemical group within the data set. It is separated due to comparatively high concentrations of the rare earth elements Sm, Lu, Yb, La and Ce, as well as Hf, and by low Ca concentrations (Table 9.1). This group almost exclusively contains utilitarian vessels made with sandy clays deriving from areas of Pleistocene alluvium (Quartz Sand fabrics) (Table 9.3). The exception is the incense burner with the same fabric type. This group's distinct chemical profile, especially with regard to the rare earth element concentrations, indicates significant chemical differences in the clay component of this

Table 9.1: A summary of the average compositions and standard deviations (actual and percent) for the main chemical groups identified. Concentrations are given in ppm apart from Ca, Na and Fe. High element values are marked in light grey and low values are marked in dark grey

	Group 1			Group 1 without Outliers			Group 2			Group 3			Group 4			Group 5			Group 6			Group 7		
	Ave. Value	S.D.	% S.D.	Ave. Value	S.D.	% S.D.	Ave. Value	S.D.	% S.D.	Ave. Value	S.D.	% S.D.	Ave. Value	S.D.	% S.D.	Ave. Value	S.D.	% S.D.	Ave. Value	S.D.	% S.D.	Ave. Value	S.D.	% S.D.
Sm	5.84	1.05	17.91	5.7	0.99	17.39	7.2	1.85	25.73	6.57	1.32	20.08	5.03	1.53	30.41	2.73	0.4	14.72	3.31	0.77	23.18	5.45	1.48	27.21
Lu	0.41	0.07	16.79	0.39	0.05	13.96	0.47	0.08	17.83	0.43	0.07	15.44	0.33	0.07	20.64	0.25	0.03	13.57	0.25	0.06	22.25	0.33	0.06	19.13
U	3.25	1.34	41.28	2.91	0.95	32.5	3.25	0.93	28.62	3.02	1.56	51.73	2.86	1.65	57.71	1.8	0.28	15.4	1.15	0.37	32.13	1.56	0.57	36.48
Yb	3.28	0.55	16.71	3.13	0.45	14.51	3.8	0.77	20.25	3.52	0.61	17.38	2.6	0.63	24.15	2.01	0.28	14.08	2.07	0.51	24.72	2.83	0.55	19.47
As	8.52	2.46	28.89	8.05	1.48	18.4	2.81	1.27	45.33	4.48	2.71	60.46	3.53	2.57	72.75	2.29	1.17	51.16	2.79	2.12	76.19	4.06	2.52	61.97
Sb	1.5	0.35	23.34	1.38	0.2	14.67	1.07	0.12	11.23	1.16	0.25	21.47	1.03	0.24	23.42	0.96	0.07	7.2	0.96	0.26	27.6	1.06	0.32	30.7
Ca	6.05	2.63	43.48	7.12	1.7	23.86	1.23	0.37	30.3	13.05	3.77	28.9	18.23	5.99	32.83	9.88	3	30.39	22.12	4.53	20.46	19.56	3.45	17.63
Na	1.06	0.25	23.38	1.13	0.15	13.03	0.14	0.07	48.73	0.14	0.04	30.6	0.13	0.06	50.96	0.13	0.04	27.32	0.11	0.04	36.98	0.13	0.04	27.99
La	30.1	5.82	19.33	29.03	4.3	14.81	36.37	9.64	26.51	33.71	6.86	20.35	26.09	9.09	34.84	16.25	2.98	18.31	16.09	3.55	22.06	28.85	8.33	28.87
Ce	65.72	12.62	19.21	63.07	9.05	14.35	81.99	22.54	27.49	80.92	31.11	38.44	55.88	18.13	32.45	34.62	5.48	15.84	38.14	9.59	25.15	72.03	18.35	25.48
Th	12.54	3.91	31.17	11.44	3.01	26.28	12.7	1.83	14.39	13.15	2.33	17.75	10.05	2.31	22.98	10.65	0.99	9.32	8.38	1.98	23.69	10.89	3.11	28.52
Cr	37.37	20.47	54.78	29.96	2.64	8.8	55.73	9.81	17.61	59.99	12.99	21.65	58.78	17.09	29.08	42.84	6.8	15.88	47.77	14.26	29.85	55.05	12.58	22.85
Hf	5.65	0.75	13.25	5.27	0.35	6.67	8	1.51	18.87	6.74	1.51	22.45	5.06	2.37	46.86	7.98	1.19	14.92	4.84	0.94	19.44	6.68	1.92	28.68
Cs	8.78	2.95	33.57	8.83	3.2	36.25	7.24	3.33	46	5.7	2.24	39.34	4.95	2.56	51.73	3.79	1.44	38.06	1.38	0.61	43.97	1.41	0.39	27.32
Tb	0.73	0.14	18.95	0.7	0.14	19.37	0.9	0.2	22.66	0.83	0.17	20.62	0.62	0.18	29.43	0.38	0.06	14.55	0.45	0.12	26.57	0.69	0.17	24.87
Sc	9.13	2.91	31.91	7.87	0.92	11.74	11.29	2.71	23.96	11.75	2.04	17.38	10.7	2.26	21.12	8.62	0.7	8.08	7.76	1.33	17.09	9.06	1.68	18.55
Rb	138.95	32.28	23.23	142.68	35.86	25.13	68.2	31.89	46.76	57.47	17.7	30.81	55.1	24.05	43.64	36.16	8.96	24.78	25.78	10.44	40.52	32.69	13.91	42.55
Fe	2.17	0.53	24.32	1.93	0.17	8.61	2.37	0.41	17.2	2.65	0.48	18.2	2.09	0.43	20.46	1.98	0.21	10.77	1.66	0.31	18.67	2	0.42	21.15
Ta	1.07	0.29	26.8	0.98	0.22	22.56	1.56	0.28	17.71	1.38	0.3	22.12	0.92	0.19	21.09	1.15	0.18	15.31	0.72	0.11	15.13	0.94	0.15	16.11
Co	5.91	1.23	20.73	5.5	0.63	11.53	5.83	1.43	24.43	6.07	1.7	28.05	4.19	0.75	17.84	4.32	0.9	20.94	2.78	0.72	25.92	3.72	1.23	33
Eu	0.89	0.18	20.21	0.82	0.06	7.45	1.31	0.38	29.48	1.2	0.27	22.19	0.9	0.29	32.45	0.48	0.07	14.07	0.63	0.15	24.68	1.02	0.3	28.91

Table 9.2: Stylistic and petrographic characteristics of the vessels comprising Group 1. Chemical outliers are marked with an asterisk.

Stylistic Categories Represented	Compositional Categories Represented	No.
Polychrome Slipped and Painted (*vases, flaring bowls, deep bowl, tripod deep bowls*)	Volcanic Glass Group A	6
Monochrome Orange Slipped and Groove-Incised (*vases, rounded bowl*)		4
Monochrome Orange Slipped (*vase, flaring bowl*)		2
*Buff/Cream Slipped (plain) - slate ware (*rounded bowl*)	Volcanic Glass Group C	1
*Monochrome Red Slipped (*flaring tripod bowl*)	Volcanic Glass Group B	1
*Polychrome Slipped and Painted (*flaring bowl*)	Volcanic Glass Group D	1
*Monochrome Orange Slipped (*rounded bow*l)	Grog-Mixed Carbonate Class	1
	Total	**12** + **4OL**

Table 9.3: Stylistic and petrographic characteristics of the vessels comprising Group 2. Chemical outliers are marked with an asterisk.

Stylistic Categories Represented	Compositional Categories Represented	No.
Unslipped Utilitarian (*=1 *jars, plate/lids, comal,* * *pedestal-based censer*)	Quartz Sand Class	11
Unslipped and Incised (appliqué impressed fillet) (*strap handle jars*)		3
Unslipped/Smoothed – Perforated (*incense burner*)		1
*Washed/Smoothed (*jar*)		1
	Total	**16**

utilitarian pottery. The high levels of Hf most likely reflect the presence of zircon in the clay, occurring as aplastic inclusions, which are too small in size to permit identification through thin section analysis (c.f. Buxeda et al. 2003a). The presence of zircon might be expected given the siliceous mineralogy of the aplastic component of these fabrics (V. Kilikoglou, personal communication 2004). Whilst the low Ca content of the samples comprising Group 1 also relates, in part, to the geological parentage of these particular clays, which are associated with deep deposits of siliceous sand deriving ultimately from the Maya Mountains, it also reflects the fact that these pastes do not contain carbonates. Interestingly, discrete chemical subgroups are not apparent within Group 2, even though thin section analysis suggested that different Pleistocene-alluvium-related clay resources are represented within the Quartz Sand class. Nonetheless, the results of the chemical analysis would seem to be in keeping with the fact that mineralogical variation within the Quartz Sand class is continuous rather than discrete.

The Main Cluster -

GROUP 3

Group 3 is the largest of the five chemical groups that form the main cluster within the dendrogram. This group contains 46 samples and is separable into four sub-groups (3A, 3B, 3C and 3D) of varying size. Subgroup 3A is the largest, containing 22 samples, and 3D, with just 6 members, is the smallest. On the whole, Group 3 is characterized by high mean values for Cs, Th, Cr, Sc and Fe, although the concentrations of these elements vary considerably (see the standard deviations listed in Table 9.1). The chemical differences apparent among the subgroups comprising Group 3 relate to the relative concentrations of particular rare earth elements. For example, the samples comprising subgroup A contain lower concentrations of La and Eu than those comprising subgroup D, even when taking into account the standard deviations of these elements for both subgroups. Similarly, subgroup C is clearly differentiated from subgroups A and B by comparatively high Th levels.

With few exceptions, Group 3, as well as its component subgroups, predominantly consists of grog-tempered Monochrome Orange Slipped and Monochrome Orange Slipped and Incised vessels (Table 9.4). Vessel forms are

Table 9.4: Stylistic and petrographic characteristics of the vessels comprising Group 3. Chemical outliers are marked with an asterisk.

	Stylistic Categories Represented	Compositional Categories Represented	No.
Subgroup A	Monochrome Orange Slipped and Incised (*jar rounded, composite silhouette and out-curving bowls, tripod bowls with segmented basal flange, tetrapod bowl, pedestal-based jars (censer)*)	Grog-Mixed Carbonate Class (main group & subgroup A) (N=7) Sascab-Quartz A subgroup (grog-tempered variant) (N=3) Crystalline Calcite-Tempered C subgroup (grog tempered variant) (N=1)	11
	Monochrome Orange Slipped (*various jar forms, drum, chalice, composite silhouette and out-curving bowls*)	Grog-Mixed Carbonate Class (N=7) Sascab-Quartz A subgroup (grog-tempered variant) (N=1)	8
	Unslipped/Smoothed-Buff (*jar*)	Crystalline Calcite-Tempered Group H	1
	*Red-Orange-Black Resist (*composite silhouette dish*)	Crystalline Calcite-Tempered Group C	1
	*Slipped Rim/Lip - Smoothed Body (*jar*)	Crystalline Calcite-Tempered Group D	1
		Total	**22**
Subgroup B	Monochrome Orange Slipped and Incised (*chalices, drum, rounded bowl, tripod dish*)	Grog-Mixed Carbonate Class (main group)	5
	Monochrome Orange Slipped (*[=1]*chalices, out-curving tripod bowl*)	*Grog-Mixed Carbonate Class (main group) (N=1) Sascab-Quartz A subgroup (grog-tempered variant) (N=2)	3
	*Monochrome Orange Slipped (interior slipped) (*rounded bowl*)	Crystalline Calcite-Tempered Group B subgroup (grog tempered variant)	1
	*Monochrome Red Slipped (*composite silhouette bowl*)	Grog-Mixed Carbonate Class (main group)	1
	Unslipped/Smoothed-Buff (*jar*)	Crystalline Calcite-Tempered Group G	1
		Total	**11**
Subgroup C	Monochrome Orange Slipped and Incised (*chalice, tripod bowl with segmented basal flange, pedestal-based jar (censer)*)	Grog-Mixed Carbonate Class (N=2) Sascab-Quartz A subgroup (grog-tempered variant) (N=1)	3
	Monochrome Orange Slipped (*rounded bowl with strap handles, out-curving dish*)	Grog-Mixed Carbonate Class (main group)	2
	*Monchrome Orange to Red Slipped - Notched and Incised (*composite silhouette bowl*)		2
		Total	**7**
Subgroup D	Monochrome Orange Slipped (*chalices, out-curving tripod dish*)	Grog-Mixed Carbonate Class (main group) (N=2) Sascab-Quartz A subgroup (grog-tempered variant) (N=2)	4
	*Red-Orange-Black Resist (*out-curving dish*)	Dolomitic-Calcitic Marl-based Group D	1
	*Monochrome Red Slipped and Model-Carved (*flaring bowl*)	Crystalline Calcite-Tempered Group I	1
		Total	**6**

typically Early Postclassic in style, with a similar range of forms occurring within each subgroup. In addition, the large majority of samples within each subgroup correspond to the Grog-Mixed Carbonate class, although the grog tempered variants of the Sascab Quartz A fabric group and groups B and C of the Crystalline Calcite-Tempered class are also represented. In the case of Group 3, therefore, there is also a strong correspondence with particular stylistic and petrographic categories on the group level. Further distinctions among the vessels that occur within the subgroups, however, are not apparent.

GROUP 4

Group 4 is the second largest of the groups comprising the main cluster, containing 43 samples. On the whole, this group is characterized by a comparatively high average value for Ca and a comparatively low average value for Hf, but with marked variability in both cases (Table 9.1). Group 4 is separable into six discrete chemical subgroups, which are represented by less than 11 samples in each case. The differences among the subgroups are reflected in their Eu content in particular. In addition, subgroup F is clearly distinguished by higher concentrations of Sm, Lu, Yb, La and Hf. The higher concentrations of some of the rare earth elements in this subgroup most likely relates to natural differences in the clay component of these particular samples. As with Group 2, higher Hf levels may reflect the presence of zircon, which is in keeping with the siliceous mineralogy of the aplastic inclusions within these fabrics (see below). The chemical distinctiveness of subgroup F is reflected in the vessels it contains, with five of seven samples deriving from Black on Red vessels with sandy sascab-tempered fabrics (Sascab-Quartz class) (Table 9.5). Significantly, both this stylistic category and the Sascab-Quartz B fabric group are confined to this particular chemical subgroup, being entirely absent in any of the other main chemical groups. In comparison, subgroups D and E, with only two exceptions, consist of vessels with crystalline calcite-tempered fabrics and these are mainly monochrome and 'resist' composite silhouette dishes and polychrome composite silhouette bowls (subgroup D only). Subgroup C also contains composite silhouette dishes with calcite-tempered fabrics, but plain and incised orange-slipped vessels with grog tempered fabrics also occur. These later vessels, representing typical Early Postclassic forms and paste compositions, are stylistically and petrographically similar to those that occur in Group 3. The same orange-slipped, grog-tempered vessels also occur in subgroups A and B of Group 4. In both cases, however, they are accompanied by unslipped utilitarian vessels with calcite-tempered fabrics, primarily large jars that conform to later jar styles within the ceramic sequence. It is significant that, with the exception of subgroup F, Group 4 exclusively comprises locally manufactured vessels, but both early and late vessel styles and paste compositions within the ceramic sequence are represented.

The chemical separation of subgroup F and its placement virtually in the middle of the main cluster of samples requires further comment, given the provenance associations of some of the Black on Red vessels it contains. As will be recalled, the Sascab-Quartz B fabric group, which exclusively comprises Black on Red rounded dishes, was ascribed a non-local provenance on mineralogical grounds. In comparison, the vessels comprising the other subgroups within Group 4, as well as the chemical group that directly follows 4F in the dendrogram (i.e. 5A), correspond to fabric groups that were ascribed a local provenance. The occurrence of these non-local fabrics both within a subgroup that also contains local fabrics and between chemical groupings that consist of local fabrics (albeit corresponding to different fabric classes) is interesting. The positioning of the Sascab-Quartz B fabric group within the dendrogram might be partly explained by the fact that it occurs amidst samples corresponding to the Sascab-Quartz A fabric group: that is fabrics belonging to the same general fabric class. These two fabric groups share in common a particular paste recipe involving a sandy calcareous clay tempered with sascab. Hence, the close chemical relationship between Sascab A and Sascab B fabrics may reflect a fundamental similarity in the approach to paste making – i.e. the use of the same general kind of clay and temper – whereas the clear chemical separation of Sascab-Quartz B fabrics from the main group of Sascab-Quartz A fabrics (5A) indicates inherent differences related to the specific source of the raw materials used (perhaps specifically with regard to the clay component, as indicated by the different concentrations of the rare earth elements Sm, Lu, Yb, and La for 4F and 5A).

GROUP 5

Group 5 contains 25 samples and is separable into two subgroups (5A and 5B), with the large majority of samples occurring in subgroup A (Table 9.6). On the whole, Group 5 is characterized by comparatively low average concentrations of many of the rare earth elements, including Sm, Lu, Yb, La, Ce, and Eu, and comparatively high Hf, levels (Table 9.1). The two subgroups are separated due to differences in their Na, Co and Cr content, with 5A containing higher levels of Na and Co and lower levels of Cr. The chemical differences between the subgroups are mirrored in the kinds of vessels they contain. Subgroup A consists primarily of monochrome and polychrome rounded dishes and monochrome rounded bowls with sandy sascab-tempered fabrics. From a petrographic standpoint, subgroup A corresponds to the main group of Sascab-Quartz A fabrics, which was ascribed a local provenance (Table 9.6). In contrast, subgroup B consists of monochrome black deep bowls and vases with dolomitic marl-based fabrics corresponding to the non-local Dolomitic-Calcitic Marl-based A fabric group. In both cases, however, the chemical subgroups correspond to early vessel styles and paste compositions within the sequence.

As with subgroup F of Group 4, it is interesting that 5B, corresponding to a fabric group ascribed a non-local provenance based on mineralogical characteristics, occurs within the same chemical group as locally produced vessels, albeit forming a distinct chemical subgroup. It is also noteworthy that 5A comprises the large majority of

Table 9.5: Stylistic and petrographic characteristics of the vessels comprising Group 4. Chemical outliers are marked with an asterisk.

	Stylistic Categories Represented	Compositional Categories Represented	No.
Subgroup A	Monochrome Orange Slipped (*chalice, composite vessel and rounded bowl, plate/lid*)	Grog-Mixed Carbonate Class (main group) (N=3) Sascab-Quartz A subgroup (grog-tempered variant) (N=1)	4
	Monochrome Orange Slipped and Incised (*tripod bowls with segmented basal flange, chile grinder*)	Grog-Mixed Carbonate Class (main group)	3
	Unslipped/Smoothed-Buff (*jar*)	Crystalline Calcite-Tempered Group H	1
		Total	**8**
Subgroup B	Unslipped/Smoothed-Buff (*jars, plate/lid*)	Crystalline Calcite-Tempered Group A (N=1) Crystalline Calcite-Tempered Group B (N=2)	3
	Monochrome Orange Slipped and Incised (*chalice, out-curving bowl*)	Grog-Mixed Carbonate Class (main group)	2
	Monochrome Orange Slipped (*chalice*)	Sascab-Quartz A subgroup (grog-tempered variant)	1
		Total	**5**
Subgroup C	Monochrome Orange Slipped and Incised (*chalice, frying pan censer*)	Grog-Mixed Carbonate Class (main group) (N=1) Sascab-Quartz A subgroup (grog-tempered variant) (N=1)	2
	Monochrome Orange Slipped (*drum*)	Grog-Mixed Carbonate Class (main group)	1
	Monochrome Red Slipped (*composite silhouette dish*)	Crystalline Calcite-Tempered Group C	1
	Red-Orange-Black Resist (*composite silhouette dish)*		1
		Total	**5**
Subgroup D	Red-Orange-Black Resist (*composite silhouette dish*)	Crystalline Calcite-Tempered Group C	4
	Polychrome Slipped and Painted *(*[=1]composite silhouette bowls*)	Sascab-Quartz A (main group) (N=2) *Crystalline Calcite-Tempered Group C(N=1)	3
	Monochrome Red Slipped (**[=1]composite silhouette dish*)	Crystalline Calcite-Tempered Group C	2
	Monochrome Orange Slipped and Incised (*jar*)		1
	*Monochrome Red Slipped, Stuccoed and Painted (*vase*)		1
		Total	**11**
Subgroup E	Monochrome Orange Slipped (*composite silhouette dishes and a bowl*)	Crystalline Calcite-Tempered Group C (N=2) Crystalline Calcite-Tempered Group D (N=1)	3
	Monochrome Red Slipped (**[=1]composite silhouette dishes*)	Crystalline Calcite-Tempered Group C	2
	Monochrome Black (groove-incised and punctated) (*vase*)		1
	*Red-Orange-Black Resist (*composite silhouette dish*)		1
		Total	**7**
Subgroup F	Black on Red Slipped and Painted (*rounded dishes, massive bowl*)	Sascab-Quartz A (main group) (N=2) Sascab-Quartz B (N=3)	5
	Polychrome Slipped and Painted (*composite silhouette dish*)	Sascab-Quartz A (main group)	1
	*Unslipped/Smoothed-Buff (*jar*)	Crystalline Calcite-Tempered Group B	1
		Total	**7**

Table 9.6: Stylistic and petrographic characteristics of the vessels comprising Group 5. Chemical outliers are marked with an asterisk.

	Stylistic Categories Represented	Compositional Categories Represented	No.
Subgroup A	Polychrome Slipped and Painted (interior slipped) (*rounded dishes*)	Sascab-Quartz A (main group)	8
	Monochrome Red Slipped (interior slipped) (*rounded dishes and a bowl*)		4
	Monochrome Orange Slipped (interior slipped) (*rounded dishes and a bowl*)		3
	Monochrome Orange Slipped (*rounded bowl*)		1
	Polychrome Slipped and Painted (*composite silhouette dish*)		1
	Buff/Cream Slipped (plain) (*rounded bowl*)		1
	Monochrome Orange Slipped (*composite silhouette dish*)		1
	Monochrome Black (*rounded bowl*)	Dolomitic-Calcitic Marl-based Group A (subgroup)	1
		Total	**20**
Subgroup B	Monochrome Black - plain and decorated (*deep bowls, vases*)	Dolomitic-Calcitic Marl-based Group A	5
		Total	**5**

vessels analysed from the main group of Sascab-Quartz A fabrics (i.e. the main group of vessels not containing grog temper), but the grog-tempered variant of this fabric group is entirely absent, occurring primarily in Group 3.

GROUP 6

Group 6 contains 33 samples and comprises 2 distinct chemical subgroups (6A and 6B). Subgroup 6A is the largest of the two, containing 28 samples. As with Group 5, Group 6 is characterized generally by comparatively low average concentrations for many of the rare earth elements, including Sm, Lu, Yb, La, Ce, and Eu (Table 9.1). In Group 6, however, the average values of Hf, Cs and Co are also comparatively low and average Ca content is very high, but variability is high in all cases. Subgroup 6A separates chemically from 6B due to higher concentrations of the rare earth elements Sm, Yb, La, and Eu, indicating an important compositional difference in the clay component of thesamples comprising the two subgroups. It is also significant that the samples comprising Subgroup 6B have a lower Ca content, which is in keeping with the fact that this subgroup nearly exclusively consists of vessels with dolomitic marl-based fabrics, corresponding to Dolomitic-Calcitic Marl-based group A (Table 9.7). These vessels are stylistically and petrographically indistinguishable from the Monochrome Black vessels that comprise subgroup B of Group 5. Hence, the separation of these vessels into subgroups of two different chemical groups is rather interesting, especially considering the high level of stylistic and petrographic homogeneity they exhibit.

In comparison, the samples comprising subgroup A predominantly derive from calcite-tempered vessels, with various local and non-local fabric groups of the Crystalline Calcite-tempered class represented. This internal variation in the fabric groups that occur in subgroup A is mirrored in the diversity of vessel styles it contains, with stylistically different fine ware *and* coarse ware vessels present. It is interesting that 6A corresponds to a range of local and non-local fabrics groups and that the non-local fabric groups represent two distinct approaches to paste-making, as reflected by the Crystalline Calcite-Tempered and the Dolomitic-Calcitic Marl-based fabric classes. Nonetheless, subgroup A can be considered as strongly associated with local calcite-tempered vessels, with the majority of samples deriving from groups A, B and C of the Crystalline Calcite Tempered Class.

GROUP 7

Cluster 7 includes 22 samples and chemical subgroups are not apparent. Positioned at the end of the main cluster, this chemical group exhibits the highest level of

Table 9.7: Stylistic and petrographic characteristics of the vessels comprising Group 6. Chemical outliers are marked with an asterisk.

	Stylistic Categories Represented	Compositional Categories Represented	No.
Subgroup A	Monochrome Red Slipped (interior slipped) (*rounded bowls*)	Crystalline Calcite-Tempered Group A (N=3) Crystalline Calcite-Tempered Group B (N=1) Crystalline Calcite-Tempered Group C (N=2) Crystalline Calcite-Tempered Group D (N=1)	7
	Polychrome Slipped and Painted (**vase, composite silhouette bowls*)	*Sascab-Quartz A (main group) (N=1) Dolomitic-Calcitic Marl-based Group D (N=4)	5
	Monochrome Red Slipped (*$^{=2}$*jars, composite silhouette dishes*)	*Crystalline Calcite-Tempered Group C (N=2) Crystalline Calcite-Tempered Group D (N=1) *Dolomitic-Calcitic Marl-based Group B (N=1)	4
	*$^{=1}$Slipped Rim/Lip - Smoothed Body (*jars*)	Crystalline Calcite-Tempered Group A (N=1) *$^{=1}$Crystalline Calcite-Tempered Group B (N=2)	3
	*Unslipped/Smoothed-Buff (*jars*)	Crystalline Calcite-Tempered Group H	2
	Monochrome Brown (*torch*)		1
	Buff/Cream Slipped (plain) (*rounded dish*)	Crystalline Calcite-Tempered Group A	1
	Monochrome Red Slipped and Incised (*vase*)		1
	Monochrome Black (raised band with a central groove) (*vase*)	Crystalline Calcite-Tempered Group C	1
	Red-Orange-Black Resist (*composite silhouette dish*)		1
	Monochrome Grey (Plumbate ware) (*pedestal base*)	Crystalline Calcite-Tempered Group E	1
	Monochrome Orange Slipped (*jar with strap handles*)	Sascab-Quartz A - main group	1
		Total	**28**
Subgroup B	Monochrome Black - decorated (*flaring and deep bowls, vases*)	Dolomitic-Calcitic Marl-based Group A	4
	Monochrome Red Slipped (*jar*)	Dolomitic-Calcitic Marl-based Group B	1
		Total	**5**

internal variation, with standard deviations well above 20% for most elements (Table 9.1). The high level of chemical variability within this group is mirrored in the fact that it contains samples that correspond to a wide range of stylistic categories, as well as multiple local and non-local fabric groups (Table 9.8). Nonetheless, discounting the numerous chemical outliers within Group 7, the majority of samples derive from locally produced, crystalline calcite-tempered vessels.

9.3. Relationships Between the Chemical Groups and Subgroups

Further inspection of the structure of the Lamanai data set with principal components analysis provides additional insight into the relationships between the different chemical groups and subgroups identified, as well as the elements most responsible for the differences among them. The first four components derived by PCA account for account nearly 85% of the total variation within the data set. As shown in Table 9.9, component 1 is loaded mainly by the rare earth elements Lu, Yb, Sm, La, Ce and Eu, whilst component 2 is characterized by high loadings for the alkali metals Na, Rb and Cs. Together, these two components account for approximately 64% of the total variation. Component 3 is loaded mainly by the transition metals Sc, Cr and Fe, and component 4, by Hf and Ca. Figure 9.2 presents of plot of components 1 and 2 with Varimax rotation. As can be seen, there is a clear separation of Group 1, which consists of volcanic ash-

Table 9.8: Stylistic and Petrographic characteristics of the vessels comprising Group 7. Chemical outliers are marked with an asterisk.

Stylistic Categories Represented	Compositional Categories Represented	No.
Monochrome Orange Slipped (*=1 *jars*, **drum, tripod bowl*)	*Crystalline Calcite-Tempered Group A (N=1) Grog-Mixed Carbonate Class (subgroup B)(N=1) Dolomitic-Calcitic Marl-based Group B subgroup (N=1) *Dolomitic-Calcitic Marl-based Group C (N=2)	5
Slipped Rim/Lip - Smoothed Body (*jars*)	Crystalline Calcite-Tempered Group A	5
Monochrome Red Slipped (*rounded bowls, composite silhouette dishes*)	Crystalline Calcite-Tempered Group A (N=1) Crystalline Calcite-Tempered Group A subgroup (grog tempered variant) (N=1) Crystalline Calcite-Tempered Group C (N=1)	3
*Red-Brown-Striated (*jars*)	Crystalline Calcite-Tempered Group G	3
*Monochrome Orange Slipped (interior slipped) (*rounded bowl*)	Crystalline Calcite-Tempered Group A subgroup (grog tempered variant)	1
Monochrome Red Slipped (interior slipped) (*rounded bowl*)	Crystalline Calcite-Tempered Group B	1
Monochrome Black (*flaring bowl*)	Dolomitic-Calcitic Marl-based Group A	1
Monochrome Orange Slipped and Incised (*stand*)	Dolomitic-Calcitic Marl-based Group B	1
*Unslipped/Smoothed-Buff (*jar*)	Quartz Sand Class subgroup (calcite-tempered variant)	1
	Total	**21**

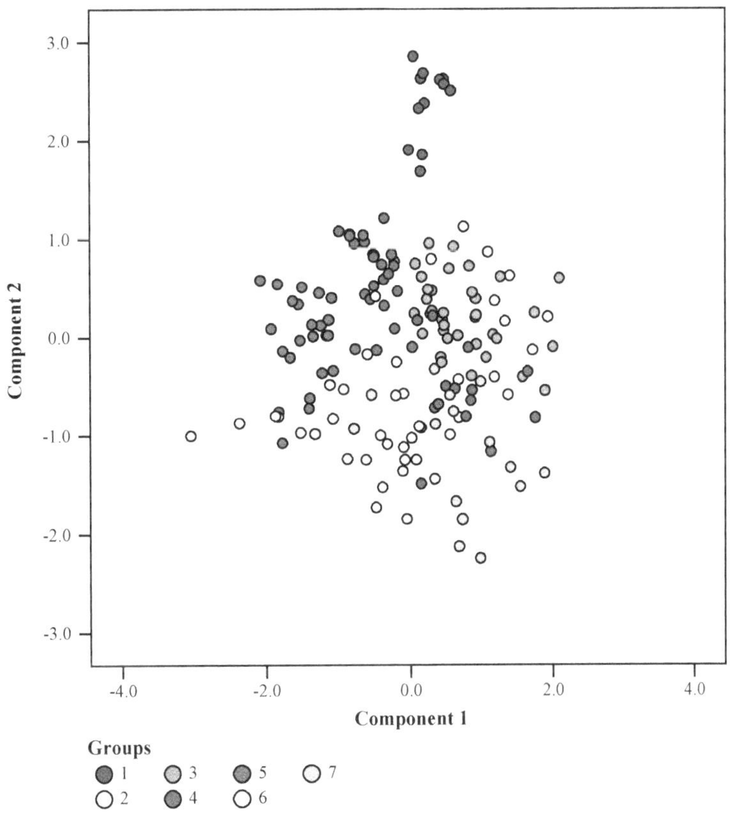

Figure 9.2: A Plot of components 1 and 2 of the Lamanai data set with Varimax rotation based on Sm, Lu, Yb, Ca, Na, La, Ce, Th, Cr, Hf, Cs, Sc, Rb, Fe, Co and Eu. Chemical groups identified through cluster analysis are differentiated with coloured symbols. (see colour version on accompanying CD)

Table 9.9: Rotated component matrix showing loadings of the components plotted in Figures 9.2 and 9.3.

	Component			
	1	2	3	4
Eu	**.961**	.065	.210	.034
Sm	**.952**	.204	.183	.072
La	**.925**	.199	.135	.191
Ce	**.914**	.138	.146	.182
Yb	**.891**	.197	.161	.266
Lu	**.846**	.294	.238	.252
Rb	.203	**.908**	.098	-.097
Cs	.213	**.809**	.294	.108
Na	.108	**.743**	-.366	.169
Co	.413	.584	.233	.319
Sc	.319	.243	**.859**	.020
Cr	.132	-.258	**.851**	.016
Fe	.254	.286	**.671**	.280
Hf	.128	-.117	.116	**.931**
Ca	-.255	-.252	.012	**-.776**
Th	.298	.266	.530	.572

tempered ceramics, from the remaining chemical groups, which comprise fabric groups from northern Belize and the Yucatan. The position of Group 1 at the top centre of the plot indicates that its discrimination is dependent upon component 2, which is mainly loaded by the alkali metals. As already has been discussed, these elements reflect the presence of alkali feldspar, which occurs in volcanic ash and tuff. This plot also reveals that although all of the chemical groups, except group 1, overlap to form a large, dispersed cluster situated at the centre of the plot, the different chemical groups separate to occupy particular areas. of the cluster. The tendency of these groups to form a single dispersed cluster may reflect the inherent overlap in the mineralogical characteristics of the different fabric groups deriving from northern Belize and the Yucatan. Nonetheless, the internal structure of the cluster reflects chemical differences at the group level, and its existence suggests that a clear separation of the main chemical groups could be achieved through detailed statistical analyses with measures of variance such a Malhalanobis distance and best relative fit (c.f. Glascock et al. 2004 and Arnold et al. 1999).

As illustrated in Figure 9.3, when components 1 and 4 are plotted, Group 2 and Group 5 are discriminated from the other main chemical groups. These groups form disparate but separate clusters just to the right and left of centre, at the top of the plot, indicating that the discrimination is mainly dependent upon component 4, which is loaded primarily by Hf and Ca, but also relates in part, to

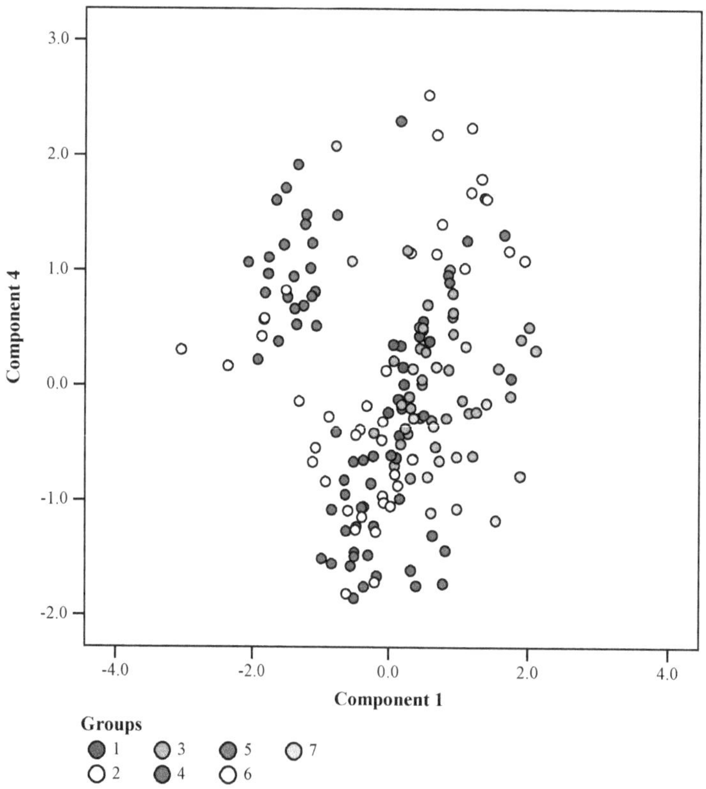

Figure 9.3: A Plot of components 1 and 4 of the Lamanai data set with Varimax rotation based on Sm, Lu, Yb, Ca, Na, La, Ce, Th, Cr, Hf, Cs, Sc, Rb, Fe, Co and Eu. Chemical groups identified through cluster analysis are differentiated with coloured symbols. (see colour version on accompanying CD)

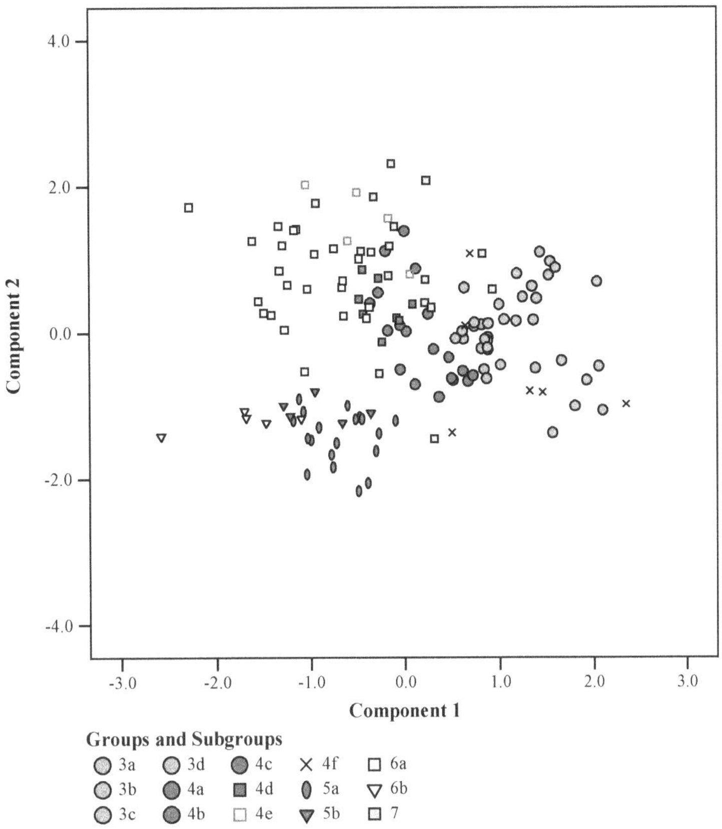

Figure 9.4: A Plot of components 1 and 2 of the main cluster within the Lamanai data set (Groups 3-7) based on Sm, Lu, Yb, Ca, Na, La, Ce, Th, Cr, Hf, Cs, Sc, Rb, Fe, Co and Eu. Chemical groups and their associated subgroups are differentiated with coloured symbols: Sub groups marked with circles are strongly associated with the Grog-Mixed Carbonate class, subgroups marked with squares are primarily associated with calcite-tempered fabrics, subgroups corresponding to Dolomitic-Calcitic Marl-based A are marked with triangles. (see colour version on accompanying CD)

component 1, which reflects rare earth element concentrations. As will be recalled, both of these groups are characterised by comparatively high concentrations of Hf, which most likely reflects the presence of zircon, and they correspond to fabric groups characterized by abundant inclusions of siliceous rocks and minerals, deriving from sandy clays. Accordingly, the discrimination of these Groups 2 and 5 from the others, reflects a distinction between ceramics containing a significant siliceous aplastic component and those with a predominantly calcitic mineralogy.

The clear separation of these groups from each other reflects the fact that Group 2 is characterized by comparatively high rare earth element concentrations, whereas Group 5 is characterized by low concentrations. The discrimination of these groups, therefore, reflects fundamental chemical differences in the clay components of the respective fabric groups with which Group 2 (Quartz Sand) and Groups 5 (Sascab-Quartz A and Dolomitic-Calcitic Marl-based A) are associated. It is also significant that the Group 4 samples occurring within the loose cluster formed by Group 2 are those from 4F with Sascab-Quartz B fabrics, which also derive from sandy clays.

The relationships among the groups and subgroups comprising the main cluster within the dendrogram were examined by performing a separate principal components analysis of the main cluster alone. A plot of components 1 and 2 is presented in Figure 9.4. Once again, component 1 is mainly loaded by rare earth elements (Lu, Yb, Sm, La, Ce and Eu), whereas component 2 is mainly loaded by Ca and Hf (Table 9.10), with the first four components accounting for approximately 84% of the total variation within the data set. Although a close chemical relationship among the different chemical subgroups is clearly demonstrated, the three local paste-making traditions (i.e. crystalline calcite-tempered, Sascab-Quartz A and Grog-Mixed Carbonate) are clearly discriminated from each other, occupying different areas of the scatter. For example, Group 7 and the subgroups consisting of calcite tempered vessels (marked with squares) cluster together at the top of the scatter, with the different chemical groupings occupying different areas of this cluster. The separation of the calcite tempered fabrics is mainly dependent upon component 2, undoubtedly reflecting the high crystalline calcite content of these pastes, primarily due to tempering. In comparison, the Grog-Mixed Carbonate Class (brown and orange circles) dominates the right side of the scatter, with the associated chemical subgroups of Groups 3 and 4 forming separate

Table 9.10: Component matrix showing loadings of the components plotted in Figures 9.4 and 9.5.

	Component			
	1	2	3	4
Lu	**.925**	.214	-.079	.023
Yb	**.888**	.297	-.204	.040
Sm	**.882**	.447	-.078	.033
La	**.881**	.344	-.197	-.093
Ce	**.844**	.374	-.214	-.075
Eu	**.841**	.504	-.110	.024
Sc	.792	-.226	.405	.228
Th	.713	-.455	-.127	.280
Co	.703	-.208	.122	-.528
Fe	.682	-.341	.105	-.173
Cs	.590	-.286	**.533**	-.264
Ca	-.403	**.723**	.401	.139
Na	.425	-.447	-.033	.033
Rb	.538	-.138	**.716**	-.073
Hf	.355	**-.578**	**-.681**	.016
Cr	.493	-.177	.203	**.766**

subclusters. Finally, Sascab-Quartz A fabrics cluster at the left hand side of the scatter. These results indicate that the different local paste-making traditions are characterized by distinct chemical profiles.

Principal components analysis also discriminates two of the non-local fabric groups from the main scatter of local fabrics. As can been seen in Figure 9.4, the samples corresponding to Dolomitic-Calcitic Marl-based A (represented by triangles), cluster together at the left edge of the scatter, but overlap slightly with subgroup 5A. The position of the Dolomitic-Calcitic Marl-based A fabrics towards the left side of the scatter reflects a discrimination heavily dependent on component 1, and thus, elements related to clay minerals. Chemical differences related to source can, therefore, be inferred. The apparent chemical overlap between this fabric group and locally produced Sascab-Quartz A fabrics (blue circles) would appear to reflect a fundamental similarity in clay chemistry. However, since the sascab temper in the Sascab-Quartz A fabrics is known to contain clay minerals (see Chapter 5), and this tempering material derives from similar deposits as the base clay of the Dolomitic-Calcitic Marl-based A (both derive from deposits formed through the deep weathering of the underlying bedrock), it is possible that similar clay minerals occur in these different raw materials. As an example, this kind of chemical overlap occurs among geographically separated deposits of Neogene clay in Crete (Hein et al. 2004a).

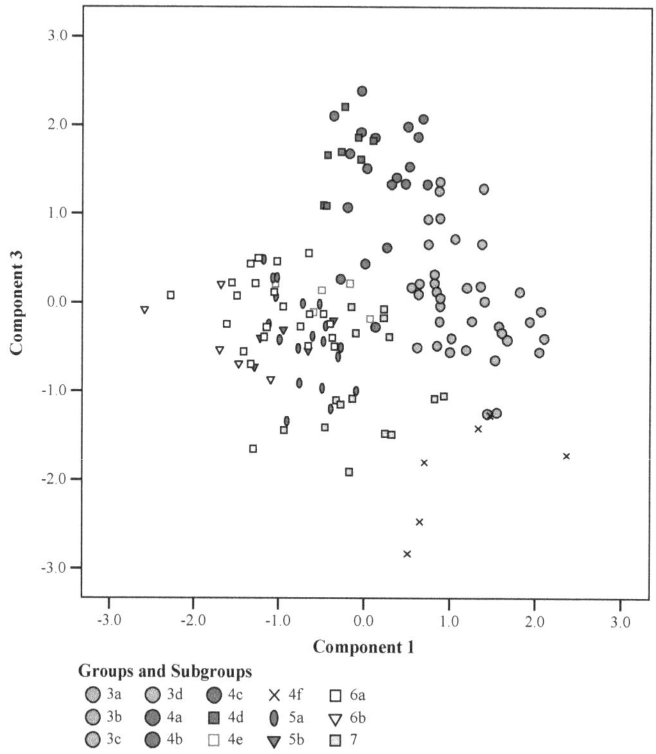

Figure 9.5: A Plot of components 1 and 3 of the main cluster within the Lamanai data set (Groups 3-7) based on Sm, Lu, Yb, Ca, Na, La, Ce, Th, Cr, Hf, Cs, Sc, Rb, Fe, Co and Eu. Chemical groups and their associated subgroups are differentiated with coloured symbols: Sub groups marked with circles are strongly associated with the Grog-Mixed Carbonate class, subgroups marked with squares are primarily associated with calcite-tempered fabrics, subgroups corresponding to Dolomitic-Calcitic Marl-based A are marked with triangles. (see colour version on accompanying CD)

As shown in Figure 9.5, which presents a plot of components 1 and 3, Sascab-Quartz B fabrics, which are also non-local, are discriminated from: 1) the other non-local fabric group, Dolomitic-Calcitic Marl-based A, 2)their local technological counterpart Sascab-Quartz A and 3) the other two local paste-making traditions. Scattered in the lower right corner of the plot, the separation of these samples from the main cluster is dependent on both component 1, which relates to rare earth concentrations, and component 2, which reflects the alkali metals Rb, and Cs, and Hf. Chemical differences related to source can therefore be inferred.

9.4. Local Fabric Groups and Modern Clay Resources

As was discussed in the previous chapter, comparison of the petrographic characteristics of fired samples of the modern clays prospected from different geological and environmental contexts in and around Lamanai to those of the ceramic fabric groups suggested that certain fabric groups were connected to particular, and sometimes different, kinds of local clay deposits. To examine the chemical relationship between different modern clays available in the local area and the chemical groups corresponding to the different local paste-manufacturing traditions, a data set comprising the modern clay samples and archaeological samples comprising chemical groups 2, 3, 4A-E, 5A, 6A and 7 was subjected to principal components analysis. Since many of the ceramic samples were known to contain carbonate temper, Ca was not included in the analysis, so as to avoid dilution/enrichment effects (see Bishop et al. 1982 and Arnold et al. 1999 for a discussion).

The first two components derived by PCA account for approximately 82% of the total variation within the data set. As shown in Table 9.11, component 1 is characterized by high loadings for many elements, including the rare earth elements Lu, Yb, Sm, La, Eu, Ce, Th, as well as Sc and Co, whereas component 2 is loaded primarily by Fe and the alkali metals Na and Rb. When these two components are plotted, the result is a clear separation of the ceramic specimens from the clay samples (Figure 9.6). The ceramic samples form a tight cluster at the top of the plot, however, the fine-structuring of this cluster is quite similar to that resulting from PCA of the ceramic samples alone. In comparison, the clay samples are spread out across the bottom half of the plot, forming loose clusters that to correspond to differences in geological context (see Chapter 5). This discrimination of the ceramic samples from the clay samples is mainly dependent upon component 2, which reflects Na, Rb and Fe content. In addition, the differentiation of the different kinds of local clay deposits is strongly dependent on component 1, which is primarily loaded by rare earth elements, suggesting that the discrimination relates to clay minerals as opposed aplastic content.

The fact that the chemical compositions of the ceramic and clay samples do not coincide is not surprising or troublesome, since experimental studies have

Table 9.11: Rotated component matrix showing loadings of the components plotted in Figures 9.6.

	Component	
	1	2
Lu	**.975**	-.105
Yb	**.966**	-.157
Sm	**.938**	-.144
La	**.929**	-.141
Eu	**.921**	-.151
Ce	**.910**	-.162
Th	**.908**	.186
Sc	**.842**	.359
Co	**.817**	-.223
Cr	.785	.253
Hf	.751	-.030
Na	-.183	**.921**
Fe	-.139	**.920**
Rb	.382	**.829**
Cs	.511	.629

demonstrated that it is rarely possible match raw clays chemically with finished ceramic products (e.g. Kilikoglou et al 1988). A similar chemical study conducted at K'axob (located just north of Lamanai), of Preclassic ceramics and an extensive sample of modern clays and rocks prospected from the surrounding area also found a lack of correspondence between pottery and raw material compositions, even when the chemical contribution of particular carbonate tempers, such as sascab, were taken into account (Bartlett et al. 2000).

It is well recognized, and has been demonstrated experimentally, that paste preparation techniques such as the purification of clays through levigation or sieving and the addition of temper can cause a chemical distinction between raw clays and 'prepared' ceramics bodies (e.g. Kilikoglou et al. 1988; Neff et al. 1988, 1989). In the case of temper, if the source material contains clay minerals, as with sascab, the effect on trace element concentrations can be quite significant (Arnold et al. 1999). Another intervening factor is chemical alteration during burial (Franklin and Vitali 1985). With regard to the present study, however, there is some evidence to suggest that paste-preparation techniques, in particular, play an important role in the distinction between the modern clays and the ceramic specimens. For example, in their natural state, raw clays from the Lamanai area, particularly those from the west side of the lagoon, contain large rock fragments. The absence in the ceramic fabrics of inclusions larger that the size of very coarse sand, suggests that clay component was processed, by some means prior to use, to remove at least some of the coarse fraction. In addition, local fabrics most often represent complex mixtures of raw material ingredients. For example, they are heavily tempered and often contain multiple tempers, including grog and sascab, which can greatly effect elemental concentrations. It does not seem

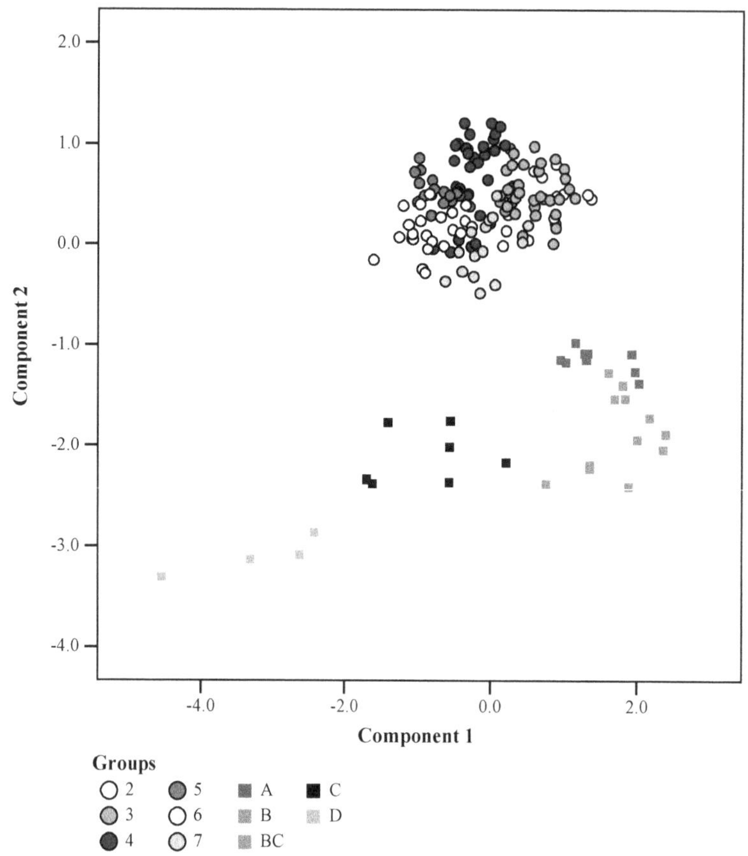

Figure 9.6: A Plot of components 1 and 2 of the chemical groups corresponding to local fabric groups (Groups 2, 3, 4[A,B,D,E], 5A, 6A, and 7) and modern clay samples prospected from the Lamanai area based on Sm, Lu, Yb, Na, La, Ce, Th, Cr, Hf, Cs, Sc, Rb, Fe, Co and Eu. Chemical groups are differentiated with coloured symbols. Ceramic groups are marked with circles and clay groups are marked with squares. For the clays: red square = samples deriving from the east of the New River Lagoon, green squares = samples deriving from the west side of the New River Lagoon, from clay horizons directly below the ground's surface, the pink square = the sample taken from the Barber Creek deposit in the area of modern wash deposits of Pleistocene alluvium, black squares = samples deriving from the west side of the New River Lagoon, from clay deposits that form in association with horizons of weathering limestone, and , turquoise squares = samples deriving from river and creek deposits. (see colour version on accompanying CD)

unreasonable, therefore, to suggest that the combined effect of these practices alone would result in a chemical distinction between local clay resources and locally produced ceramic products.

9.5. SUMMARY

With regard to the questions posed at the beginning of this chapter the following observations can be made:

1. The results show that the relationship between geochemical, mineralogical and stylistic variation within the Lamanai assemblage is very complex. In some cases, there is a direct correspondence between chemical groups and fabric groups, particularly on the class level. Such is the case for Groups 1 and 2, since the associated fabric classes (Volcanic Glass and Quartz Sand, respectively) are confined to these chemical groups. In contrast, the Grog-Mixed Carbonate, the Crystalline Calcite-Tempered and the Dolomitic-Calcitic Marl-based classes occur in different chemical groups, but showing a strong association with only one or two of these groups. Moreover, with regard to the Crystalline Calcite Class, there is a complete lack of correspondence between the different fabric groups within this class and the chemical groups and subgroups identified.

2. In some cases, statistical analysis only discriminates non-local fabrics from local fabrics. For example, Sascab-Quartz B, Dolomitic-Calcitic Marl-based A and the Volcanic Glass fabric groups are differentiated chemically, both from the local fabric groups and from each other. Nonetheless, several non-local fabric groups are not discriminated, especially in cases where mineralogy is calcareous and the number of samples is small (e.g. the non-local Crystalline Calcite-Tempered

fabric groups and Dolomitic-Calcitic Marl-based B, C and D).

3. The different local paste-making traditions are discriminated chemically with PCA. In the case of the Grog-Mixed Carbonate fabrics and the local crystalline calcite-tempered fabrics, however, they separate into multiple chemical groups and subgroups and the significance of this fine-structuring is not apparent based on the current body of stylistic, petrographic and chronological evidence. With regard to the Grog-Mixed Carbonate this fine-structuring might be considered to resolve the internal heterogeneity of this fabric group, and perhaps lends support to the suggestion that compositional variation within the Early Postclassic component of the assemblage is partly due to the activities of contemporaneous potters/groups of potters (i.e. the use of slightly different paste recipes).

4. A chemical correspondence between local pottery and the clay resources analysed is not demonstrated.

5. As a final comment, it is noteworthy that the results of the current study demonstrate that the ability to interpret chemical data sets is greatly enhanced by a comprehensive understanding of the technological and provenance characteristics, relationships and inter-relationships of the pottery under study. In the present study, it would have been difficult, if not impossible, to discriminate among source-related chemical distinctions relating to the activities of local potters vs. non-local producers, for mineralogically similar pottery, in the absence of detailed petrographic and geological information.

CHAPTER 10
VARIABILITY IN DECORATIVE AND SURFACE TREATMENTS AND FIRING OF TERMINAL CLASSIC AND EARLY POSTCLASSIC FINE WARE

Surface finishing and decorative techniques and firing procedures, are two fundamental aspects of Maya pottery production that have seldom been the focus of systematic scientific investigation. With the exception of Neff's (2003) detailed examination of Plumbate ware with laser ablation-inductively coupled plasma-mass spectrometry, scientific enquiry into these aspects of Maya ceramic technology have not advanced substantially beyond Shepard's (e.g. 1939) pioneering investigations using a binocular microscope, re-firing experiments and microchemical analyses (e.g. Rice 1980, 1987b; Reents-Budet 1994). Scanning electron microscopy (SEM) offers a means of examining, in detail, specific points of technological similarity and difference, elucidating information regarding the technical procedures involved in the manufacture of different kinds of pottery.

The present study focuses on the fine ware component of the Lamanai assemblage, using SEM to investigate variation in the nature of the finished surfaces, slips and paints, as well as the microstructural characteristics of the ceramic body, within and among the numerically dominant stylistic categories identified in Chapter 7. A central interest was examining the relationship between the decorative and firing technology of fine ware vessels and their paste technology and provenance, in order to identify additional technological similarities and differences between and among locally produced and imported pottery, as well as changes over time in local technical practices. A total of 31 vessels were included in the study, representing six different stylistic groups. The decorative and morphological characteristics of these vessels, as well as their fabric group and provenance associations are given in Table 10.1. The results of the analyses are organized chronologically for monochrome, bichrome and polychrome vessels and the data discussed is summarized in Table 10.2. Table 10.3 presents the standard that was used to assess body microstructures and estimated firing temperatures following Maniatis and Tite (1981:68).

10.1. TERMINAL CLASSIC MONOCHROME VESSELS

Monochrome Black Vases and Deep Bowls

Of the five vessels analysed from the Monochrome Black stylistic group, two were ascribed a local provenance based on their geological characteristics, and three were determined to be imports, likely deriving from coastal areas of northeastern and eastern northern Belize. Both of these fabric types are characterized by dark grey surfaces and dark coloured firing horizons (black cores with a brown subsurface zone). In both cases the clay used is calcareous (>10% CaO), but EDAX analysis shows that the imported fabrics contain magnesium, which is absent in the local fabric, as well as comparatively lower levels of iron. The presence of magnesium in the imported fabrics is in keeping with the fact that they derive from dolomitic marl.

Firing - The presence in these fabrics of a brown subsurface zone or horizon, intervening between a black core and dark grey surfaces, suggests a succession of firing conditions (Rye 1981:118). This cycle appears to have involved a period of reduction, followed by a period of mildly oxidizing conditions and ending with a final period of reduction. An intervening period of mildly oxidizing conditions is also suggested by the occurrence of orange splotches on the finished surfaces of two the vessels, representing areas that remained oxidized due to an uneven distribution of air. This apparent lack of control over the firing atmosphere would be expected in conjunction with open firing.

Although the local and imported vessels appear to have been fired under a similar sequence of conditions, the microstructural characteristics of their fabrics suggest differences in firing temperature. The microstructure of the imported fabric suggests a firing temperature of <750°C (NV stage) (see Table 10.3), since there are no signs of sintering or vitrification (Figure 10.1). In comparison, the local fabric appears to have been fired at a slightly higher temperature. These microstructures are characterized by fairly extensive sintering but isolated glassy areas, which are indicative of Initial Vitrification (IV) are absent (Table 10.2). An estimated firing temperature approaching the range of 750-800°C for IV microstructures is, therefore, suggested.

The distinctive microstructure observed in the local fabrics appears to be characteristic of local calcite-tempered fabrics in general, as identical microstructures were also observed in temporally equivalent (Terminal Classic) monochrome red and orange and bichrome resist vessels corresponding to the three local crystalline calcite-tempered fabric groups (Groups A, B, and C) associated with fine ware production. As will be discussed below, however, a different a firing regime at least involving a period of oxidation at the end of the cycle would have been required to produce the final colour of the later vessels.

Surface Finishing Techniques - Further technological differences between the local and imported Monochrome Black vessels relate to the techniques employed to achieve their black surfaces. At the surface of the imported vessels, a discrete layer approximately 2-5μm in thickness can be discerned (Figure 10.3). This layer is

Table 10.1: Stylistic characteristics and petrographic group and provenance associations of vessels included in the SEM study. Samples are organized chronologically for monochrome, bichrome and polychrome vessels.

Sample No.	Vessel/Lot No.	Decorative Treatment	Form	General Provenance	Fabric Group
Terminal Classic Vessels					
Monochrome Black					
125	LA508/7	groove-incised and punctated	vase, cylindrical	local	Calcite C
126	LA508/6	raised band with a central groove	vase, barrel-shaped	local	Calcite C
124	LA658/1	groove-incised and stamped	vase, cylindrical	non-local	Marl-Based A
39	LA814/1	groove-incised with false gardooning	vase, cylindrical	non-local	Marl-Based A
40	LA706/1	groove-incised	deep bowl	non-local	Marl-Based A
Monochrome Red And Orange Slipped					
281	LA717	Orange	composite silhouette dish	local	Calcite C
23	LA872/4	Red	composite silhouette dish with a ring base	local	Calcite C
136	LA630/7	Red	composite silhouette dish with a ring base	local	Calcite C
401	LA701	Orange	composite silhouette dish	non-local?	Calcite D
254	LA717	Orange (interior slipped)	bowl, rounded	local	Calcite B (subgroup)
257	LA717	Red (interior slipped)	bowl, rounded	local	Calcite A
252	LA717	Red (interior slipped)	bowl, rounded	local	Calcite B
261	LA717	Orange (interior slipped)	dish, rounded	local	Sascab Quartz A
144	LA693/1	Red (interior slipped)	dish, rounded	local	Sascab-Quartz A
Red-Orange-Black Resist					
289	LA717		composite silhouette dish	local	Calcite C
291	LA717		composite silhouette dish	local	Calcite C
295	LA717		composite silhouette dish	local	Calcite C
296	LA717		composite silhouette dish	local	Calcite C (subgroup)
Black On Red Slipped And Painted					
48	LA244/2		massive bowl	local	Sascab-Quartz A
118	LA240/1		dish, rounded	non-local	Sascab-Quartz B

Table 10.2: Summary of macroscopic and microstructural characteristics of vessels analyzed by SEM.

Sample No.	Colour of Fabric	Surface Colour	Slip Colour	Paint Colour	Vitrification Stage	Estimated Firing Temp.	Firing Atmosphere
125	black core bordered by very thin brown horizons and very thin black margins	dark grey	black	—	NV+	approaching 750-850	R-N/O-R
126	black bordered by thin light brown horizons and a very thin black exterior margin	dark grey	black with orange mottles	—	NV+	approaching 750-850	R-N/O-R
124	black core with and very thin brown layers and very thin black margins	dark grey to brown	black	—	NV	>750	R-N/O-R
39	black core with dark brown margins	dark grey to brown	black with orange mottles	—	NV	>750	R-N/O-R
40	black core with and very thin brown layers and very thin black margins	dark grey to brown	black with orange mottles	—	NV	>750	R-N/O-R
281	brownish-buff core with light orange margins	light orange	reddish-orange	—	NV+	approaching 800-850	?-O
23	black core with orangish-buff margins	buff	red	—	NV+	approaching 800-850	?-O
136	dark grey core with pinkish-orange margins	light orange	red	—	NV+	approaching 800-850	?-O
401	grey core bordered by thin brownish-buff horizons and light orange margins	buff	reddish-orange	—	NV+	approaching 800-850	?-O
254	black core with brownish-buff margins	buff	reddish-orange	—	NV+	approaching 800-850	?-O
257	buff throughout	buff	red	—	NV+	approaching 800-850	?-O
252	black core with a thin dark brown exterior margin and a brownish-buff interior margin	grey to buff	red	—	NV+	approaching 800-850	?-O
261	grey core with brownish-buff margins	buff	reddish-orange	—	IV	800-850	?-O
144	grey core with brownish-buff margins	buff	red	—	IV	800-850	?-O
289	dark grey core with orangish-brown margins	light orange	burgundy and black	—	NV+	approaching 800-850	?-O
291	greyish-buff core with pinkish-orange margins	light orange	reddish-orange and black	—	NV+	approaching 800-850	?-O
295	black core with thin buff margins	buff	reddish-orange and black	—	NV+	approaching 800-850	?-O
296	black core with orangish-brown margins	light orange	reddish-orange and black	—	NV+	approaching 800-850	?-O
48	grey core with very thin light orange margins	light orange	red	black	NV	<800	?-O
118	reddish-brown throughout	reddish-brown	red	black	IV	800-850	O

Table 10.2: Summary of macroscopic and microstructural characteristics of vessels analyzed by SEM (continued).

Sample No.	Colour of Fabric	Surface Colour	Slip Colour	Paint Colour	Vitrification Stage	Estimated Firing Temp.	Firing Atmosphere
54	dark greyish-brown core with thin orange margins	orange	orange	red and black	NV+	approaching 800-850	O
707	dark grey core with thin brownish-buff margins	buff	N/A	orange and black	IV	800-850	O
202	grey core with light brown margins	brownish-buff	orange	red and black	IV+	around 850	O
207	light grey with a thin light orange interior margin	buff	orange	red and black	IV+	around 850	O
335	orangish-buff throughout	orangish-buff	orange	—	NV+	approaching 750-800	R-O
333	very thin grey core with grayish-buff margins	buff	orange	—	NV+	approaching 750-800	R-O
76	black core bordered by very thin light grey horizons and very thin light orange margins	buff	orange	—	VI+	around 850	R-O
327	dark grey core with a thin brownish-buff exterior margin and greyish-buff interior margin	light orange to light grey (interior)	orange	—	VI+	around 850	R-O
81	buff orange throughout	orangish-buff	orange	—	NV+	approaching 800-850	O
99	grey core bordered by light grey horizons and with a light orange exterior margin	light grey(interior) to buff	orange	—	NV+	approaching 750-800	R-O
152	light orange throughout	orangish-buff	reddish-orange	—	NV+	approaching 800-850	O

unvitrified and the clay particles are compacted and aligned, suggesting that is was produced through intense burnishing (c.f. Kilikoglou 1994:70-72). EDAX analysis of this layer and the underlying ceramic body indicated no appreciable differences in the levels of the different oxides present. Hence, there is no conclusive evidence (based on the small number of samples analysed) that this surface treatment involved the application of a slip. Rather, the highly glossy black surface was achieved through intense burnishing and firing in a reducing atmosphere at low temperatures.

In contrast, a thin layer of slip about 5-8μm in thickness is present on both of the local vessels (Figure 10.4). The slip is unvitrified with extensive sintering and appears to have been made using a fine-grained suspension of a clay similar to that used to make the ceramic body. This is suggested by the similarity of compositional readings obtained for the slip layers and bodies, apart from elevated levels of potassium and iron in the slip, which would result from levigation (Aloupi and Maniatis 1990:461). Also present in the slip layers are irregular-shaped voids and spherical inclusions, which were identified as iron oxide conglomerates (also see Noll 1975). The presence of the voids is due to the uneven distribution of iron oxide inclusions. Burnishing marks are readily observable on the exterior of the local vessels, suggesting some compaction of the surface, likely to improve the lustre of the slip. The local vessels, although still glossy in appearance, are characterized by a comparatively dull surface finish. The porosity of the slip is undoubtedly a contributing factor to this aspect of the visual appearance of the local vessels (c.f. Aloupi and Maniatis 1990:468).

To improve understanding of the microstructures that were observed for this stylistic group, a sample taken from one of the imported vessels (sample 40) was refired in an oxidizing atmosphere at 850°C for one hour. Analysis in the SEM showed a substantial change in the microstructure of both the compacted surface layer and the body. Both were vitrified and a microstructure equivalent to the extensive stage of vitrification (V) had developed in the body (Figure 10.5). In addition, the slip had turned orange. These results confirm that the original firing temperatures were below 850°C and that reduction firing was a key element in producing the visual properties of the surface treatment.

Monochrome Red and Orange Dishes and Bowls

The SEM analysis of Terminal Classic red and orange slipped dishes and bowls focussed on three form categories: composite silhouette dishes,

Table 10.3: Vitrification stages and firing temperature ranges following the standard presented in Maniatis and Tite (1981: 68).

		Firing Temperature Ranges (°C)	
	Vitrification Stage	**Oxidizing Atmosphere**	**Reducing Atmosphere**
NV	No Vitrification	>800	>750
IV	Initial Vitrification	800-850	750-800
V	Extensive Vitrification	850-1050	850-1050

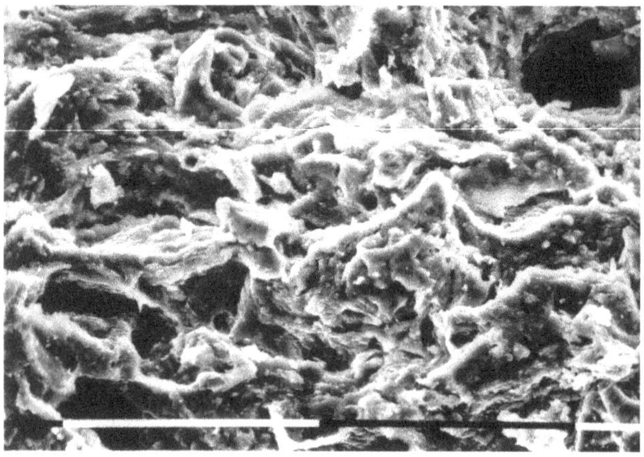

Figure 10.1: Imported Monochrome Black vessel (sample 40) displaying an NV microstructure (x500). Bar scale = 10μm.

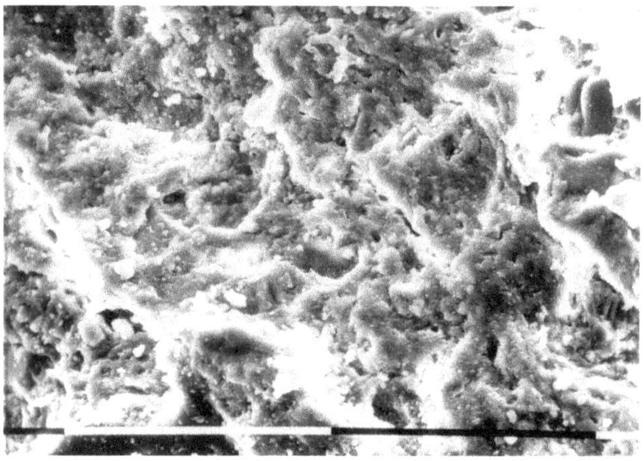

Figure 10.2: Local Monochrome Black vessel (sample 125) displaying an NV microstructure with extensive sintering (NV+), which is typical of all monochrome and bichrome vessels with calcite-tempered fabrics (x500). Bar scale = 10μm.

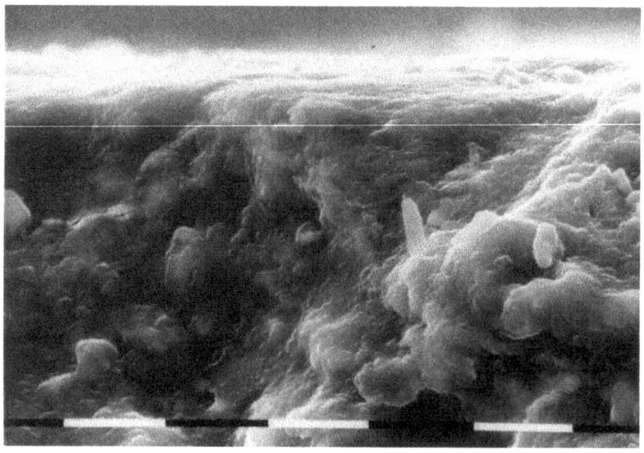

Figure 10.3: Imported Monochrome Black vessel (sample 39) with an unvitrified, compacted surface layer (x2000). Bar scale = 10μm.

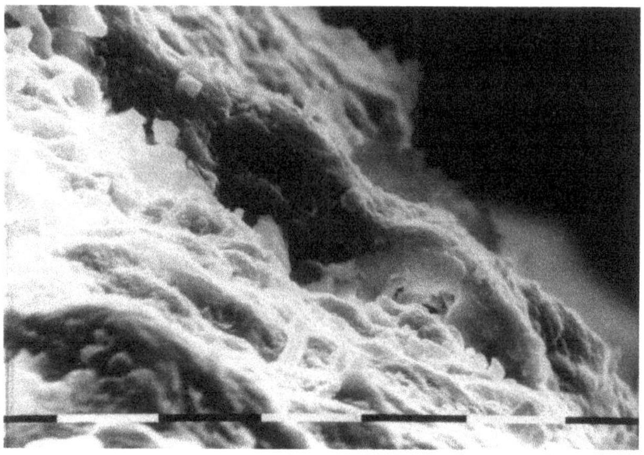

Figure 10.4: Local Monochrome Black vessel (sample 125) displaying an unvitrified layer of slip with extensive sintering and containing irregular-shaped voids and spherical iron conglomerates (x2000). Bar scale = 10μm.

rounded dishes and rounded bowls. A total of four composite silhouette dishes were selected for analysis: two with a red slip and two with an orange slip. All of these dishes have calcite-tempered fabrics, and with the exception of sample 401, which is possibly non-local, all were ascribed a local provenance, deriving from group C of this fabric class. Of the three rounded bowls that were analysed, two have a red-coloured slip and one has an orange slip. On all of these bowls, the slip is confined to the interior surface and the rim area of the exterior surface, distinguishing them stylistically from bowls on which the slip completely covers both surfaces. As with the composite silhouette dishes, these bowls have calcite-tempered fabrics but they derive from different local fabric groups within this fabric class (i.e. groups A and B). With regard to the rounded dishes that were analysed, both derive from the Sascab-Quartz A fabric group and, therefore, are associated with a different local tradition of paste-making. One of these dishes has a red slip and the

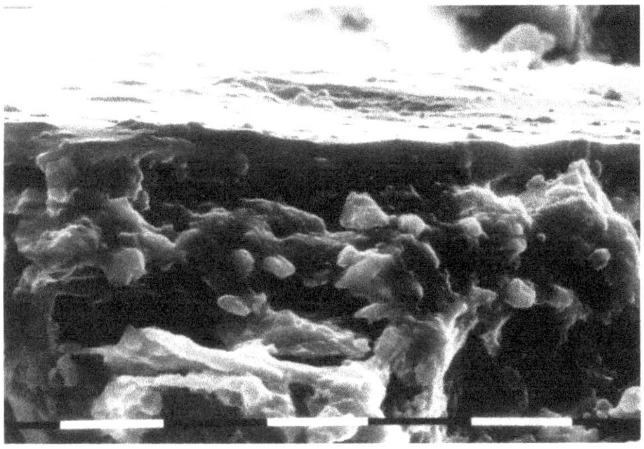

a)

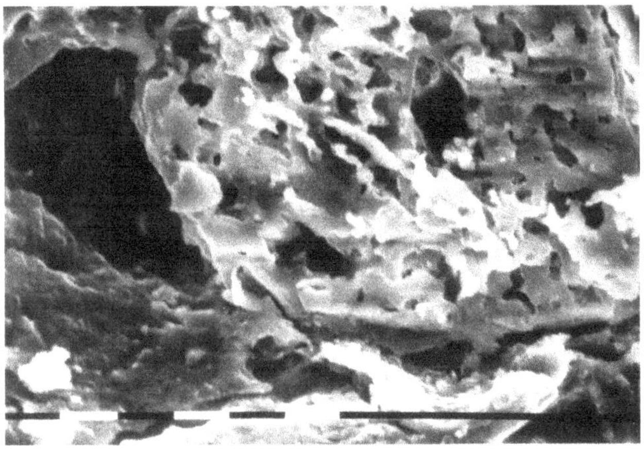

b)

Figure 10.5: a) vitrified surface layer on sample 40 after refiring at 850°C under oxidizing conditions for 1 hour (x2000); b) body of sample 40 after refiring showing and extensively vitrified microstructure (V) (x1000). Bar scale = 10μm.

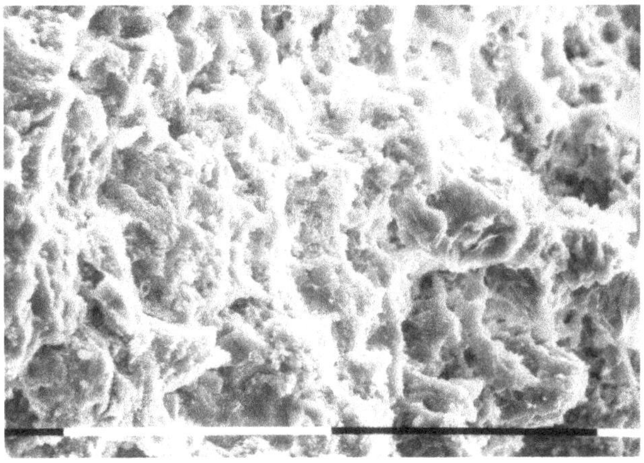

a)

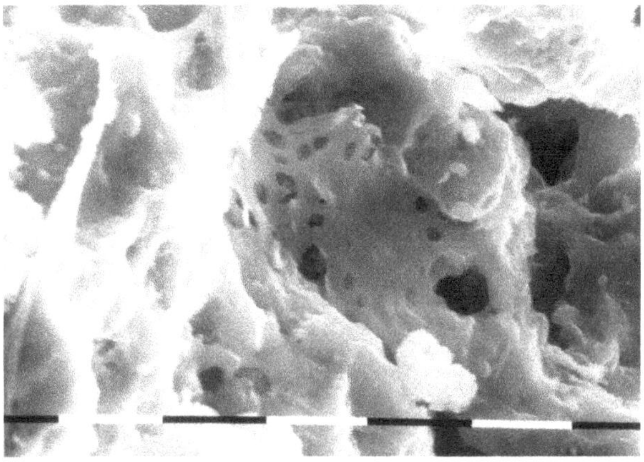

b)

Figure 10.6: a) body of a locally produced monochrome rounded dish with a Sascab-Quartz A fabric (sample 261) displaying an IV microstructure (x500); b) vitrified area (x2000).

other, an orange slip, and in both cases, the slip is confined to the interior surface and rim area of the exterior.

The SEM analysis showed there to be significant differences between the calcite-tempered vessels and the rounded dishes with Sascab-A fabrics, in terms of both their firing and slip technology. Among the calcite-tempered vessels deriving from different fabric groups, body microstructures were found to be strikingly similar, suggesting similar firing temperatures. With regard to the nature of the slip layer, however, differences were observed among the three form categories with regard to the thickness of the slip layer and its iron content.

Firing - With the exception of one of the rounded bowls (sample 257), the fabrics of all of the vessels described above display firing horizons, with light brown or grey to black cores bordered by lighter-coloured margins (light orange to buff). In all cases, however, the clay used to make the ceramic body is calcareous (>10% CaO). The microstructure of the calcite-tempered fabrics is strikingly similar to that observed in local monochrome black vessels, characterized by extensive sintering, but lacking areas of vitrification. However, the light colour of the body surface, together with the red to orange colour of the slip, indicates that oxidizing conditions prevailed, during the final stages of the firing regime. The darker-coloured cores that are present in all but one of the calcite-tempered vessels are due to incomplete oxidation. Since isolated glassy areas were not observed in these fabrics, an estimated firing temperature approaching the range of 800-850°C for bodies displaying Initial Vitrification (IV) is suggested. Refiring experiments also indicate an original firing temperature below 850°C since samples refired in an oxidizing atmosphere at 850°C (for 1 hour) showed an extensively vitrified body microstructure (stage V) and slip layer. Also noteworthy is that the microstructure of the composite silhouette dish that is possibly of non-local origin is virtually identical to the local calcite-tempered vessels.

The examination of the microstructures of the rounded dishes with Sascab-Quartz A fabrics showed that these vessels were fired at a higher temperature than the calcite-tempered vessels. The microstructure of the rounded dishes corresponds to Initial Virification (IV) since glassy areas occur throughout the analysed samples (Figure 10.6). An estimated firing temperature within the range of 800-850°C is therefore indicated.

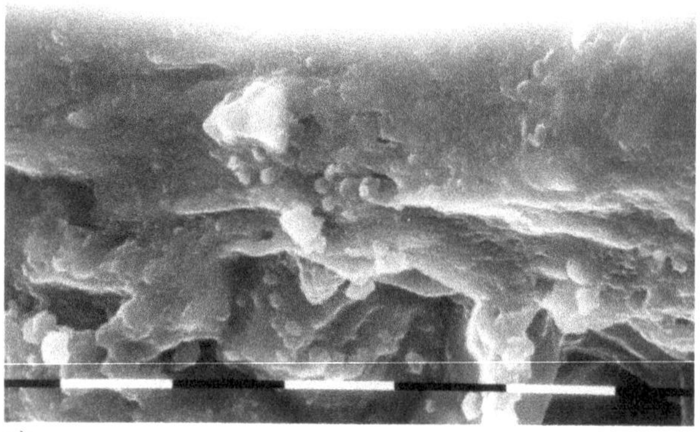

a)

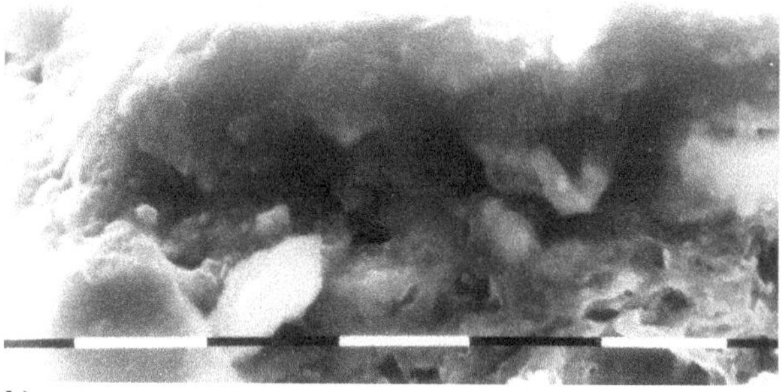

b)

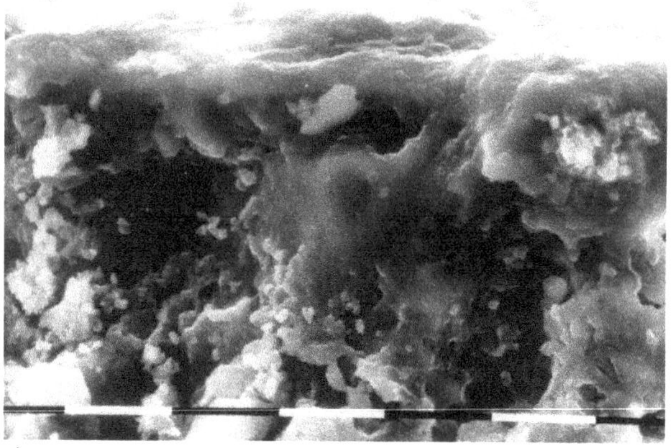

c)

Figure 10.7: a) Red-slipped calcite-tempered rounded bowl (sample 252) displaying a very thick (15-25μm) layer of slip (x2000); b) red-slipped calcite-tempered composite silhouette dish displaying a thick slip layer (10-15μm) (x2000); orange slipped rounded dish with a Sascab-Quartz A fabric displaying a comparatively thin (5-10μm) layer of slip (x2000). Bar scale = 10μm.

Surface Finishing Techniques - A discrete layer of slip is present on all of the monochrome and red and orange vessels and the there were no differences observed in its microstructure. In all cases, the slip is fine-grained and unvitrified but with extensive sintering. iron-rich conglomerates are present in all samples. The similarity of the compositional readings obtained for the slip layers and bodies of individual vessels, apart from elevated levels of potassium and iron in the slip, suggests that the slips were produced through refinement of clays similar to those used to make the ceramic bodies. Both the slips and bodies of the calcite-tempered vessels (the composite silhouette dishes and the bowls), however, contain higher levels of iron than the rounded dishes, pointing to a chemical difference in the clay used in each case. In addition, there appears to be no apparent difference in the relative amount of iron present in the red and orange slips in either case. This suggests that different slip colours were produced through manipulation of the firing atmosphere. The addition of a small quantity or iron to the slip would have helped to promote intense red colours. Shepard (1939:264-267) reports, however, that localized deposits of iron rich clay at San Jose (located southwest of Lamanai) fire to the same colour as the red slips on vessels dating to the Classic period. Further analyses involving experimental work on local clay resources would help to clarify slip preparation procedures.

A main difference in the slips that occur on the different form categories that were focussed upon is the thickness of the slip layer (Figure 10.7). The slip layer on the calcite-tempered bowls and dishes is comparatively thick, but it is thickest on the bowls. On the bowls it is 15-25μm in thickness, whereas on the composite silhouette dishes it is 10-15μm. The slips on the rounded dishes are appreciably thinner in comparison, measuring 5-10μm. The thickness to which the slip was applied appears to have had a direct affect on the visual and tactile properties of the surface finish. The vessels with a comparatively thick slip layer have a glossier surface finish and a waxy feel.

10.2. Terminal Classic Bichrome and Polychrome Vessels

Red-Orange-Black Resist Composite Silhouette Dishes

A total of four bichrome composite silhouette dishes were selected for detailed analysis with SEM. On three of these vessels the slipped surface has black decorative elements on a red to orangish-red background and on the remaining vessel, the background colour is burgundy. These dishes are morphologically identical to their monochrome counterparts and they are indistinguishable petrographically from the monochrome composite silhouette dishes that occur in the same fabric group (Group C of the Crystalline Calcite-Tempered class).

Firing - The fabrics of the bichrome composite silhouette dishes display similar firing horizons to those that occur in their monochrome counterparts, with a light grey to black core bordered by lighter coloured subsurface and surface zones. The SEM showed that the microstructure of the bichrome dishes is also similar to that of the

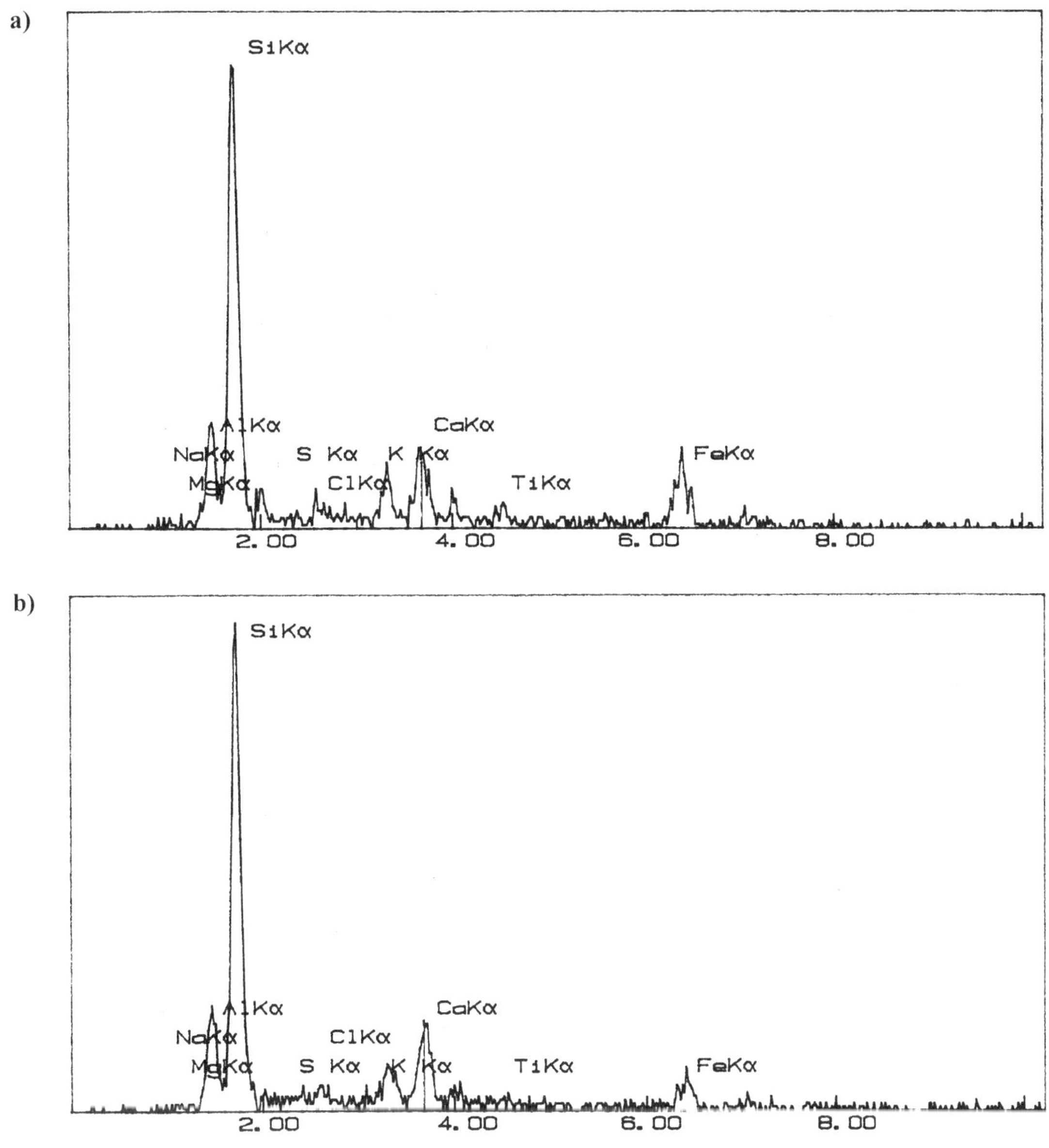

Figure 10.8: Spectra of the red (a) (FSD=181) and black (b) (FSD=202) areas of the slip on a composite silhouette dish with a resist decorative treatment (sample 296) showing elevated levels of iron oxide in the red areas.

monochrome dishes, indicating they were fired within the same temperature range and under similar conditions.

Surface Finishing Techniques - No apparent differences were observed in the microstructural characteristics of the slip layer on the bichrome composite silhouette dishes and those on other locally produced vessels of the Terminal Classic period. In addition, slip thickness was found to be comparable to the monochrome composite silhouette dishes. The absence of microstructural inconsistencies between the red and black coloured areas of the slip on the bichrome vessels provides some evidence that the contrasting colours were in fact produced through a resist technique. This method of colour decoration involves the application of a protective material to select areas of the slipped surface prior to firing to produce localized areas of reduction. The protective material is subsequently removed from the blackened areas, leaving a smooth finished surface (see Shepard 1956:206-213 for a discussion). The light colour of the surface and subsurface zones of the firing horizons indicate that the black decoration was produced whilst firing in an oxidizing atmosphere and the absence of a colorant such as manganese that remains black when oxidized confirms that the black decoration is not simply paint. This interpretation is supported by the results of a refiring experiment, which showed that that the black decoration turns the same red as the background colour when fired under oxidizing conditions (850°C for 1 hour). Interestingly, EDAX analysis of the red and black coloured areas on individual vessels showed that iron levels were consistently lower in the black areas (Figure 10.8). This was the only significant chemical variation observed among the vessels analysed. Based on the current evidence, however, it is unclear what specifically might have caused this chemical difference. The composition of the protective material used to produce negative decorations might have been a contributing factor.

Red Slipped Vessels with Black Painted Decoration

Two vessels from this stylistic group were analysed by SEM: a rounded dish and a massive bowl. Although the fabrics of both of these vessels correspond to the Sascab-Quartz class, the massive bowl corresponds to the local fabric group (Sascab-Quartz A), whilst the dish, connected geologically to areas bordering the Rio Hondo in the Corozal District, is non-local. Both of the fabrics were made using calcareous clay, but spectral analysis indicates that calcium levels are significantly lower in the imported fabric. A further chemical difference between the two fabrics is that the local fabric contains relatively higher levels of iron.

Firing - The imported fabric is completely oxidized, displaying a uniform reddish-brown colour throughout the cross-section. The red colour of both the slip and the body, as well as and the absence of firing horizons, suggests that this vessel was fired under oxidizing conditions. In comparison, the local fabric has a thick grey core that is bordered by thin light orange margins. The light coloured margins, together with the red slip, indicate that the firing cycle ended with a period of oxidation. Whether this was preceded by an initial period of reduction is unclear, as the darker-coloured core might have been caused by incomplete oxidization or fast firing (see Buxeda et al. 2003b).

SEM analysis of the bodies of the imported and local vessels indicates that they were fired at different temperatures. The microstructure of the local fabric is unvitrified, corresponding to the NV stage of vitrification. A firing temperature below 800°C is therefore indicated. A slightly higher firing temperature of 800-850°C can be estimated for the imported fabric since its microstructure, containing glassy areas, corresponds to Initial Vitrification (IV).

Surface Finishing Techniques - Significant differences were observed in the nature of the slip layer and black paint on the local and imported vessels. In both cases, a thin layer of slip approximately 5-10μm in thickness can be distinguished. On the local vessel, the slip is unvitrified (Figure 10.9a) and its micromorphology is inconsistent, exhibiting areas in which the clay particles are compacted and sintering is more extensive. This non-uniform microstructure is caused by an uneven burnishing of the vessel's surface. The general composition of the slip on the local vessel is similar to that of the body, except that iron and potassium levels are higher in the slip. Elevated iron levels in conjunction with potassium suggest that the slip was produced through levigation of clay similar to that used to make the body. In comparison, the slip layer on the imported vessel is uniform with extensive sintering (Figure 10.9b). Burnishing marks occur on this vessel, but differential compaction of clay particles was not observed with the

a)

b)

Figure 10.9: Slip layers on Black on Red vessels; a) unvitrified slip on the local vessel (sample 48) (x2000); b) extensively vitrified slip on the imported vessel (sample 118) (x2000). Bar scale = 10μm.

SEM, suggesting a more careful treatment of the surface. EDAX analysis indicates that the composition of the slip is clearly different from the body, due to the presence of sulphur in the slip. The spectra obtained for the slip are also characterized by a low peak:background ratio, which may suggest the presence of an organic or carbon-based component. The black painted areas of the imported vessel produce similar readings, but with even higher levels of 'noise' (Figure 10.10b).

On both vessels, a distinct layer of black paint sitting atop the slip layer cannot be distinguished. The presence of aluminium and silicon in the black painted areas of both vessels suggests a clay-based paint. With regard to the local vessel, the spectra obtained for the painted area were not substantially different from those obtained for the slip layer, suggesting that accurate readings were not obtained for the paint. Nonetheless, the painted areas on the local and imported vessels are clearly compositionally different (Figure 10.10). The black painted areas of the imported vessel are characterized by a low count rate and consequent high levels of 'noise'. This 'noise', together with the absence of manganese, suggests that the black paint is carbon-based, being made from a charred organic material.

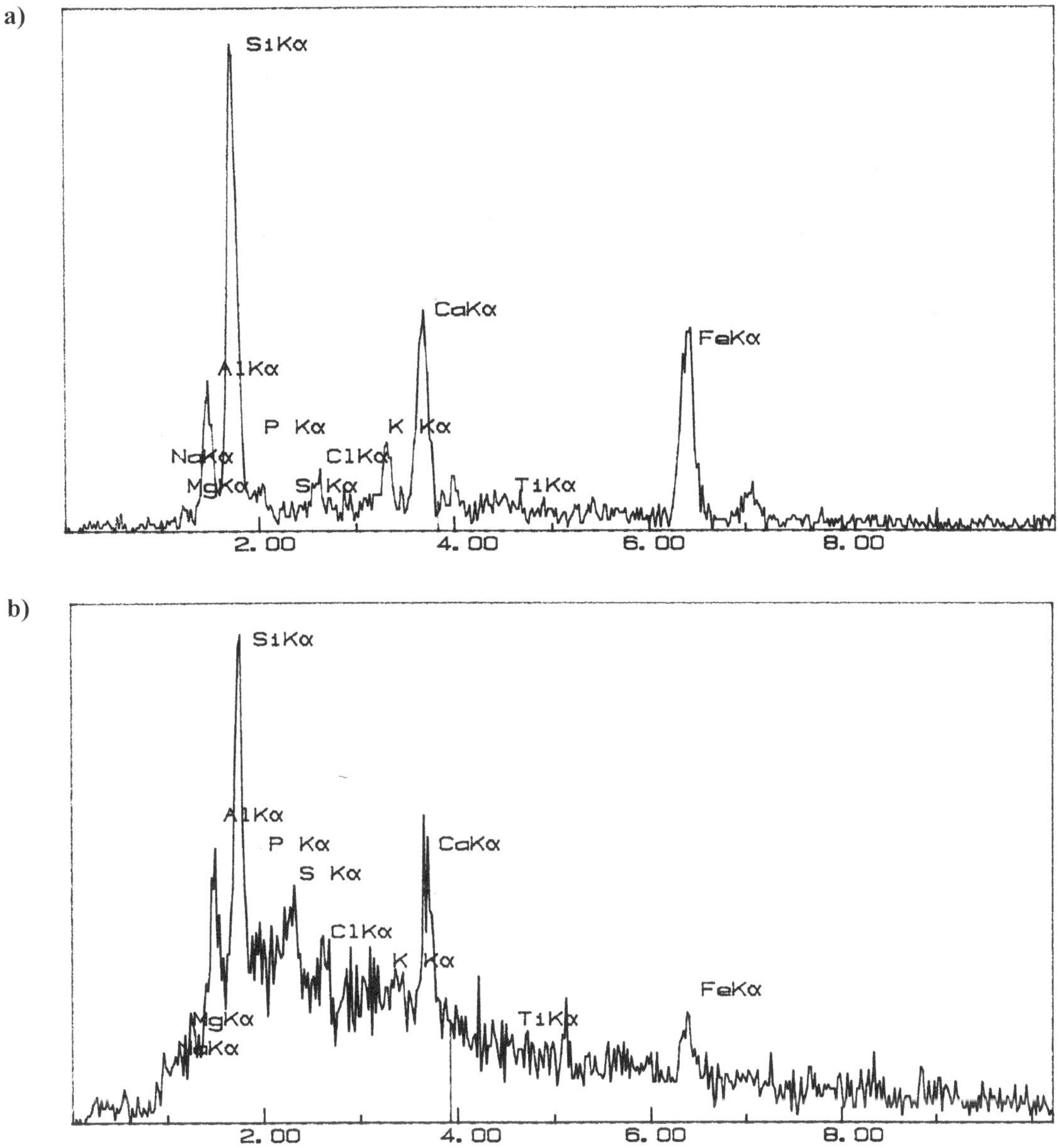

Figure 10.10: Spectra of the black painted areas on the local (a) (FSD=235) and imported (b) (FSD=127) vessels showing significant compositional differences.

Polychrome Dishes and Bowls

The polychrome vessels included for detailed examination by SEM derive from three different fabric groups. Three of the vessels, a bowl and both dishes, derive from local fabric groups, corresponding to two different local traditions of paste-making. The bowl has a calcite-tempered fabric (Group C) and the dishes have Sascab-Quartz A fabrics. The remaining vessel, although similar in form to the local bowl, was ascribed a non-local provenance, connected geologically to coastal areas of northeastern northern Belize, extending into southern Yucatan. In all cases, the clay used to make the ceramic body is calcareous (>10% CaO), and the imported vessel is distinguished by the presence of titanium, which is absent in the local vessels. The primary focus of the SEM analysis was to compare the vitrification microstructures and paints on the polychrome vessels and to compare the vitrification microstructures of the local polychrome vessels to those of monochrome vessels with equivalent fabrics.

Firing - The fabrics of all of the polychrome vessels exhibit firing horizons, with light to dark grey cores bordered by orange to buff subsurface and surface zones. The colour of the later zones indicates oxidizing conditions and the darker-coloured cores were most likely caused by incomplete oxidation. Although there appears to be no appreciable difference in firing atmosphere, the microstructural characteristics of the fabrics suggest variation in firing temperature. The microstructure of the local calcite-tempered bowl is similar to that observed in the oxidized monochrome and bichrome vessels that were examined from the same fabric group. Thus, a firing temperature approaching 800-850°C can be estimated. In comparison, the microstructure of the local dishes

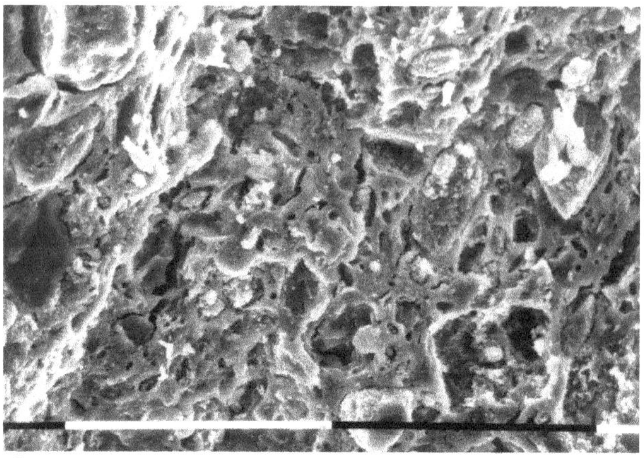

a)

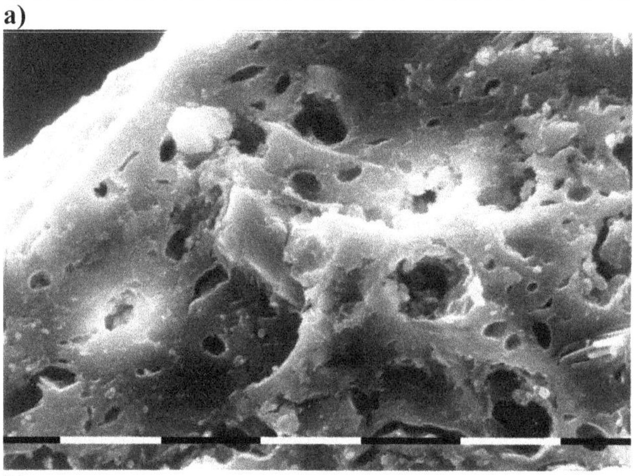

b)

Figure 10.11: a) body microstructure of a polychrome rounded dish with a Sascab-Quartz A fabric (sample 202) displaying a vitrified microstructure (IV+) that falls between the Initial Vitrification and Extensive Vitrification (x500); b) red painted decoration on an orange-slipped polychrome rounded dish with the same fabric (sample 207). Note that the surface layer is vitrified and discrete paint and slip layers can not be distinguished (x2000). Bar scale = 10μm.

appears to fall between Initial Vitrification (IV) and Extensive Vitrification (V), indicating an estimated firing temperature of 800-900°C (Figure 10.11). Since these fabrics have sascab temper, however, the temperature could not have been much higher than 850°C so as to avoid the decomposition of calcium carbonate. Significantly, the polychrome dishes appear to have been fired at a higher temperature than conformal monochrome dishes with the same fabric. The microstructure of the imported vessel is different from the local vessels, equivalent to Initial Vitrification (IV), which corresponds to an estimated firing temperature of 800-850°C.

Decorative Techniques - A discrete layer approximately 5-10μm can be distinguished at the surface of all of the polychrome vessels. This layer displays extensive sintering in all cases, but vitrified areas were only observed in the local dishes (Figure 10.11b). EDAX analysis showed that orange slips that occur on three of the vessels contain higher concentrations of Fe- and potassium than the respective bodies, suggesting the use of a refined suspension of the body clay for the slip.
Painted decoration occurs on all of the vessels analysed, although it was not possible to get accurate readings for the black paint on the local calcite-tempered vessel and the imported vessel. For all the vessels, no differences were observed in the microstructure of the painted and undecorated areas of the surface layer (Figure 10.11b). In most cases, however, the presence of paint could be detected through EDAX analysis. Comparison of the upper and lower regions of the surface layer showed differences in the concentrations of iron and manganese, indicating the presence of the paint.

In all cases, the paints used appear to be clay-based since aluminium and silicon were consistently found to occur in the painted areas that were distinguished chemically with EDAX analysis. The red paints, which range from deep red to orangish-red, are characterized by high levels of iron in comparison to the slip, indicating that these paints were made using an iron-based pigment. In comparison, the black paints, which could only be distinguished on the local dishes, were found to contain both iron and manganese, suggesting the use of a manganese-rich compound as the colourant. Since the dishes were fired in an oxidizing atmosphere, the presence of the manganese is the key factor in producing the black painted decoration (Noll 1975:608-610).

10.3. EARLY POSTCLASSIC MONOCHROME ORANGE VESSELS

A total of seven vessels representing vessel styles typical of the Early Postclassic period were selected for detailed analysis by SEM. A range of vessel forms were included as well as vessels with incised and modelled decorative elements and undecorated surfaces. All but one of the vessels (152) derive from local fabric groups (Grog-Mixed Carbonate class and Sascab-Quartz A). The imported vessel is connected geologically to coastal areas of northeastern northern Belize, extending into southern Yucatan. In all cases, the clay used for the body is calcareous (>10%CaO), but the presence of magnesium in the imported fabric clearly distinguishes it from the local fabrics. The presence of magnesium the imported fabric reflects the fact that this vessel was made using a dolomitic marl. Among the local vessels, there is some variation in iron, titanium and potassium levels.

Firing - The fabrics within this group exhibit a range of different firing horizons. Three of the fabrics (samples 335, 81 and 152) are completely oxidized, displaying a uniform orangish-buff colour throughout the cross-section. The colours of the slip and the body, as well as the absence of firing horizons, suggests that these vessels were fired under oxidizing conditions. In the case of sample 335, however, the microstructural characteristics of the fabric are most similar to those displaying firing horizons, suggesting that this vessel was fired under different conditions than the other two vessels with oxidized fabrics (see below).

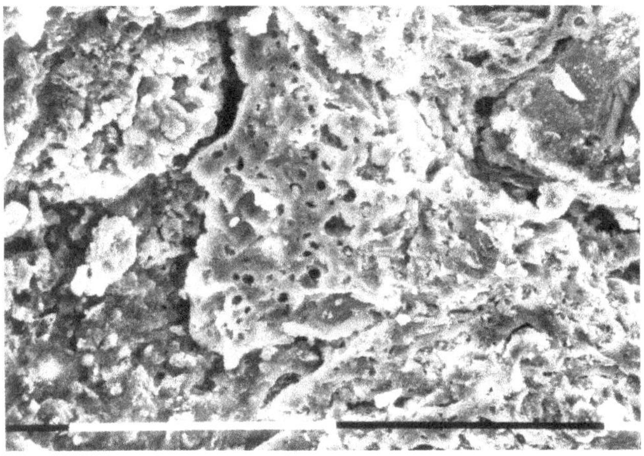

Figure 10.12: Body microstructure of a Early Postclassic chalice (sample 327) subjected to reduction-oxidation firing regime with a maximum firing temperature of 800-900°C, showing abundant fine bloating pores in a vitrified microstructure (x500). Bar scale = 10μm.

Firing horizons are present is four of the fabrics, which exhibit grey to black cores, bordered by lighter-coloured subsurface and/or surface zones. A distinctive characteristic of the fabrics displaying a thin light orange oxidized zone adjacent to the surface is that the interior margin of this zone is sharply defined. Rye (1981:118) suggests that the presence of a sharply defined oxidized zone is indicative of rapid oxidation or cooling in air and is characteristic of open firing. The microstructural characteristics of these fabrics indicate that the firing regime involved a period of reduction. Reduction firing is indicated by the presence and abundance of fine bloating pores in extensively sintered and vitrified areas, which gives these fabrics a 'spongy appearance' (Maniatis and Tite 1981:65-66) (Figure 10.12) Such microstructures are also considered to be typical of a fast-firing technique (Buxeda et al. 2003b). This microstructure is distinctive of these particular fabrics and was not observed in any of the other vessels analysed in this study. In this instance, therefore, the presence of the grey to black cores is due to reduction firing. The lighter-coloured subsurface zone, together with the orange colour of the slip, indicates that reduction firing was followed by a period of oxidation. In the case of sample 335, which lacks a dark-coloured core, it would appear that this period was of sufficient duration to remove any carbon present. In addition, the presence of a thin, sharply defined, oxidized surface zone in some fabrics, together with the 'spongy' appearance of the microstructure, suggest that the whole processes happened quite quickly, indicating 'fast-firing' under open-firing conditions (Buxeda et al 2003b).

Differences in firing temperature were also observed among the Early Postclassic vessels. The microstructures observed in samples 81 and 152, both of which are oxidized fabrics, exhibit extensive sintering but isolated glassy areas were not observed. Since this microstructure falls between No Vitrification (NV) and Initial Vitrification (IV), a firing temperature approaching the range of 800-850°C for IV microstructures can be estimated. Among the fabrics subjected to a reduction-oxidation firing cycle, the microstructure of three of the samples (99, 333 and 335) is extensively sintered and contains abundant fine bloating pores. As with the two oxidized vessels, this microstructure falls between No Vitrification (NV) and Initial Vitrification (IV). However, since these vessels were partly fired in a reducing atmosphere, a slightly lower temperature approaching the range of 750-800°C (IV and a reducing atmosphere) can be estimated. In comparison, the microstructure of samples 76 and 327, which is also characterized by the presence of fine bloating pores, appears to fall between Initial Vitrification (IV) and Extensive Vitrification (V), indicating an estimated firing temperature of 800-900°C. Since these fabrics contain inclusions of micrite and crystalline calcite, however, the temperature could not have been much higher than 850°C so as to avoid the decomposition of calcium carbonate.

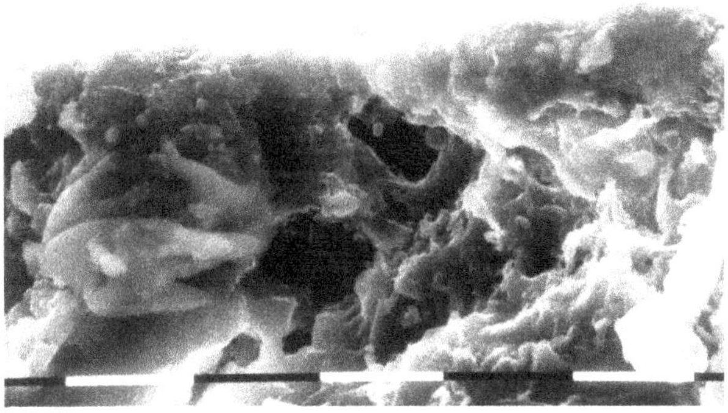

a)

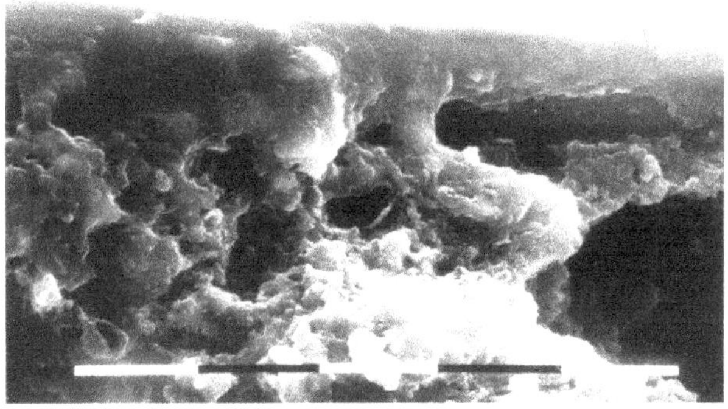

b)

Figure 10.13: Slip layer on Early Postclassic vessels; a) unvitrifed layer with extensive sintering (sample 333) (x2000); b) vessel 99, displaying a compacted and extensively sintered layer due to intense burnishing (x2000). Bar scale = 10μm.

Surface Finishing Techniques - A thin, fine-grained layer of slip, approximately 5-10μm, can be distinguished at the surface of all of the vessels analysed (Figure 10.13Figure). This layer is unvitrified, but sintering is more extensive in some cases (e.g. 76, 327 and 99), which appears, in most instances, to relate directly to slight differences in firing temperature. Iron-rich conglomerates are present in all cases. The slip layer on vessel 99, which is a tripod bowl, however, is extensively sintered and compacted, and was likely produced through intense burnishing of the vessel's surface (Figure 10.13b). Although burnishing marks are present on the other Early Postclassic vessels, comparably compacted slip layers were not observed. EDAX analysis of the slips and bodies of individual vessels suggests that the slips were produced through levigation of a clay similar to the body clay, producing higher iron and potassium levels in the slip. Comparison of the spectra obtained for the slip layers indicated differences in iron levels. Iron levels are comparatively low in samples 335 and 76 (both are chalices). Interestingly, the spectra obtained for these two vessels are strikingly similar to those obtained for the Terminal Classic monochrome rounded dishes that derive from the same local fabric group. In comparison, iron levels in slips of the other Postclassic vessels, are similar to those indicated for monochrome and bichrome Terminal Classic vessels with calcite-tempered fabrics.

10.4. SUMMARY

SEM analysis has revealed several interesting points of technological similarity and difference within and among the different pottery styles analysed with regard to decorative and surface finishing techniques and firing.

1) All of the vessels analysed share two main technological similarities, which can be considered to represent regional commonalities in Maya ceramic technology. Firstly, with the possible exception of the imported Black on Red vessel, slips appear to have been made through the refinement of clay that is compositionally similar to that used for the ceramic body, indicating that similar, or possibly, the same clay was used for both purposes. Secondly, although differences in firing temperature and regime are evident, the maximum temperature to which vessels were fired rarely exceeded 850°C. It is likely that firing temperatures were purposely kept low due to the nature of the clay bodies, which in all cases contain inclusions of either calcite or dolomite. Since calcium carbonate starts to decompose at around 850°C, it would have been necessary to carefully monitor and control firing temperature to avoid failures.

2) The differences observed in surface finishing and decorative techniques correlate with differences in vessel composition. For example, the locally produced vessels that are associated with different paste-making traditions are also discriminated by differences in slip thickness, and in the case of calcite-tempered fine ware, by the compositional characteristics of the slip. Similarly, the imported Monochrome Black and Black on Red vessels are distinguished from local stylistic equivalents by significant differences in surface treatment and decorative techniques. SEM analysis, therefore, has provided further evidence of technological variation among vessels already distinguished by differences in paste technology and provenance.

3) The results of the study indicate that during the Terminal Classic Period there is greater variability in surface finishing and decorative techniques and firing procedures than in the Early Postclassic period. Nonetheless, SEM analysis has documented further aspects of technological variation among Early Postclassic ceramics, indicating fundamental differences in the manufacture of vessels that share a common and distinctive set of stylistic characteristics. For example, the macroscopic and microstructural evidence distinguishes three different types of fired bodies, reflecting differences in firing procedures. It is also significant that the results of the present study suggest that only Early Postclassic vessels were produced using a reduction-oxidation firing sequence that resulted in a substantially different body microstructure. This body microstructure suggests fast-firing under open-firing conditions. The emergence of this new microstructure is suggestive of a significant shift in local firing practices, representing a new method of producing vitrified bodies at relatively low firing temperatures.

Although the sample size is admittedly small, when the results of the SEM analysis are taken together with the findings of the petrographic and chemical analyses, it would seem that the scientific examination of the Lamanai fine ware pottery has documented an important diachronic change in the relationship between stylistic and technological variability within the assemblage. During the Terminal Classic period, fine ware vessels made using the same set of raw material ingredients also display a high level of consistency in terms of the methods used to slip, decorate and fire these vessels. Generally speaking, stylistic diversity correlates with technological diversity. By Early Postclassic times, however, fine ware pottery exhibits a high level of stylistic consistency, but is characterized by compositional and technological heterogeneity. The nature of this aspect of ceramic change is explored further in the following chapter through a consideration of the wider social setting within which ceramic production and consumption took place.

CHAPTER 11

CERAMIC PRODUCTION AND CONSUMPTION AT LAMANAI DURING THE CLASSIC TO POSTCLASSIC TRANSITION: EMERGENT PATTERNS AND CONTRIBUTING FACTORS

The preceding chapters have focussed on documenting the nature and temporal dimensions of variability in the stylistic, technological and provenance characteristics of the pottery comprising the Terminal Classic to Early Postclassic assemblage. The data generated through this detailed physical analysis of the pottery provides a basis for the reconstruction of community-level patterns of ceramic production and consumption. Through the identification of instances of continuity and change in manufacturing procedures and ceramic depositional patterns, this chapter investigates shifts in local demand and manufacturing priorities, the nature of ritual and ceremonial practice, and the range of pottery used within different contexts of religious and ceremonial activity. There are two main focal points of the discussion. The first concerns the compatibility of the patterns that have emerged at Lamanai with the conventional characterizations of ceramic economic activities that form the basis of current explanatory models of ceramic change during the Classic to Postclassic transition. The second relates to the contributing role of factors external and internal to the community in shaping local patterns in ceramic production and consumption during this period.

11.1. CONTINUITY AND CHANGE IN LOCAL POTTERY PRODUCTION

As with other community-based activities at Lamanai during the Terminal Classic to Early Postclassic period, local pottery production shows evidence of both continuity and change. Local potters continued to be the primary producers of ceramic items destined for final deposition as part of ritual and ceremonial events that took place within the community, demonstrating a sustained self-sufficiency in terms of fulfilling the community's ceramic requirements in this regard. Nonetheless, changes in the kinds of vessels used in theses activities, as well as their visual appearance and technological characteristics, indicates that ceramic needs and tastes had changed by Early Postclassic times, reflecting an important shift in demand as well as manufacturing priorities. The technological data presented in the preceding chapters provide abundant evidence of the kinds of changes that occurred in local manufacturing procedures, and yet it is also evident that the different choices that potters were making by Early Postclassic times were informed by a local tradition of technical and environmental knowledge that had also guided earlier potters. For example, although paste preparation methods for fine ware production had changed significantly by the Early Postclassic period, potters continued to exploit the same local raw material resources as their predecessors. Similarly, there is good evidence that certain surface-finishing techniques and firing methods continued to be used in the manufacture of local fine ware, although alongside new techniques and a significantly different approach to decorative embellishment. It is also evident that changes in the morphology of the predominant fine ware forms produced locally (i.e. bowls and dishes) document a continuous development from earlier conventions. Accordingly, the ceramic evidence indicates a complex developmental process characterized by the blending and merging of new ideas and approaches with established conventions (Table 11.1). Central to this dynamic was the ongoing dictates of the community's ceramic needs, potters' practical responses to changes in demand and an expanding stock of technical knowledge and 'know-how'.

Temporal Trends in Fine Ware Production

The early part of Terminal Classic to Early Postclassic ceramic sequence is characterized by the existence of several different local approaches to pot-making, distinguished by a particular set of choices with regard to raw material selection, vessel forms, surface-finishing and decorative technology and firing methods (Table 11.2). These approaches are quite consistent, especially with regard to paste and slip technology and firing methods. The stylistic data indicate notable variation and overlap in vessel forms, slip colour and other decorative characteristics within and among these 'technological groups'. Within the assemblage analysed, however, each of these different approaches is characterized by a predominance of stylistically distinctive form categories. Two of these approaches are strongly associated with the production of different fine ware dishes that have a variety of decorative treatments. The remaining two approaches are associated strongly with the production of coarse ware jars with red-slipped rims and 'interior-slipped' bowls, which seems to represent an original service set comprising vessels for drinking or eating and a storage container. Nonetheless, whether the predominance of these particular vessel forms within the different technological groups can be considered to reflect consumption tendencies within the contexts analysed, or actual production specialization, cannot be judged based on the current body of evidence due to the limited range of depositional contexts considered.

By Early Postclassic times, the multiple, internally-consistent, early approaches to fine ware production, give way to a single, more general approach to vessel manufacture which is characterized by a high level of internal variability (Table 11.2). Unlike earlier approaches to pottery manufacture, this later approach

Table 11.1 : Local production patterns during the Terminal Classic to Early Postclassic period, according to stages of the production process.

Selection of Manipulation of Raw Materials

Fine ware

- Multiple local approaches to paste manufacture that exhibit a high level of internal consistency give way to a single general approach that is characterized by a high level of internal heterogeneity (as shown by petrographic and chemical analysis).
- A shift in practice away from the use of a single tempering material, primarily different forms of calcium carbonate, and towards the simultaneous use of multiple tempering materials including carbonates, grog and sand.
- A shift towards a more careful or rigorous processing of raw materials resulting in comparatively fine-textured pastes.
- Continuity in the local clay resources chosen for fine ware production.

Coarse Ware

- The emergence of Quartz Sand fabrics, which represent an approach to paste manufacture not prevalent at Lamanai in preceding time periods.
- A continuation of the local tradition of crystalline-calcite-tempered coarse wares, although with some changes in raw material ingredients and perhaps their processing.

Forms and Morphology

- The emergence of new vessel forms, which in many cases represent new functional categories of pottery – e.g. a range of incense-burning paraphernalia, drums and comals.
- Particular forms of service and drinking vessels – e.g. the vase – cease to be produced.
- Vessel morphology increases in complexity, in terms of body geometry and incorporating hand-modelled and possibly mould-made decorative embellishments (e.g. effigy appliqués and foot supports and flanges) and appendages. Dependent composite forms increase in prevalence, often consisting of base, body and connective components and with enlarged or exaggerated rims.
- Bowls and dishes remain the dominant fine ware forms produced and the development of vessel morphologies within these shape classes appears to be continuous in nature, indicating an indigenous development out of earlier conventions.
- Crystalline calcite-tempered coarse ware jars continue to exhibit a comparatively standardized morphology, but with changes over time in surface treatment and specific morphological attributes. In comparison, jars manufactured within the Quartz-Sand tradition exhibit a high level of morphological heterogeneity.
- A loss of diversity in surface treatment and decorative technology.
- A shift in decorative technology away from painting and the manipulation of slip colour through firing methods and a resist technique to produce to different colours and colouring effects, and towards oxidized colours and surface penetrating decorative techniques (incision and gouging). By the Early Postclassic period, vessels exhibiting more than one colour are quite rare, and coloured decoration is achieved applying a coat of stucco to selected areas, such as foot supports, subsequent to firing and painting it, or through the addition of Maya Blue after firing.
- A change of technique for rendering incised decoration. In the early part for the sequence decorative embellishing through stamping, incision, punctation etc, is done prior to slipping. This technique ceases and such decoration is always added to the surface after slipping by Early Postclassic times.

Table 11.1: Local production patterns during the Terminal Classic to Early Postclassic period, according to stages of the production process (continued).

Surface Finishing a Decorative Techniques
• Decorative designs become more intricate and complex. The composition of the designs rendered on individual vessels is most often unique and yet derives from a limited repertoire of fairly standardized motifs. Decorative designs characteristic of the different surface treatment groups dating to the early part of the sequence are comparatively consistent in terms of their composition.
• Decorative embellishment becomes confined almost exclusively to the exterior surface by the end of the sequence. In addition, the scale of the decoration is generally larger and there is a tendency to cover all the available space.
• Continuity in slip manufacturing and application techniques. Techniques used in the Early Postclassic period derive from earlier precedents.
• A simplification of surface treatments accorded to coarse ware jars. By the end of the sequence these vessels are rarely accorded slips or other decorative embellishments.
Firing Methods
• Variability in firing methods is generally greater in the early part of the sequence. In addition, early in the sequence differences in firing technology coincide with differences in other aspects of vessel technology as well as stylistic characteristics. By Early Postclassic times, firing methods still vary but the general style and decorative technology of the vessels being produced are similar.
• The emergence of a firing regime, involving a period of reduction followed by a final oxidation stage, which produces vitrified ceramic bodies with oxidized surfaces at low temperatures.
• A continuation of some firing methods into the Early Postclassic period.

cannot be characterized as expressing a particular set of choices indicative of consistency in technical practices. Rather, variation in paste technology, surface-finishing and decorative treatment and firing methods indicates different technical choices and, thus, the existence of different, but technically overlapping, manufacturing processes. As a result, Early Postclassic pottery exhibits a superficial stylistic and technological uniformity, sharing common stylistic and decorative elements and petrographic characteristics, but in terms of technique, is quite variable, due to differences in the specific set of technical procedures employed in its production.

Paste technology is characterized by the use of multiple tempers, including grog, and a more rigorous or careful processing of raw material ingredients to produce fine-textured fabrics. In comparison, earlier paste technologies are characterized by comparatively coarse-textured fabrics, and the use of tempering materials other than sascab and calcite is rare. It is also apparent that design elements and decorative technology of Early Postclassic pottery exhibit a higher level of complexity than is seen in earlier approaches to fine ware manufacture. Early Postclassic vessels display a comparatively complex body geometry and dependent composite forms, consisting of high pedestal bases, bodies characterized by a complex curvature including enlarged or exaggerated rims, and connective elements are common. Basic forms are made more elaborate through the addition of hand-modelled and, possibly, moulded decorative embellishments such as segmented flanges and anthropomorphic and zoomorphic appliqués, handles and foot supports. Additional decorative embellishments most often include the rendering of intricate designs through incision and gouging. In contrast to the earlier technique of incising prior to slipping (which, incidentally, appears to have been rarely employed by local potters in general), the decorative designs on Early Postclassic vessels are always rendered subsequent to slipping, most likely to ensure their visibility. In addition, they tend to cover the majority of the available space on the exterior surface. This focus on extensive decoration of the exterior surface, specifically with regard to open forms, also represents a departure from earlier conventions. The composition of the incised designs on individual Early Postclassic vessels is most often unique, and yet appears to derive from a repertoire of standardized motifs that are repeatedly encountered on different vessels. The net effect of this high level of variation in decorative embellishment is that Early Postclassic vessels often have an individual character that is not prevalent among earlier vessel styles.

Another important aspect of this later approach to local fine ware manufacture is that it is accompanied by an expansion of the local ceramic repertoire to include new kinds of pottery vessels. Of particular significance is the inception of ceramic musical instruments, particularly

Table 11.2: Varied local approaches to pottery manufacture during the Terminal Classic to Early Postclassic period.

		Paste Making Tradtion	Paste Technology	Predominant Vessel Forms	Surface Treatment	Firing Methods (atmosphere/temp.)
Terminal Classic	Fine Ware	Crystalline Calcite-Tempered (Group C)	Yalbac clays (weathered limestone horizon) tempered with finely-crystalline calcite and sascab	composite silhouette dishes	thick slip layer, monochrome red, orange and bichrome resist	oxidizing conditions/approaching 800-850°C
	Fine Ware	Sascab-Quartz (Group A)	Filipe clay tempered with sascab	rounded dishes	thin slip layer ('interior-slipped'), monochrome red, orange, buff, bichrome black on red and polychrome	oxidizing conditions /800-850°C
	Fine Ware and Coarse Ware	Crystalline Calcite-Tempered (Group A)	Yalbac clays (subsurface) tempered with coarsely-crystalline calcite	rounded bowls and large storage jars	bowls - very thick slip layer ('interior-slipped') jars - red slip on lip or rim to neck area, well smoothed	bowls - oxidizing conditions /approaching 800-850°C
	Fine Ware and Coarse Ware	Crystalline Calcite-Tempered (Group B)	Yalbac clays (weathered limestone horizon) (subsurface) tempered with finely-crystalline calcite	rounded bowls and large storage jars	bowls - very thick slip layer ('interior-slipped') jars - red slip on lip or rim to neck area, well smoothed	bowls - oxidizing conditions /approaching 800-850°C
	Coarse Ware	Quartz Sand	Boom clays, possible sand temper in some cases	large storage jars	unslipped, crudely smoothed	—
	Coarse Ware	Crystalline Calcite-Tempered (Group H)	Yalbac clays (subsurface) tempered with coarsely-crystalline calcite	large storage jars	unslipped, well smoothed	—
Early Postclassic	Fine Ware	Grog-Mixed Carbonate	Yalbac clays (various) tempered with grog, and varying amounts of sascab, calcite and, in some cases, sand	a variety of forms	thin slip layer, orange slip, commonly with incision	various

different drum forms, and a range of distinctive vessel forms, which presumably had specialized functions, but appear to be associated primarily with burning incense and other substances based on the occurrence of carbon deposits (e.g. incense burners, frying pan censers and sieves). These new vessel forms represent functional categories of pottery that were not produced earlier.

Temporal Trends in Coarse Ware Production

Local coarse ware production follows a significantly different developmental trajectory than that of fine ware. The earliest stratigraphic contexts examined in this study are dominated by crystalline calcite-tempered coarse ware vessels, primarily jars. Their prevalence in these contexts suggests a continuation of a well-established local tradition of coarse ware manufacture. These vessels are well made and exhibit a high level of stylistic homogeneity, including a fairly standardized morphology, but were made using two distinct paste recipes, suggesting the existence of two slightly different approaches to calcite-tempered coarse ware manufacture. The results of the petrographic study indicated that this local coarse ware tradition continued into the Early Postclassic period, but with changes in jar morphology and surface treatment, as well as the raw material ingredients of the paste. For example, early jars are slipped to just below the rim or neck area whereas later jars are unslipped. By Early Postclassic times, calcite-tempered coarse ware exhibits greater stylistic and compositional variability, which presents a direct contrast to the high level of stylistic consistency that characterizes the early jars. Nonetheless, the different jar styles that are typical of the Early Postclassic period are still well made and display a fair degree of morphological standardization, with at least three distinct jar styles represented (see Appendix II, Unslipped/Smoothed Buff stylistic group). Occurring alongside these later jar styles in later stratigraphic contexts are new vessel forms, such as plates/lids and incense burners, with identical fabrics. The appearance of these new forms indicates an expansion in the range of vessel forms produced within this local tradition. The absence of slip on the later jars suggests a simplification of surface treatment.

Some time during the Terminal Classic period, a significantly different approach to coarse ware production emerges. This approach appears to lack a prior local precedent, but continues into the Late Postclassic period (see Chapter 8), suggesting that its appearance in the Terminal Classic period represents the inception of a new local tradition of coarse ware manufacture. This new approach continues alongside the calcite-tempered coarse ware tradition into at least the Early Postclassic period.

The new approach to coarse ware manufacture is based on a different set of raw materials, including sandy clays and, possibly, sand that derive from the east side of the New River Lagoon. The vessels produced are comparatively poorly made and are characterized by a high level of morphological heterogeneity, which gives individual vessels a unique character. In addition, a greater range of functional classes of pottery appears to have been produced within this tradition, although large storage jars are most frequent in the assemblage analysed here. Many of the vessels forms with utilitarian functions are also produced within the calcite-tempered coarse ware tradition. It is noteworthy that the jars produced within the new tradition bear stylistic and petrographic similarities to Late Classic jars at Altun Ha, but with mineralogical differences relating to the use of local clays (see Howie-Langs 1999). Similarly, the results of the petrographic analysis suggested that the local jar fabrics reflect an approach to coarse ware manufacture based on a particular choice of raw material ingredients (clays formed in areas of Pleistocene alluvium and sand) that also occurred elsewhere in Northern Belize. A significant aspect of the emergence of this tradition at Lamanai is that it involves the use of raw material resources that had not been habitually exploited previously by local potters.

Local Production Patterns in Comparison to Conventional Perspectives

Conventional perspectives on ceramic economic patterns during the Classic to Postclassic transition emphasize two main dichotomies: 1) that between the level of technological sophistication and standardization exhibited by polychrome pottery and trade wares in comparison to common service and utilitarian wares; and 2) that between the Classic period pattern of variability in pottery styles and technology and the Postclassic pattern of standardization. Taken to imply fundamental differences in organizational aspects of production, these characterizations form the basis of current explanatory models of ceramic change during the Terminal Classic to Early Postclassic period. Of particular relevance to the present study, therefore, is whether or not these characterizations and, thus, associated inferences relating to Maya economic infrastructures, can be considered to accurately reflect the situation at Lamanai. The evidence that has been revealed through detailed scientific examination and comparison of the technological characteristics of locally produced fine ware and coarse ware, suggests that they cannot. For example, the results of the analysis indicate a high level of technological similarity between polychrome rounded dishes, the predominant form of polychrome pottery manufactured locally, and bichrome, and monochrome rounded dishes, especially in terms of paste and slip technology, with no apparent difference in the level of technical consistency exhibited among the polychrome vessels. Just as these vessels are conformal, they are most often virtually indistinguishable petrographically. In addition, the SEM results suggest similar firing regimes, except that the polychrome dishes appear to have been fired at slightly higher temperatures, as indicated by their different body microstructures. The implication is that these vessels, regardless of the colour of their slips or the presence/absence of painted decoration, share important technological similarities reflective of commonalities in manufacturing procedures. In addition, there is no evidence to suggest that the locally-manufactured polychrome vessels are more 'standardized' technologically than their bichrome and monochrome counterparts.

Comparison of the two distinct local approaches to coarse ware production has also revealed a more complex pattern than the conventional characterizations imply. For example, the different styles of storage jars produced within the calcite-tempered tradition tend to exhibit a high level of morphological consistency and a uniform paste technology, reflecting a degree of technical standardization that stands in direct contrast to contemporaneous sandy-pasted jars. These jars exhibit greater variation with regard to their morphology and the petrographic characteristics of their fabrics, indicating an inconsistency in specific technical practices. These different patterns may relate, in part, to differences in organizational aspects of production, suggesting that it cannot merely be assumed that pottery corresponding to the same functional class was produced through similar means and methods. The implication is that there is some evidence that there may have been greater variation in organizational aspects of production on the community level than has been traditionally recognized and, consequently, that the proposed two-tiered system revolving around elite-sponsored workshops and part-time rural producers may be an over-simplistic characterization.

A related issue concerns the absence of slip on calcite-tempered coarse ware jars by Early Postclassic times, which might be considered to reflect a trend towards simplification of the surface treatment of utilitarian vessels. Such changes are often viewed as cost-saving measures related to a shift towards the mass production of ceramics in workshops (e.g. Masson and Rosenswig n.d.). A concomitant trajectory towards increased homogeneity in other aspects of vessel technology is assumed, considered as a natural by-product of potters working closely together in a workshop environment. Interestingly, at Lamanai, this change in the surface finishing techniques used in the production of coarse ware jars does not appear to have been accompanied by an increased homogeneity or standardization of other aspects of vessel technology. The different styles of calcite-tempered jars typical of the Early Postclassic period exhibit a similar level of morphological standardization as the earlier calcite-tempered jars with slipped rims. In addition, they are made to the same high standard, in terms of the time invested in achieving a uniform, high quality surface finish. Nonetheless, the results of the petrographic analysis indicate that the later jar styles exhibit greater compositional variability, in terms of the textural properties and aplastic inclusion content of their pastes. At least some of this variation appears to be chronological, but later jars still exhibit a more variable paste technology. Accordingly, there is no evidence that the change in the surface treatment of calcite-tempered coarse ware jars is somehow linked to a broader change in production methods. Rather, it would seem to relate, at least in part, to demand, reflecting changing ceramic tastes and preferences.

The loss of diversity in surface treatment among fine ware pottery by the Early Postclassic period has also been conventionally interpreted as reflecting a standardization and simplification of vessel styles, which is most often taken to imply technological standardization. Standardized production methods are generally assumed, which is perceived as symptomatic of the inception of mass-production. Analysis of Early Postclassic pottery from Lamanai, however, has revealed, that stylistic and technological similarities are perhaps more superficial than has been recognized previously, as this pottery is characterized by a high level of internal technological and stylistic heterogeneity. In addition, a simplification of elements of design associated with vessel morphology and decorative embellishment is not demonstrated. It could be argued that the stylistic and technological variation within the Lamanai assemblage is just as prevalent in the later part of the ceramic sequence as it is earlier on. In fact the high level of internal variability that characterizes Early Postclassic pottery stands in direct contrast to the high level of consistency exhibited by earlier approaches to fine ware manufacture. Accordingly, whilst it is apparent that stylistic and raw material characteristics of locally produced fine ware become more uniform by the Early Postclassic period, a shift towards technical standardization, that could be taken as indicative of mass-production, is not demonstrated.

11.2. CONTINUITY AND CHANGE IN CONSUMPTION PATTERNS

The Consumption of Pottery in Community-Based Ceremonial and Ritual Activities

The stylistic characteristics, functional associations and provenances of pottery provide important evidence of similarities and differences between the patterns of consumption reflected in ceramic deposits that arise out of different kinds of human activity. This information is fundamental to our understanding of the kinds of activities that took place in different physical and social settings, as well as to the identification and delineation of cultural practices and conventions.

A prominent characteristic of the different types of ceremonial and ritual deposits analysed in this study is the striking overlap in the kinds of pottery they contain. For example, the ceramic components of the midden deposits associated with structures N10-27 and N10-9 are similar in terms of the proportions of fine ware and coarse ware vessels, as well as the range of functional classes of pottery. From this we might surmise a general similarity in the patterns of human activity behind these deposits and, perhaps, the kinds of activities that took place at these different ceremonial structures. Since the midden deposits encompass a sequence of human activity that spans the Terminal Classic to Early Postclassic period, they are also indicative of the longevity of such behaviour. It is also evident that the fine ware pottery used and deposited as part of activities and events that took place at the ceremonial structures was also used in mortuary and offertory ceremonies. The fact that these latter types of ceremonies did not involve different or special kinds of pottery, suggests a basic congruency in the objects used in these different contexts and, perhaps, in certain aspects of ritual practice. Burials and offerings, however, rarely contain coarse ware pottery, particularly large storage jars, whereas these vessels are abundant in the midden assemblages. The presence and abundance of storage jars and other utilitarian pottery in the temple middens, points to an important difference in the nature of the activities that lead to the accumulation of these deposits, perhaps relating to the scale or functional aspects of associated activities and events. This general pattern of similarities and differences in the kinds of ceramics used in different religious and ceremonial contexts continues into the Early Postclassic period, despite the dramatic changes in vessel styles, indicating continuity in certain aspects of local customs or perceptions relating to the kinds of pottery deemed appropriate or required for different ceremonial and ritual events.

The provenance of the pottery comprising the burial, offertory and temple midden assemblages provides insight into the social and economic networks that underlie its use in associated ceremonies and events. The petrographic data indicate that throughout the Terminal Classic and Early Postclassic periods the large majority of the pottery used and deposited as part of these activities continued to be supplied by local producers. In addition, there is no evidence that consumption of non-local or 'imported'[1] pottery substantially increased by Early Postclassic times, as predicted by conventional explanatory models of the development of the lowland ceramic economy during the Classic to Postclassic transition. The petrographic data also indicate that, by the Early Postclassic period, non-local pottery came from a more restricted geographical region of the lowland area than previously. In the early part of the Terminal Classic period, non-local pottery derives from multiple points of origin within several different areas of the lowland region, including northern Belize and adjacent areas of the Yucatan Peninsula of Mexico. By Early Postclassic times, however, non-local ceramics appear to have originated exclusively from localities within this particular geographic area. Also significant is that, throughout the time period, drinking vessels and jars remained the primary types of non-local pottery brought to the site and subsequently used in different ceremonial and ritual activities and events. These ceramics include a number of different styles of service ware and utilitarian jars, which, in most cases, are visually distinct from the locally-manufactured functional equivalents. However, the relative frequencies of non-local fine ware and coarse ware pottery in different stratigraphic contexts within the temple middens provide some evidence that the amount of non-local fine ware, and particularly drinking vessels, used in local ceremonial and ritual events decreased significantly by Early Postclassic times. In contrast, the amount of non-local storage jars brought to the site and deposited as part of these activities appears to have increased over time.

Consideration of the specific archaeological contexts within which non-local pottery occurs yields additional insight into the consumption patterns associated with different kinds of ceramic deposits. Table 11.3 gives a breakdown of the occurrence of local and non-local ceramics in the different types of ritual/ceremonial deposits according to the type of architectural structures from which individual deposits were recovered. As can be observed, whilst the large majority of pottery deposited as part of different kinds of ceremonial and ritual activity was of local manufacture, non-local pottery also occurs in each of the different kinds of ceramic deposits. The temple middens contain the largest amount of non-local pottery, and it is also evident that the proportion of non-local ceramics is greater in the temple middens than in any of the other deposit types. With regard to burials and offerings, it can be observed that non-local ceramics tend to occur in burials associated with residential structures, whereas offerings containing non-local ceramics tend to be associated with public architecture, including elite residential and administrative complexes. This apparent difference in the kinds of structures with which burials and offerings containing non-local ceramics are associated may relate to a variety of social and ideological factors. But considering the nature of depositional contexts concerned, the association of these 'exotic' items with particular kinds of architecture might be considered to reflect material expressions of wealth, prestige and identity within the context of different kinds of religious and ceremonial events.

Changes in Modes of Consumption

Starting in the Terminal Classic period, the appearance of extensive midden deposits containing large amounts of pottery, perhaps purposefully broken prior to discard, in public areas of the central precinct, not only provides evidence of different depositional practices, but also changes in consumption patterns. It has been argued in previous chapters (Chapters 2, 6 and 7) that the contents of these middens, as well as their pattern of accumulation in relation to the history of maintenance of the ceremonial structures with which they are directly associated, suggests that these garbage dumps are the product of ongoing ritual and ceremonial activities, perhaps partly of a communal nature, that took place in these settings. The fact that at least some of the pottery used in these

[1] I am hesitant to use the term 'imported' as it implies a specific kind of economic behaviour that I am unconvinced can be assumed as primarily responsible for the occurrence of non-local pottery within these depositional contexts. In Maya archaeology, this term is often taken to imply that the movement of pottery from different manufacturing loci to Lamanai is rooted in market-based exchange and there is no evidence to suggest that other distributive mechanisms, such as gift-giving, tribute, or simply pottery being brought to the site with its owners, are not involved, given the archaeological contexts of the material.

Table 11.3: Frequencies of locally produced and imported vessels summarised by context.

	Archaeological Context	Provenance				
		Local	Local/ Non-local	Non-local	Non-local?	Total
Burials	ceremonial structure	13	0	0	0	13
	residential structure	26	0	9	1	36
	residential/admin. structure	20	0	0	0	20
	temple	2	0	0	0	2
	post-abandonment interment	2	0	0	0	2
	miscellaneous burial	1	0	0	0	1
	Total	**64**	**0**	**9**	**1**	**74**
Offerings	cache – ceremonial structure	1	0	0	0	1
	cache – residential structure	0	0	1	0	1
	cache – residential/admin. structure	11	0	0	0	11
	cache – temple	25	1	2	0	28
	general – ceremonial structure	0	0	1	0	1
	general – residential structure	4	0	1	0	5
	general – residential/admin. structure.	11	1	4	0	16
	general – temple	5	0	3	0	8
	Total	**57**	**2**	**12**	**0**	**71**
Temple Middens	N10-27	213	83	42	17	355
	N10-9	96	23	5	1	125
	Total	**309**	**106**	**47**	**18**	**480**
Refuse Deposits	residential structure	3	0	0	1	4
	residential/admin. structure	10	2	4	0	16
	miscellaneous refuse deposit	1	0	0	0	1
	Total	**14**	**2**	**4**	**1**	**21**
Grand Total		**444**	**110**	**72**	**20**	**646**

activities appears to have been purposefully broken during or at the conclusion of these events (as indicated by fragmentation patterns) is indicative of a particular mode of consumption that might not have been as prevalent in preceding time periods. The same observation can be made with regard to the appearance of smash-and-scatter offerings, and the shift towards the interment of purposely broken pottery in burials, albeit with greater certainty, since these practices represent clear departures from earlier patterns. When considering that the midden pottery constitutes ceramics permanentlyremoved from use as part of possibly large scale ceremonial and ritual events, it seems likely that this different mode of consumption may have had a significant effect on demand as well as manufacturing priorities. The deliberate breakage of pottery as part of such activities could have accelerated demand for such pottery since new vessels would have been needed for subsequent events and occasions, leading to an increase in production (Arnold 1985:163). In addition, assuming that the timing of such events would have been known to members of the community, including potters, the demand for ceramics might have increased at certain times of the year, thereby stimulating increased production, or the production of particular kinds of pottery in anticipation of particular ceremonial occasions or festivities. Arnold (1985:160) describes a similar situation in conjunction with festivities associated with All Saints Day for contemporary communities in Yucatan, Mexico and Guatemala.

The Local Ceramic Repertoire and Shifts in Demand

Additional changes in consumption patterns are indicated by the appearance of new functional classes of pottery vessels in burial, offertory and midden deposits, as well as by the loss of stylistic diversity in the fine ware pottery occurring in these deposits by Early Postclassic times. The appearance of ceramic drums, whistles, and a range of new vessel forms that appear to have functioned, at least in part, as receptacles and implements for burning incense and other substances, is reflective of a new demand for particular kinds of ceramic items. This new demand relates directly to the use of musical instruments and incense-burning paraphernalia in ritual and ceremonial activities and hence, is suggestive of innovations or changes in ritual and ceremonial practices, which may have initially stimulated the production of these items. The disappearance of vases from burial, offering and temple midden assemblages provides further evidence of changes in demand and production associated with the use of particular drinking vessels in certain rites and ceremonial events. The loss of stylistic diversity in the repertoire of fine ware vessels used in ritual and ceremonial activities is indicative of a trend towards the consumption of pottery conforming to a particular and distinctive set of stylistic traits. Within the context of ceremonial and ritual activities ceramics act as vehicles for the expression of cultural values and ideological beliefs and ideas (Arnold 1985:158-159). Accordingly, changes over time in their stylistic characteristics, decorative embellishment and the nature of design elements and decorative treatments reflect changes in the manner in which this information is communicated and perhaps, the kinds of ideas being expressed. Viewed from this perspective, the preference for particular vessel shapes, as well as for pottery of a particular colour and with particular kinds of decorative embellishment, suggests a change in the ideological information conveyed by the pottery used in local religious and ceremonial events. It might also be considered to reflect an increased concern for consistency and order, at least in terms of the nature of the pottery used in these behavioural contexts.

11.3. Factors Contributing to Ceramic Change During the Terminal Classic to Early Postclassic Period

Maya scholars have identified four regional-level developments that contributed to the dramatic changes in the material and archaeological manifestations of Maya culture that occurred during the Terminal Classic to Early Postclassic period:

1. Military pressures caused by the predatory expansion of northern lowland City states, particularly Chichen Itza.
2. The disruption of networks of exchange, communication and political affiliation as a result of the decline and failure of elite culture and socio-political institutions in the Maya heartland and the emergence of a new system of trade and commerce as part of the expanding influence of northern lowland hegemonies such as Chichen Itza.
3. Population migrations stimulated by ongoing conflicts and instability in areas of the southern and northern lowlands.
4. Ideological shifts relating to the expansion of a regional religious cult, driven by elites, that emphasized militaristic proselytism.

The evidence from sites in northern Belize suggests that individual communities were affected differentially by these developments, such that the contributing role these factors played in shaping local-level trajectories, either individually or in combination, appear to have varied from site to site. As a result, individual community histories document different local experiences of internal and external pressures, degrees of prosperity, actions and reactions.

The extent to which these regional-level developments affected community life at Lamanai is particularly ambiguous. The archaeological evidence indicates that the community was a hive of activity during a time when nearby city centres such as Altun Ha and Colha were experiencing decline, abandonment and in the case of the latter, militaristic invasion. Construction activities dedicated to the improvement and expansion of the ceremonial precinct suggest a time of continued prosperity and socio-political stability. This general pattern of stability and continuity, however, stands in direct contrast to the often striking changes that occur with regard to other aspects of day-to-day life such as building procedures, offertory and burial practices, the use of public areas associated with large ceremonial structures and material culture, including the stylistic and technological characteristics of locally manufactured pottery. It is also clear that the link between these significant changes and certain regional-level developments is less than direct. For example, the structural remains and artifact inventories of the time offer no indication that Lamanai ever underwent a militaristic invasion, whether as part of the expansion of centres of power based in Northern Yucatan, a religious cult driven by the political ambitions of elites, or otherwise. In fact, there is no compelling evidence of a defensive community posturing starting in the Terminal Classic period, as a response to a perceived external threat, apart from the possibility that the new residential structures built in the vicinity of public areas of the central precinct represent local families that had moved to this area of the settlement for safety reasons. In addition, the large scale construction projects undertaken in the ceremonial precinct starting as early as the Late Classic period, together with the continued local production of material goods, such as pottery, and subsistence products (particularly maize), attest to the ongoing stability of the community's economic and socio-political infrastructure. Accordingly, it would appear that any disruptions in traditional networks of exchange and political affiliation did not have a particularly significant impact on local economic patterns. The question that emerges, therefore, is that if it can be demonstrated that certain external pressures appear to have played a limited role in the community's developmental trajectory (at least in some respects) then what factors did contribute to the changes in local conventions, in both material and behavioural sense, during this period? A closer examination of local patterns of ceramic production and consumption, in light of certain regional-level developments, provides some interesting insight into the kinds of factors involved, from a ceramic point of view.

Evidence of a Reorientation of Politico-Economic Relationships

Local ceramic economic patterns at sites in northern Belize are considered to have been altered significantly during the Terminal Classic to Early Postclassic period due to two developments on the regional level: 1) the loss of trading partners as a direct result of the economic decline of southern lowland polities situated in the heartland area and 2) the expansion of maritime networks of exchange driven by Chichen Itza (Masson and Mock 2004). The absence of ceramics deriving from areas of the southern lowlands to the south and west of Lamanai in later stratigraphic contexts within the Terminal Classic to Early Postclassic sequence offers some evidence that contact and interaction with sites situated in this geographic area broke down during the Terminal Classic period. Based on the paucity of ceramics deriving from this area in general, however, it would appear that the flow of southern lowland pottery into the local community (at least into the central precinct) was quite limited to begin with. Consequently, a disruption or cessation of economic relations with sites situated in and bordering the heartland area would have had little impact on the local ceramic economy. In comparison, the presence of ceramics deriving from areas to the north of Lamanai in both early and late stratigraphic contexts within the ceramic sequence indicates that economic contacts or interaction with sites situated in this particular area of the lowland region continued into the Early Postclassic period. Nonetheless, as with pottery that originated from areas to the south and west of Lamanai, the evidence suggests that the volume of ceramics coming into the community from sites in the north was quite small, with no evidence of a substantial increase by Early Postclassic times. When considering Lamanai's potential involvement in trade networks stimulated by the expansion of northern lowland hegemonies such as Chichen Itza, the ceramic evidence has two major implications. Firstly, whatever contacts and interaction the community had with areas to the north were firmly in place by the early part of the Terminal Classic period. Secondly, there is no indication that economic interaction with this area of the lowland region became more intensive by Early Postclassic times. Hence a reorientation of the local ceramic economy towards the north in conjunction with a growing participation of the local community in maritime trade networks, as has been suggested for other sites in northern Belize (e.g. Progresso Lagoon, Caye Coco and Saktunja) (Masson and Mock 2004), is not demonstrated based on the ceramic evidence. The rarity of trade wares such as Slate ware, Fine Orange ware and Plumbate ware at Lamanai, the occurrence of which is generally seen as direct evidence of participation in trade networks driven by interests based in the northern lowlands, lends additional support to this conclusion. In fact, the ceramic evidence might also be viewed as a sign that Lamanai's commercial connections to the north operated on a relatively limited scale. What seems to be fairly clear is that the emergence and expansion of these new trade networks had little effect on local patterns of ceramic consumption and, hence, demand, at the very least as concerned the central precinct and adjacent residential areas.

Evidence of the Presence of Immigrants

The population movements that would have been triggered by ongoing conflicts and the economic decline of polities and city centres situated to the north, south, east and west of Lamanai, begs the question of whether the community accepted refugees. The strongest ceramic evidence that suggests the presence of immigrants at the site is the sudden appearance of a new tradition of coarse ware manufacture during the Terminal Classic period. There are two factors that suggest that the emergence of this new tradition relates to the influx of immigrants into the local community. Firstly, this approach to coarse ware manufacture involves raw material resources that had not been exploited previously by local potters, but similar clays were apparently used for coarse ware manufacture in other areas of northern Belize, particularly in inland areas adjacent to the Caribbean coast (as indicated by petrographic analysis). Secondly, the location of these clays in inland areas on the opposite side of the New River Lagoon, poses certain logistical challenges to their use. The successful manufacture of pottery using these new clays would have required knowledge of their behaviour at different stages of the manufacturing process, as well as how their natural properties could be altered and manipulated to best advantage. It is entirely possible that potters already working in the vicinity of Lamanai could have acquired this knowledge through experimentation. However, they would have had to have expended a considerable amount of time and effort to exploit these sources, when clays that they were already familiar with were abundant in the immediate vicinity of the settlement. It seems more likely, therefore, that the potters who used these particular clays chose them based on prior knowledge and experience with similar raw materials. Since these clays only occur in areas of Pine Ridge savannah, it would have been readily apparent where they occurred in the local area. It is well documented ethnographically that potters moving into a new area tend to seek out clay resources similar to those with which they are most familiar (e.g. Day 2004). Given the possibility that during the Terminal Classic period potentially large groups of people, which undoubtedly included potters, were in search of stable surroundings in which to re-establish themselves, the emergence of a significantly different way of producing utilitarian pottery at Lamanai might reflect this type of behaviour. This change in local patterns of pottery production, therefore, may reflect the influx of immigrants who brought with them a substantially different way of making coarse ware pottery.

Changes in Religious and Ceremonial Practice

The different ceramic depositional patterns that arise during the Terminal Classic period provide compelling evidence of a significant change in religious and ceremonial practices, particularly as concerned the appropriate treatment and function of the pottery used within different contexts of ritual and ceremonial activity. The rarity of the various trade wares postulated as material markers of participation in a regional cult driven by a network of elites dedicated to the proliferation of a new, formalized system of beliefs, practices, imagery and religious paraphernalia, suggests that these changes cannot be simply put down to a strong influence that came from outside of the local community. Rather, there are many aspects of the material record that suggest that this is a local phenomenon. For instance, the general focus of construction activities centred in the central precinct, as well as the accompanying changes in the patterns of use of communally-built structures and public areas, appear to have been predicated on the establishment and maintenance of physical links between the past and present, as well as the affirmation of traditional religious beliefs. The placement of new ceremonial structures and facilities, including a ball court, literally in the shadows of buildings that were major sites of ceremonial activity in the Classic period and earlier, would have generated a strong visual and psychological sense of continuity in the built setting of religious and ceremonial events, despite changes in the actual physical surroundings and specific aspects of ritual practice. Likewise, the interment of numerous burials within newly built ceremonial and administrative structures might be considered to reflect a particularly public form of ancestor veneration that, apparently, was not as prevalent in preceding time periods (c.f. McAnany 1995). Occupying a prominent place within the built environment, these interments are strong public expressions of the bonds between past and present members of the community, as well as the active role of the ancestors in perpetuating prosperity among their living descendants (Marcus 2000). The active creation and maintenance of such links through the built settings in which ceremonial and ritual activities took place can be viewed as a material manifestation of Lamanai's individual community history, fostering a sense of community identity deriving from shared experiences (past and present) and a common social memory (Joyce and Hendon 2000).

As the physical remains of the events and occasions through which shared experiences were created, ceramic depositional patterns provide additional insight into the motivations and concerns that lay behind certain changes in ritual practice. Viewed from this perspective the deliberate breakage of pottery used in religious and ceremonial events might be interpreted as acts of 'enchainment' – i.e. a means of creating and maintaining an active bond between people, places and a shared experience predicated on material culture (Chapman 2000:226). There is good evidence to suggest that within the contexts of burial and offertory ceremonies participants in these events retained or removed fragments of the pottery deliberately broken as part of associated rites. This practice might be viewed as a practical attempt to permanently rid the objects involved in religious acts of any potentially harmful powers they possess when complete. However, an integrating function would appear to be more in keeping with the considerable effort expended in establishing physical expressions of the important link between activities and people of the past and present through the manipulation of public architecture. Following Chapman's argument for the centrality of such acts in the creation and maintenance of a shared sense of community identity and history, the sudden appearance of large midden deposits dominated by deliberately broken pottery on the physical boundary between ceremonial structures and public plazas capable of accommodating large assemblies of people, might be viewed as another instance of enchainment. In this case, however, bonds are created through the intentional deposition of the remains of activities in which the wider community participated (either directly or indirectly) at strategic, highly visible locations. As permanent fixtures within public areas of the settlement set aside for ceremonial and ritual activity, these deposits would have served as a constant visual reminder of the shared experience of participating in religious and ceremonial events or, alternatively, of active efforts undertaken to protect or ensure the interests and wellbeing of the community.

Changes in local patterns in the production of the fine ware pottery used in ritual and ceremonial activities also suggest changes in practice, which might be considered to reflect a new emphasis or increased concern relating to the creation and active maintenance of a shared community identity. If manufacturing procedures can be regarded as material expressions of group identity, then changes in local patterns of fine ware production would appear to suggest a shift towards a community focus. The replacement of multiple distinct local approaches to fine ware production by a single, more generalized approach characterized by high internal variability is suggestive of a subsuming of individual expressions of group identity in favour of a shared community-level identity. From the standpoint of vessel design, the development of a more generalized approach to pottery production by Early Postclassic times would appear to reflect the emergence of a shared set of ideas not only with regard to what fine ware pottery ideally should look like, in terms of its colour and basic morphological and decorative characteristics, but also what their fabrics should contain. The inclusion of grog temper in all fine ware pottery produced locally by the Early Postclassic period might be considered to reflect new ideas concerning a vessel's 'substance'. Whilst the use of grog has several material advantages (e.g. if made from the same clay as the ceramic body it has the same thermal expansion rate, which has benefits during firing) (Rice 1987c:407;

Arnold 1985: 24; Rye 1976: 114-115), it is also recognized as an important social practice. Chapman (2000: 54) describes the use of grog as a means of "presencing the past through the incorporation of old or ancestral material culture into currently used ceramics" (see also Rainbird 1999). The inclusion of grog, therefore, acts as another form of enchainment. The fact that grog does not replace other forms of temper, but instead, was added alongside tempering materials that have a long history of use at the site, provides some evidence that the choice to include grog may reflect ideological, as opposed to purely practical considerations. Observations made by early Spanish explorers that visited the Maya area lend some support to this interpretation. For example, Chuchiak (2004) reports that Captain Ruiz de Arce observed that Maya potters incorporated crushed fragments of censers deliberately broken as part of ancestor veneration rituals into the new censers they made for subsequent ceremonies. Chuchiak also reports that Captain Ruiz de Arce observed that potters did this because they believed that the new censers would not function properly in these rituals if they did not contain pieces of vessels used in previous rituals. Whilst it can not be demonstrated that this actual practice has roots in the Terminal Classic period, the fact that grog tempered fine ware becomes prevalent during the same time period that the deliberate breakage of pottery included in offerings and burials become the standard practice suggests a strong connection between this technological innovation and this change in ritual practice.

A final issue concerns the inception of new ideas surrounding the decorative treatment of fine ware vessels used in religious and ceremonial events, particularly those that took place at ceremonial structures. The shift towards the manufacture of orange-slipped fine ware with elaborate decorative embellishments marks the movement of decorative imagery from the interior surface of the vessel to the exterior surface. When decorative images are placed on the inside of a vessel they can only be fully appreciated by those who handle the vessel directly, whereas decoration placed on a vessel's exterior can be seen and appreciated by a wider audience. Accordingly, this change in the placement and nature of decorative images may reflect an important change in the nature of audience that was meant to see these images when the vessels were used in particular social and/or ritual contexts. The high visibility of the ideological information conveyed on these vessels, together with the formalized nature of decorative elements and their content, may suggest that an underlying intention (conscious or subconscious) was to enable a wider segment of the population to share and participate in the beliefs and ideas communicated through the use of pottery in particular events and activities, as well as the messages it communicated, both literally and symbolically. The introduction of ceramic instruments and a range of incense burning paraphernalia into these same contexts of religious and ceremonial activity might also be considered to suggest a new emphasis on large communal gatherings, reflecting an elaboration of ritual performance aimed at enhancing the personal and sensory experience of a wider audience of participants and observers.

11.4. Intra-Community Dynamics Contributing to Continuity and Change in the Local Material Record

When the ceramic evidence is situated within a broader framework and community-based activities and regional-level developments and potential pressures, there are two main factors that can be identified as contributing to changes in local patterns of production and consumption at Lamanai: 1) the movement of immigrants into the local area and 2) an intra-community dynamic relating to an ideological shift on the local-level. Changes in patterns of local pottery production and the use and depositional patterns of ceramics within contexts of ritual and ceremonial activity suggest an increased emphasis on creating and establishing bonds between the past and present, as well as between places, people and their shared experiences and social linkages. Material-human interactions, particularly with regard to ceramics, formed an important conduit through which these ideas and concern were expressed. It would appear that at the root of these community patterns lay a new concern for fostering and maintaining a shared community identity, history and social memory, which was enacted through both repeated practices and the manipulation of material culture and the built environment. The ceramic evidence suggests that pottery constituted an important medium through which these ideas were conveyed and perpetuated, and that large, public gatherings were perhaps a means through which they were communicated. Chapman (2000) and others (e.g. McAnany 1995, 1998; Brumfiel and Fox 1994) have observed that these kinds of emphases or strategies often come into view archaeologically at times of crisis and socio-political change. The function or intention of these acts and endeavours is to foster community integration and solidarity, constituting a symbolic breaking down of social differences (real or perceived) in favour of a common and shared identity. In state-level societies, such strategies are often employed by the ruling class or competing factions to legitimate, maintain and expand claims to authority, power, privilege and influence (c.f. Pohl and Pohl 1994). At Lamanai, however, the evidence suggests that the new emphasis on enchainment and community integration relates to other factors, which are at least partly tied to changes in the local population. For example, there is no compelling evidence that these efforts and concerns were motivated specifically by elite interests, as indicated by the striking pattern of socio-political and economic stability over the course of the Terminal Classic to Early Postclassic period. On the other hand, the ceramic data provides evidence of immigrants moving into the local area, as well as ongoing contacts with, if not the physical presence of, people/groups from other areas, as suggested by the occurrence of non-local pottery, particularly storage jars within the temple middens. The presence of outsiders, at least some of whom were new residents, suggests that the situations of crisis, instability, economic decline etc. faced by other communities would have been well known to the people living under comparatively stable and prosperous conditions at Lamanai. One way of responding to any perceived threat to the local community would have been to ensure a continuing mutual support of the gods, the ancestors and other members of the community (new and old) through ritual performance and through acts and endeavours directed at solidarity and enchainment. Set within this broader frame of reference, therefore, local shifts in ideology, ritual practice and the nature of associated material culture might be seen as a practical local response to changing community dynamics and shifting world conditions.

CHAPTER 12
CONCLUSIONS AND FUTURE RESEARCH

12.1. A SUMMARY OF RESEARCH AND MAIN CONCLUSIONS

This study has assessed the nature and interrelationships of physical, technological and functional variation among Terminal to Early Postclassic ceramics at the site of Lamanai, as well as their depositional patterns within contexts of ceremonial and ritual activity. It has examined the technological and provenance characteristics of different stylistic and functional categories of pottery comprising the Terminal Classic to Early Postclassic ceramic assemblage and has documented temporal trends in the stylistic and technological characteristics of locally manufactured and non-local pottery. This detailed examination of the physical evidence has been situated alongside a contextual analysis, on both the local and regional levels, of the wider cultural and environmental setting within which Terminal Classic to Early Postclassic pottery was made, circulated, used, and finally, deposited as part of religious and ceremonial activities and events that took place within the community at Lamanai. This contextualization of the material evidence within a wider framework of material-human interactions has enabled a comprehensive reconstruction of community-level patterns of ceramic production and consumption that is mindful of the synergistic relationship between creative processes, as learned practices, the material and social necessities of society and practical and environmental considerations and constraints. Integrating the physical, archaeological, environmental and cultural evidence has enabled a consideration of the material, economic and symbolic representational aspects of emergent patterns and variation in the local material record.

This study has demonstrated the high level of insight and interpretative resolution that can be achieved when multiple analytical techniques and independent lines of evidence are combined within a single research programme brought to bear on a specific archaeological problem. The results of the study have revealed that ceramic economic patterns at Lamanai during the Classic to Postclassic tradition depart significantly from conventional characterizations. Moreover, the material evidence has brought to light a significantly different picture of the nature of and factors contributing to ceramic change than is accounted for in current explanatory models.

The specific findings of different aspects of this study have been reported at the conclusion of each chapter. The following presents a summary of particular points of interest.

Pot-making within the Local Environment

The results of the geological survey of raw material resources available in the local area has revealed that significant compositional variation exists among local clays, as well as the existence of different forms of calcite and limestone. Clays formed in different geological and environmental contexts within the local area can be discriminated both petrographically and chemically and, therefore, provide a basis for differentiating ceramic fabrics based on their raw material ingredients. A thorough review of the geological literature concerning northern Belize and adjacent areas, viewed in concert with the results of the local geological survey and the thin-section analysis of the pottery, has demonstrated the potential for ceramics made using different raw materials to be discriminated, and their provenances, either identified or approximated through thin section analysis.

Considered in light of the information generated by the local geological survey, the results of the petrographic analysis demonstrate that ancient potters chose specific raw materials for pottery manufacture, indicating that they had considerable knowledge of environmental resources. It is also evident that they possessed in depth knowledge of the physical properties of local clays and how they could be manipulated and used successfully in combination with tempering materials, despite their more problematic natural properties.

Stylistic Trends within the Terminal Classic to Early Postclassic Ceramic Sequence

Among the fine ware, there is a general loss of diversity in surface treatment. The early part of the sequence comprises multiple surface treatment groups, characterized by different slip colours and decorative techniques. By Early Postclassic times, however, all pottery is orange-slipped with primarily incised and hand-modelled decorative embellishments. There is also a general increase in the morphological complexity of fine ware pottery. By Early Postclassic time vessel forms often display a complex body geometry, further elaborated through the addition of appliqués, supports, appendages and flanges.

New vessel forms are introduced into to the local ceramic repertoire (both fine ware and coarse ware) during the Terminal Classic period.

Paste Technology and Provenance

The majority of vessels comprising the assemblage are of local manufacture. Four different local traditions of paste-making were identified: Crystalline Calcite-Tempered, Sandy-Sascab Tempered, Grog-Mixed Carbonate and Quartz-Sand, and the Crystalline Calcite-tempered tradition contains at least four slightly different approaches to paste preparation based on different raw material ingredients. Early pastes contain rock temper, whereas by Early Postclassic times they contain both rock temper and grog. Later pastes are also comparatively fine-textured, suggesting a more careful or rigorous processing of raw material ingredients by this time.

Only a small proportion of the ceramic comprising the assemblage analysed was found to be of non-local origin. During the early part of the sequence non-local pottery can be connected geologically to several different areas of the lowland region, including the Maya Mountains area, inland areas of northern Belize, extending into southern Yucatan, coastal areas to the north of the site and, possibly, Central Peten. By early Postclassic times non-local ceramics derive exclusively from areas to the north of the site, extending into southern Yucatan.

Locally-manufactured and non-local pottery can be discriminated petrographically and, in some cases, chemically. The results of the study show, however, that the relationship between mineralogical, geochemical and stylistic variation within the assemblage is very complex. Chemical variability within the local ceramic assemblage can be attributed to both provenance and technology. Discrimination of these two factors in the interpretation of the chemical data set would have been difficult, if not impossible, without a detailed knowledge of the geological and technological characteristics of the pottery, especially as concerns mineralogically similar fabrics.

Decorative and Firing Technology

There is a general shift in the technology of decoration of fine ware pottery away from painting and the manipulation of slips to achieve desired colouring effects and towards a technology centred on the use of one colour and decorative embellishments achieved through post-slip-pre-firing incision and the addition of hand-modelled appliqués and appendages.

All of the pottery analysed by SEM share two technological characteristics: 1) similar clays were used to make the ceramic body and the slip and 2) maximum firing temperatures rarely exceeded 850°C.

Differences in decorative technology and firing methods tend to correlate with differences in composition, but often not with stylistic differences.

Shifts in Local Patterns of Ceramic Production and Consumption

Multiple local approaches to pottery manufacture characterized by a high level of internal consistency give way to a more general approach that is characterized by a specific set of general stylistic and compositional traits, but also a high level of technological inconsistency.

The Terminal Classic period marks the appearance of large midden deposits containing deliberately broken pottery in the immediate vicinity of ceremonial structures, indicating a significant change relating to modes of pottery consumption. This pattern of behaviour might have had a significant effect on demand and manufacturing priorities at certain times of the year. There is also A shift towards the inclusion of deliberately broken pottery in burials and offerings as opposed to whole vessels, indicating changes in ritual practice.

The emergence of new vessel forms during the Terminal Classic period and the predominance of fine ware pottery displaying a particular set of stylistic and, possibly, compositional characteristics by Early Postclassic times point to a significant shift in patterns of production and demand. Nonetheless, there is a basic congruence in the fine ware pottery used in different kinds of religious and ceremonial activities and events. In addition, nearly all of the non-local vessels used in local ceremonial and ritual activities are drinking vessels and storage jars.

The results of the study reveal a significantly different picture of local ceramic economic activities and the factors influencing ceramic change during the Terminal Classic to Postclassic transition than current models maintain. There are two primary factors that appear to have contributed to shaping local patterns of ceramic production and consumption: firstly, the movement of immigrants into the local area and, secondly, a shift in local ideology reflecting a new emphasis on enchainment and community solidarity and integration.

12.2. Future Research

The results of this study demonstrate the detailed information that potentially could be brought to light through similar studies of site assemblages, context-specific or otherwise, for all periods of Maya cultural development. Ceramic studies that incorporate an assessment of vessel technology using scientific techniques are slowly increasing in prevalence, and as the findings of the present study have indicated, there is still much to learn about myriad aspects of Maya ceramic technologies and manufacturing techniques. The level of insight that can be gained through such studies, however, would be enhanced greatly by a contextual approach to analysis and interpretation of the physical evidence. As this information base builds, a greater understanding of the spatial and temporal dimensions of ceramic variation at successively higher scales of analysis will be achieved, providing a powerful tool for investigating a range of issues concerning economic and social interactions among the lowland Maya, as well as the nature and technical logic of their material culture. Specific areas of research that arise directly out of the present study are as follows:

Concerning Local and Regional Patterns of Ceramic Production and Consumption

1. Detailed studies aimed at the reconstruction of local patterns of ceramic production and consumption during the preceding Classic period at Lamanai, as well as time periods following the Early Postclassic, focussing on similar cultural contexts.

2. A similarly detailed study of Terminal Classic to Early Postclassic ceramic deposits associated with elite and commoner domestic contexts at Lamanai.

3. Regional-level studies focussing on the provenance and technology of specific styles (*types*) of Terminal Classic to Early Postclassic pottery. Based on the results of the current study, particularly significant insights could emerge through studies focussing on Early Postclassic orange-slipped and incised pottery and coarse ware pottery (sandy-pasted and calcite-tempered). These studies should be undertaken in concert with a survey of the raw material resources, especially clays, available in the vicinity of the sites included in the study.

Specific Analytical and Experimental Research

1. The results of this study suggest that local paste preparation techniques used in fine ware production were highly complex, possibly involving refinement of natural clays and, in some cases, the addition of multiple tempering materials. Our current understanding of these procedures and how they relate to the appearance of ceramic fabrics under the microscope, as well as their chemistry, would be advanced substantially by experimental studies on local raw materials aimed at the reconstruction of the specific techniques involved in the manufacture of different types of local fabrics. The results of such studies would be of great value as they have a direct bearing on our ability to interpret petrographic and chemical data sets.

2. Decorative techniques and firing methods are two aspects of Maya ceramic technology that have been seldom investigated in detail using scientific techniques. SEM analysis of only a small sample of the Terminal Classic to Postclassic pottery at Lamanai has revealed that considerable variation exists in surface finishing techniques, paint and slip compositions and body microstructures, among both locally produced and non-local ceramics. In addition, our understanding of certain surface treatments, such as those involving a resist technique, would benefit greatly from more intensive and detailed examination by SEM.

BIBLIOGRAPHY

Adams, Richard E.W.

1971 *The Ceramics of Altar de Sacrificos, Guatemala.* Papers of the Peabody Museum of Archaeology and Ethnology Vol. 63, No. 1, Harvard University, Cambridge M.A.

1973a The Collapse of Maya Civilization: A Review of Previous Theories. In: *The Classic Maya Collapse*, edited by T. Patrick Culbert, pp. 21–34. School of American Research Advanced Seminar Series, Douglas W. Schwartz, general editor, University of New Mexico Press, Albuquerque.

1973b Maya Collapse: Transformation and Termination of the Ceramic Sequence at Altar de Sacrificios. In: *The Classic Maya Collapse,* edited by T Patrick Culbert, pp. 133–164. School of American Research Advanced Seminar Series, Douglas W. Schwartz, general editor, University of New Mexico Press, Albuquerque.

Adams, R.E.W., H.R. Robichaeux, Fred Valdez Jr., Brett A. Houk, and Ruth Matthews

2004 Transformations, Periodicity and Urban Development in the Three Rivers Region. In: *The Terminal Classic in the Maya Lowlands: Collapse Transition and Transformation,* edited by Arthur A. Demarest, Prudence M. Rice and Don S. Rice, pp. 324–341. University Press of Colorado, Boulder, Colorado.

Adams, Richard E.W. and Fred Valdez Jr.

1980 The Ceramic Sequence of Colha, Belize: 1979 and 1980 Seasons. In *The Colha Project: Second Season, 1980 Interim Report,* edited by Thomas R. Hester, Jack D. Eaton and Harry J. Shafer, pp. 15-40. Center for Archaeological Research, The University of Texas at San Antonio, San Antonio, Texas.

Aloupi, E. and Y. Maniatis

1990 Investigation of the Technology and Manufacture of Local LBA Theran Pottery: The Body and Pigment Analysis. In *Thera and the Aegean World III: Papers Presented at the Third International Scientific Congress, Santorini, Greece, 3–9 September 1989*, edited by D. A. Hardy, C. G. Doumas, J. A. Sakellarakis and P. M. Warren, pp. 459–469. Thera Foundation, London.

Andres, Christopher R. and Anne K. Pyburn

2004 Out of Sight: The Postclassic and Early Colonial Periods at Chau Hiix, Belize. In: *The Terminal Classic in the Maya Lowlands: Collapse Transition and Transformation,* edited by Arthur A. Demarest, Prudence M. Rice and Don S. Rice, pp. 402–421. University Press of Colorado, Boulder, Colorado.

Andrews, E. and V. Wyllys

1973 The Development of Maya Civilization After Abandonment of the Southern Cities. In: *The Classic Maya Collapse,* edited by T. Patrick Culbert, pp. 243–268. School of American Research Advanced Seminar Series, Douglas W. Schwartz, general editor, University of New Mexico Press, Albuquerque.

Andrews, E., V. Wyllys and Jeremy A. Sabloff

1986 Classic to Postclassic: A Summary Discussion. In: *Late Lowland Maya Civilization: Classic to Postclassic,* edited by Jeremy A Sabloff, V. Wyllys and E. Andrews, pp. 433–456. School of American Research Advanced Seminar Series, Douglas W. Schwartz, general editor, University of New Mexico Press, Albuquerque.

Angelini, Mary L.

1998 The Potter's Craft: A Study of Formative Maya Ceramic Technology at K'Axob Belize. Unpublished Ph.D. dissertation, Graduate School of Arts and Sciences, Boston University, Boston.

Arnold, Dean E.

1971 Ethnominerology of Ticul, Yucatan Potters: Etics and Emics. *American Antiquity* 36(1):20–40.

1978 Ethnography of Pottery Making in the Valley of Guatemala. In *The Ceramics of Kaminaljuyu, Guatemala,* edited by Ronald K. Wetherington, pp.327–400. Monograph Series on Kaminaljuyu. The Pennsylvania State University Press, Pennsylvania.

1985 *Ceramic Theory and Cultural Process.* Cambridge University Press, Cambridge.

1987 Maya Pottery After 20 Years: Archaeological Implications. In *Maya Ceramics: Papers from the 1985 Maya Ceramics Conference, Part ii,* edited by Prudence M. Rice and Robert Sharer, pp. 545–561. BAR International Series 345(i). British Archaeological Reports, Oxford.

Arnold, D. E., H. Neff and R. L. Bishop
1991 Compositional Analysis and "Sources" of Pottery: An Ethnoarchaeological Approach. *American Anthropologist* 93:70–90.

Arnold, Dean E., Hector A. Neff, Ronald L. Bishop and Michael D. Glascock
1999 Testing Interpretative Assumptions of Neutron Activation Analysis: Contemporary Pottery in Yucatan, 1964–1994. In *Material Meanings: Critical Approaches to the Interpretation of Material Culture,* edited by Elizabeth S. Chilton, pp. 61–84. The University of Utah Press, Salt Lake City.

Arnold, Dean E. and Alvaro L. Nieves
1992 Factors Affecting Ceramic Standardization. In *Ceramic Production and Distribution: An Integrated Approach*, edited by George J. Bey III and Christopher A. Pool, pp. 93–114. Westview Press, Boulder CO.

Ball, Joseph W.
1983 Notes on the Distribution of Established Ceramic Types in the Corozal District, Belize. In *Archaeological Excavations in Northern Belize, Central America,* edited by Raymond V. Sidrys, pp.203–220. Monograph XVII. University of California, Los Angeles.

1993 Pottery, Potters, Palaces, and Polities: Some Socioeconomic and Political Implications of Late Maya Ceramic Industries. In *Lowland Maya Civilization in the 8th Century,* edited by Jeremy A. Sabloff and J. F. Henderson, pp.243–272. Dumbarton Oaks, Washington DC.

Bartlett, Mary Lee and Patricia A. McAnany
2000 Crafting Communities: The Materialization of Formative Maya Identities. In *The Archaeology of Communities: A New World Perspective*, edited by Marcello A. Canuto and Jason Yaeger, pp. 102–122. Routledge, London and New York.

Bartlett, M.L., Hector A. Neff and Patricia A. McAnany
2000 Differentiation of Clay Resources on a Limestone Plain: The Analysis of Clay Utilization During the Maya Formative at K'axob, Belize. *Geoarchaeology* 15(2):95–133.

Bateson, J. H.
1972 New Interpretation of Geology of Maya Mountains, British Honduras. *American Association of Petroleum Geologists Bulletin* 55:956–963.

Bateson, J. H. and I. H. S. Hall
1971 Revised Geologic Nomenclature for Pre-Cretaceous Rocks of British Honduras. *American Association of Petroleum Geologists Bulletin* 55:529-530

Beaudry, Marilyn P.
1984 *Ceramic Production and Distribution in the South-Eastern Maya Periphery : Late Classic Painted Serving Vessels*. BAR International Series 203. British Archaeological Reports, Oxford.

1991 New World Paste Compositional Investigations. In *The Ceramic Legacy of Anna O. Shepard*, edited by Ronald L. Bishop and Frederick W. Lange, pp. 224–256. University Press of Colorado, Niwot, Colorado.

Beaudry-Corbett, Marilyn
2003 A Social History of Archaeological Materials Characterization Studies. In *Patterns and Process: A Festschrift in Honor of Dr. Edward V. Sayre*, edited by Lambertus van Zelst, pp. 19–26. Smithsonian Center for Materials Research and Education, Suitland, Maryland.

Becker, Marshall J.
2003 A Classic-Period Barrio Producing Fine Polychrome Ceramics at Tikal, Guatemala. Notes on Ancient Maya Firing Technology. *Ancient Mesoamerica* 14:95-112

Bishop, Ronald L.
1980 Aspects of Compositional Modeling. In *Models and Methods in Regional Exchange*, edited by Robert Fry, pp. 46–67. SAA Papers No. 1. Society for American Archaeology, Washington, D.C.

1991 Anna O. Shepard: A Correspondence Portrait. In *The Ceramic Legacy of Anna O. Shepard*, edited by Ronald L. Bishop and Frederick W. Lange, pp. 42–87. University Press of Colorado, Niwot, Colorado.

1992 Instrumentation and the Future of Archaeology. In *Quandaries and Quests: Visions of Archaeology's Future*, edited by LuAnn Wandsnider, pp.160–169. Center of Archaeological Investigations Occasional Paper No.20. Southern Illinois University at Carbondale, Carbondale.

1994 Pre-Columbian Pottery: Research in the Maya Region. In *Archaeometry of Pre-Columbian Sites and Artifacts*, edited by D.A. Scott and P. Meyers, pp. 15–65. The Getty Conservation Institute, Los Angeles.

2003 Five Decades of Maya Fine Orange Ceramic Investigation by INAA. In *Patterns and Process: A Festschrift in Honor of Dr. Edward V. Sayre*, edited by Lambertus van Zelst, pp. 81–92. Smithsonian Center for Materials Research and Education, Suitland, Maryland.

Bishop, Ronald L. and Robert L. Rands
1982 Maya Fine Paste Ceramics: A Compositional Perspective. In *Excavations at Seibal: Ceramics,* edited by Jeremy A. Sabloff, pp. 283–314. Memoirs of the Peabody Museum of Archaeology and Ethnology Vol. 15, No. 2, Harvard University, Cambridge M.A.

Bishop, R.L., V. Canouts, P.L. Crown and S.P. De Atley
1990 Sensitivity Precision and Accuracy: Their Roles in Ceramic Compositional Databases. *American Antiquity* 55:537–546.

Bishop, Ronald L., Robert L. Rands and George R. Holley
1982 Ceramic Compositional Analysis in Archaeological Perspective. In *Advances in Archaeological Method and Theory,* Vol. 5, edited by Michael B. Schiffer, pp. 275–330. Academic Press: New York and London.

Braswell, Geoffry E., Joel D. Gunn, Maria del Rosario, Dominguez Carrasco, William J. Folan, Laraine A. Fletcher, Abel Morales López, and Michael D. Glascock
2004 Defining the Terminal Classic at Calakmul, Campeche. In: *The Terminal Classic in the Maya Lowlands: Collapse Transition and Transformation,* edited by Arthur A. Demarest, Prudence M. Rice and Don S. Rice, pp. 162–194. University Press of Colorado, Boulder, Colorado.

Brumfiel, Elizabeth M. and John W. Fox
1994 Factional Competition and Polical Development in the New World. Cambridge University Press, Cambridge.

Bullard, William R. Jr.
1973 Postclassic Culture in Central Peten and Adjacent British Honduras. In: *The Classic Maya Collapse*, edited by T. Patrick Culbert, pp. 221–242. School of American Research Advanced Seminar Series, Douglas W. Schwartz, general editor, University of New Mexico Press, Albuquerque.

Buxeda I Garrigos, J., M.A. Cau Ontiveros and V. Kilikoglou
2003a Chemical Variability in Clays and Pottery from a Traditional Cooking Pot Production Village: Testing Assumptions in Pereruela. *Archaeometry* 45(1):1–18.

Buxeda I Garrigos J., R.E. Jones, V. Kilikoglou, S.T. Levi, Y. Maniatis, J. Mitchell, L. Vagnetti, K.A. Wardle and S. Andreou
2003b Technology Transfer at the Periphery of the Mycenean World: The Cases of Mycenaean Pottery Found in Central Macedonia (Greece) and the Plain of Sybaris (Italy). *Archaeometry* 45(2):263–284.

Chapman, John
2000 *Fragmentation in Archaeology. People, Places and Broken Objects in the Prehistory of South Eastern Europe*. Routledge, London.

Carmean, Kelli, Nicholas Dunning and Jeff K. Kowalski
2004 High Times in the Hill Country: A Perspective from the Terminal Classic Puuc Region. In: *The Terminal Classic in the Maya Lowlands: Collapse Transition and Transformation,* edited by Arthur A. Demarest, Prudence M. Rice and Don S. Rice, pp. 424–449. University Press of Colorado, Boulder, Colorado.

Carpenter, Andrea J. and Gary M. Feinman
1999 The Effects of Behaviour on Ceramic Composition: Implications for the Definition of Production Locations. *Journal of Archaeological Science* 26:783–796.

Chase, Arlen F.

1986 Time Depth or Vacuum: The 11.3.0.0.0 Correlation and the Lowland Maya Postclassic. In: *Late Lowland Maya Civilization: Classic to Postclassic*, edited by Jeremy A. Sabloff, V. Wyllys and E. Andrews, pp. 99–140. School of American Research Advanced Seminar Series, Douglas W. Schwartz, general editor, University of New Mexico Press, Albuquerque.

Chase, Arlen F. and Diane Z. Chase

2004 Terminal Classic Status-Linked Ceramics and the Maya "Collapse": De Facto Refuse at Caracol, Belize. In: *The Terminal Classic in the Maya Lowlands: Collapse Transition and Transformation,* edited by Arthur A. Demarest, Prudence M. Rice and Don S. Rice, pp. 642–366. University Press of Colorado, Boulder, Colorado.

Chase, Arlen F. and Prudence M. Rice (editors)

1985 *The Lowland Maya Postclassic.* University of Texas Press, Austin.

Chase, Diane Z.

1982 Spatial and Temporal Variability in Postclassic Northern Belize. Unpublished Ph.D. dissertation, Department of Anthropology, University of Pennsylvania, Philadelphia.

1985 Ganned But Not Forgotten: Late Postclassic Archaeology and Ritual at Santa Rita Corozal, Belize. In *The Lowland Maya Postclassic,* edited by Arlen F. Chase and Prudence M. Rice, pp.104–125. University of Texas Press, Austin.

1986 Social and Political Organization in the Land of Cacao and Honey: Correlating the Archaeology and Ethnohistory of the Postclassic Lowland Maya. In: *Late Lowland Maya Civilization: Classic to Postclassic*, edited by Jeremy A. Sabloff, V. Wyllys and E. Andrews, pp. 347–378. School of American Research Advanced Seminar Series, Douglas W. Schwartz, general editor, University of New Mexico Press, Albuquerque.

1988 Caches and Censerwares: Meaning from Maya Pottery. In *A Pot for All Reasons: Ceramic Ecology Revisited,* edited by Charles C. Kolb and Louana M. Lackey, pp.81–104. Papers Dedicated to Frederick R. Matson, 1986, A Special Publication of Ceramica de Cultura Maya et al., Muriel Kirkpatrick, general editor, Laboratory of Anthropology, Temple University, Philadelphia, PA.

Chase, Diane Z. and Arlen F. Chase

1982 Yucatec Influence in Terminal Classic Northern Belize. *American Antiquity* 47(3):596–613.

1988 *A Postclassic Perspective: Excavations at the Maya Site of Santa Rita, Corozal, Belize.* Monograph 4. The Pre-Columbian Art Research Institute, San Francisco..

Closs, Michael P.

1988 The Hieroglyphic Text of Stela 9, Lamanai, Belize. *Research Reports on Ancient Maya Writing* 21. Center for Maya Research, Washington, D. C.

Chuchiak, John F.

2004 De Descriptio Idolorum: An Enthnohistorical Examination of the Production, Imagery and Functions of Colonial Yucatec Maya Idols and Effigy Censers, 1540-1700. Paper presented at the 69th Annual Meeting of the Society for American Archaeology, Montreal.

Cobos Palma, Rafael

2004 Chichén Itzá: Settlement and Hegemony during the Terminal Classic Period. In *The Terminal Classic in the Maya Lowlands: Collapse, Transition, and Transformations,* edited by Arthur D. Demarest, Prudence M. Rice and Don S. Rice, pp.517–544. University Press of Colorado, Boulder, Colorado.

Coe, William R.

1965 Caches and Offertory Practices of the Maya Lowlands. In *Archaeology of Southern MesoAmerica Part 1,* edited by G. R. Willey, pp.462–468. Handbook of Middle American Indians, Vol. 2, Robert Wauchope, general editor, The University of Texas Press, Austin.

Cornec, J. H.

1985 *Note on the Provisional Geological Map of Belize at the Scale of 1:250,000.* Petroleum Office, Belmopan.

Cresswell, R.

1983 Transferts de techniques et chaînes opératoires. *Techniques et Culture* 2:145–164

Culbert, T. Patrick (editor)
1973a *The Classic Maya Collapse*. School of American Research Advanced Seminar Series, Douglas W. Schwartz, general editor, University of New Mexico Press, Albuquerque.

1973b Introduction: A Prologue to Classic Maya Culture and the Problem of Its Collapse. In: *The Classic Maya Collapse*, edited by T. Patrick Culbert, pp. 3–20. School of American Research Advanced Seminar Series, Douglas W. Schwartz, general editor, University of New Mexico Press, Albuquerque.

1977 Maya Development and Collapse: An Economic Perspective. In: *Social Processes in Maya Prehistory: Essays in Honor of Sir J Eric Thompson*, edited by Norman Hammond, pp. 509–530. Academic Press, New York.

1988 The Collapse of Classic Maya Civilization. In: *The Collapse of Ancient States and Civilizations*, edited by Norman Yoffee and George L. Cowgill, pp. 69–101. University of Arizona Press, Tuscon.

Darch, Janice P.
1981 The Characteristics and Origins of Sascab in Northern Belize, Central America. In *Zeitschrift für Geomorphologie,* edited by Göttingen H. Mortensen, pp. 400–419. Neue Folge Band 25. Gebrüder Borntraeger, Berlin, Stuttgart.

Darch, Janice P. and P. A. Furley
1983 Observations on the Nature of Sascab and Associated Soils in Cayo and Orange Walk Districts, Belize and in El Peten, Guatemala. In *Resources and Development in Belize: An Account of the University of Edinburgh Expedition to Central America, 1981*, edited by G. M. Robinson and P. A. Furley, pp. 179–222. Department of Geography, University of Edinburgh, Edinburgh.

Day, Peter M.
1989 Technology and Ethnography in Petrographic Studies on Ceramics. In *Archaeometry: Proceedings of the 25th International Symposium*, edited by Y. Maniatis, pp. 139–147. Elsevier, Oxford and Amsterdam.

1995 Pottery Production and Consumption in the Siteia Bay Area during the New Palace Period. In *Achladia: Scavi e Richerche della 'Missione Greco-Italiana in Creta Orientale (1991–1993),* edited by M. Tsipopoulou and L. Vagnetti, pp. 148–175. CNR/GEI, Rome.

2004 Marriage and Mobility: Traditions and the Dynamics of the Pottery System in Twentieth Century East Crete. In *Pseira VIII: The Archaeological Survey of Pseira Island Part 1*, edited by P.P. Betancourt, C. Davaras and R. Hope Simpson, pp. 105–142. INSTAP Academic Press, Philadelphia.

Day, P.M., E. Kiriatzi, A. Tsolakidou and V. Kilikoglou
1999 Group Therapy: A Comparison between Analyses by NAA and Thin Section Petrography of Early Bronze Age Pottery from Central and East Crete. *Journal of Archaeological Science* 26:1025–1036.

Day, Peter M. and David E. Wilson
1998 Consuming Power: Kamares Ware in Protopalatial Knossos. *Antquity* 72:2.

Demarest, Arthur A., Prudence M. Rice and Don S. Rice (editors)
2004a *The Terminal Classic in the Maya Lowlands: Collapse Transition and Transformation.* University Press of Colorado, Boulder, Colorado.

2004b The Terminal Classic in the Maya Lowlands: Assessing Collapses, Terminations and Transformations. In: *The Terminal Classic in the Maya Lowlands: Collapse Transition and Transformation,* edited by Arthur A. Demarest, Prudence M. Rice and Don S. Rice, pp. 545–572. University Press of Colorado, Boulder, Colorado.

Dickerson, R. E. and N. E. Weisbord
1931 Cretaceous Limestone in British Honduras. *Journal of Geology* 39:483–486.

Dixon, C.G.
1956 *Geology of Southern British Honduras.* Belize Government Printing Office, Belize.

Dobres, Marcia-Ann
2000 *Technology and Social Agency*. Blackwell Publishers Inc., Malden.

Eaton, J. D.
1982 Colha: An Overview of Architecture and Settlement. In *Archaeology at Colha, Belize: the 1981 Interim Report,* edited by Thomas R. Hester, Harry J. Shafer and Jack D. Eaton, pp. 11–20. Center for Archaeological Research, The University of Texas at San Antonio and Centro Studi e Ricerche Ligabue, Venezia and San Antonio.

Faber, E.W., V. Kilikoglou, P.M. Day and D.E. Wilson
2001 A Technological Study of Middle Minoan Polychrome Pottery from Knossos, Crete. In *Modern Trends in Scientific Studies on Ancient Ceramics: Papers Presented at the 5th European Meeting on Ancient Ceramics, Athens 1999,* edited by V. Kilikoglou, A. Hein and Y. Maniatis, pp. 129–142. BAR International Series 1011. BAR Publishing, Oxford.

Fash, William L., E. Wyllys Andrews and T. Kam Manahan
2004 Political Decentralization, Dynastic Collapse and, the Early Postclassic in the Urban Center of Copán, Honduras. In: *The Terminal Classic in the Maya Lowlands: Collapse Transition and Transformation,* edited by Arthur A. Demarest, Prudence M. Rice and Don S. Rice, pp. 260–287. University Press of Colorado, Boulder, Colorado.

Flores, G.
1952 Geology of Northern British Honduras. *American Association of Petroleum Geologists Bulletin* 36(2):404–409.

Foias, Antonia E. and Ronald L. Bishop
1997 Changing Ceramic Production and Exchange in the Petexbatun Region, Guatemala: Reconsidering the Classic Maya Collapse. *Ancient Mesoamerica* 8:275–291.

Folk, Robert L.
1974 *Petrology of Sedimentary Rocks.* Hemphill Publishing Company, Austin, Texas.

Ford, Anabel, Nicole Woodman and Lisa Lucero
1992 Review and Revision of Ceramic Analysis: Results of the 1991 Maya Ceramic Workshop. *Ceramicas de Cultural Maya et al.* 16:24–29.

Fox, John W.
1987 *Maya Postclassic State Formation: Sedimentary Lineage Migration in Advancing Frontiers.* New Studies in Archaeology, C. Renfrew and J.A Sabloff, general editors, Cambridge University Press, Cambridge.

Fox, Richard G.
1977 *Urban Anthropology: Cities in their Cultural Settings.* Prentice-Hall, Inc., Englewood Cliffs.

Franklin, U.M. and V. Vitali
1985 The Environmental Stability of Ancient Ceramics. *Archaeometry* 27(1):3–15.

Freidel, David A.
1985 New Light on the Dark Age: A Summary of Major Themes. In *The Lowland Maya Postclassic,* edited by Arlen F. Chase and Prudence M. Rice, pp. 285–310. University of Texas Press, Austin.

1986 Terminal Classic Lowland Maya: Success, Failure and Aftermaths. In: *Late Lowland Maya Civilization: Classic to Postclassic,* edited by Jeremy A. Sabloff and E. Wyllys Andrews V, pp. 409–432. School of American Research Advanced Seminar Series, Douglas W. Schwartz, general editor, University of New Mexico Press, Albuquerque.

Freidel, David A. and Justine Shaw
2000 The Lowland Maya Civilization: Historical Consciousness and Environment. In *The Way the Wind Blows: Climate, History, and Human Action,* edited by Roderick J. McIntosh, Joseph A. Tainter and Susan Keech McIntosh, pp. 271–300. The Historical Ecology Series, William Balée and Carole L. Crumley, general editors, Columbia University Press, New York.

Freidel, David A., Kathryn Reese-Taylor and David Mora-Marín
2002 The Origins of Maya Civilization: the Old Shell Game, Commodity, Treasure, and Kingship. In *Ancient Maya Political Economies,* edited by Marilyn A. Masson and David A. Freidel, pp. 41–86, Altamira Press, New York.

Freestone, Ian C.
1991 Extending Ceramic Petrology. In *Recent Developments in Ceramic Petrology*, edited by Andrew Middleton and Ian Freestone, pp. 399–410. British Museum Occasional Paper No. 81. British Museum, London.

Fry, Robert E.
1979 The Economics of Pottery at Tikal, Guatemala: Models of Exchange for Serving Vessels. *American Antiquity* 44:494–512.

1980 Models of Exchange for Major Shape Classes of Lowland Maya Pottery. In *Models and Methods in Regional Exchange*, edited by Robert Fry, pp. 3–18. SAA Papers No. 1. Society for American Archaeology, Washington, D.C.

1981 Pottery Production-Distribution Systems in the Southern Maya Lowlands. In *Production and Distribution: A Ceramic Viewpoint*, edited by H. Howard and E. L. Morris, pp. 145–167. BAR International Series 120. British Archaeological Reports, Oxford.

1989 Regional Ceramic Distribution Patterning in Northern Belize: The View from Pulltrouser Swamp. In *Research in Economic Anthropology, a Research Annual: Prehistoric Maya Economies of Belize,* edited by Patricia A. McAnany and Barry L. Isaac, pp. 91–111. Research in Economic Anthropology, Supplement 4, JAI Press Inc., Greenwich, Connecticut and London.

Furley, P. A. and A. J. Crosby
1974 Geography of Belize. Collins, London.

Gifford, James C.
1960 The Type-Variety Method of Ceramic Classification as an Indicator of Cultural Phenomena. *American Antiquity* 25(3):341–347.

1963 A Statement Concerning the Ceramic Group. *Ceramica de Cultura Maya et al.* 2–3:23–25.

1976 *Prehistoric Pottery Analysis and the Ceramics of Barton Ramie in the Belize Valley*. Papers of the Peabody Museum of Archaeology and Ethnology Memoir 18. Harvard University, Cambridge.

Gill, Richardson B.
2000 *The Great Maya Droughts: Water, Life, and Death.* University of New Mexico Press, Albuquerque.

Gischler, E. and A. J. Lomando
1999 Recent Sedimentary Facies of Isolated Carbonate Platforms, Belize-Yucatan System, Central America. *Journal of Sedimentary Research* 69(3 PtA):747–763.

Glascock, Michael D.
1992 Characterization of Archaeological Ceramics at MURR by Neutron Activation Analysis and Multivariate Statistics. In *Chemical Characterization of Ceramic Pastes in Archaeology*, edited by Hector Neff, pp. 11–26. Monographs in World Archaeology No. 7. Prehistory Press, Madison, Wisconsin.

Glascock, M.D., H. Neff, and K.J. Vaughn
2004 Instrumental Neutron Activation Analysis and Multivariate Statistics for Pottery Provenance. *Hyperfine Interactions* 145:95–105.

Gosselain, Oliver P.
1998 Social and Technical Identity in a Clay Crystal Ball. In *The Archaeology of Social Boundaries*, edited by M.T. Stark, pp. 78–106. Smithsonian Institution Press, Washington D.C.

Graham, Elizabeth
1985 Facets of Terminal to Postclassic Activity in the Stann Creek District, Belize. In: *The Lowland Maya Postclassic,* edited by Arlen F. Chase and Prudence M. Rice, pp. 215–230. University of Texas Press, Austin.

1987 Terminal Classic to Early Historic Period Vessel Forms from Belize. In *Maya Ceramics: Papers from the 1985 Maya Ceramics Conference, Part i*, edited by Prudence M. Rice and Robert Sharer, pp. 73–98. BAR International Series 345(i). British Archaeological Reports, Oxford.

2002 Perspectives on Economy and Theory. In *Ancient Maya Political Economies,* edited by Marilyn A. Masson and David A. Freidel, pp. 398–418. Altamira Press, New York.

2004 Lamanai Reloaded: Alive and Well in the Early Postclassic. In *Archaeological Investigations in the Eastern Maya Lowlands*, edited by J. Awe, J. Morris, & S. Jones, pp. 223-241. Research Reports in Belizean Archaeology, Volume 1. Institute of Archaeology, National Institute of Culture and History, Belmopan, Belize.

Graham, Elizabeth and David M. Pendergast

1989 Excavations at the Marco Gonzalez Site, Ambergris Cay, Belize, 1986. *Journal of Field Archaeology* 16(1):1–16.

Gregg, Jay M., Scott A. Howard and S. J. Mazzullo

1992 Early Diagenetic Recrystallization of Holocene (<3000 Years Old) Peritidal Dolomites, Ambergris Cay, Belize. *Sedimentology* 39:143–160.

Gunn, Joel D. and William J. Folan

1981 Climatic Change, Culture and Civilization in North America. *World Archaeology* 13:87–100.

1996 Tres ríos: una superficie de impacto climatológico global interregional par las tierras bajas de los mayas del suroeste. In *Los investigadores de la cultura maya.* Universidad Autónoma de Campeche, México.

2000 Three Rivers: Subregional Variations in Earth System Impacts in the Southwestern Maya Lowlands (Candelaria, Usumacinta, and Champotón Watersheds). In *The Way the Wind Blows: Climate, History, and Human Action,* edited by Roderick J. McIntosh, Joseph A. Tainter and Susan Keech McIntosh, pp. 223–270. The Historical Ecology Series, William Balée and Carole L. Crumley, general editors, Columbia University Press, New York.

Hall, I. H. S. and J. H. Bateson

1972 Late Paleozoic Lavas in Maya Mountains, British Honduras, and Their possible Regional Significance. *The American Association Of Petroleum Geologists Bulletin* 56(5):950–963.

Hamblin, Robert L. and Pitcher, Brian L.

1980 The Classic Maya Collapse: Testing Class Conflict Hypotheses. *American Antiquity* 45(2):246–267.

Hammond, Norman

1974 The Distribution of Late Classic Maya. Major Ceremonial Centres in the Central Area. In *Mesoamerican archaeology – New Approaches : Proceedings of a Symposium on Mesoamerican Archaeology held by the University of Cambridge Centre of Latin American Studies, August 1972*, edited by Norman Hammond, pp. 313–334. Duckworth, London.

1982 *Ancient Maya Civilization.* Rutgers University Press, New Brunswick, New Jersey.

Hammond, Norman and Gair Tourtellot

2004 Out with a Whimper: La Milpa in the Terminal Classic. In *The Terminal Classic in the Maya Lowlands: Collapse Transition and Transformation,* edited by Arthur A. Demarest, Prudence M. Rice and Don S. Rice, pp. 288–301. University Press of Colorado, Boulder Colorado.

Hammond, N., G. Harbottle and T. Gazard

1976 Neutron Activation and Statistical Analysis of Maya Ceramics and Clays from Lubaantun, Belize. *Archaeometry* 18(2):147–168.

Harbottle and Sayre

1975 Appendix: Current status of examination of sherds of Fine paste ceramics from Altar de Sacrificio and Seibal and their comparison with other Maya Fine paste ceramics. In *Excavation at Seibal, Vol. 2: Ceramics*, edited by Jeremy A. Sabloff, pp. 241–253. Memoirs of the Peabody Museum of Archaeology and Ethnology, Vol. 13, No. 2. Harvard University, Cambridge, Massachusetts.

Harrison, Peter D.

1977 The Rise of the Bajos and the Fall of the Maya. In *Social process in Maya prehistory : studies in honour of Sir Eric Thompson,* edited by Norman Hammond, pp. 469–508. Academic Press, London.

Hein, A., A Tsolakidou, I. Illiopoulos, H. Mommsen, J. Buxeda i Garrigos, G. Montana and V. Kilikoglou
2002 Standardization of elemental analytical techniques applied to provenance studies of archaeological ceramics: and Inter laboratory calibration study. *The Analyst* 127: 542–553.

Hein, A., P.M. Day, P.S. Quinn and V. Kilikoglou
2004a The Geochemistry of Neogene Clay Deposits in Crete and Its Implications for Provenance Studies of Minoan Pottery. *Archaeometry* 46(3): 357–384.

Hein, A., P.M. Day, M.A. Cau Ontiveros and V. Kilikoglou
2004b Red Clays from Central and Eastern Crete: Geochemical and Mineralogical Properties in View of Provenance Studies of Ancient Ceramics. *Applied Clay Science* 24: 245–255.

Hester, Thomas R., Harry J. Shafer and Jack D. Eaton
1982 *Archaeology at Colha, Belize: the 1981 Interim Report.* San Antonio: Center for Archaeological Research, The University of Texas at San Antonio and Centro Studi e Ricerche Ligabue, Venezia.

Hosler, Dorothy, Jeremy A. Sabloff and Dale Runge
1977 Simulation Model Development: A Case Study of the Classic Maya Collapse. In *Social process in Maya prehistory: studies in honour of Sir Eric Thompson,* edited by Norman Hammond, pp. 553–590. Academic Press, London.

Houston, Stephen D., Hector Escobedo, Mark Child, Charles Golden and Réne Muñoz
2001 Moral Community and Settlement Transformation Among the Classic Maya: Evidence from Piedras Negras, Guatemala. In *Social Construction of Ancient Cities*, edited by M. L. Smith. Smithsonian Institution Press, Washington DC.

Howie, Linda A., Peter M. Day and Elizabeth Graham
2004 Late Classic Maya Paste Recipes at Altun Ha, Belize and the Meaning of Paste Variability. Paper presented at the 69th Annual Meeting of The Society for American Archaeology, Montreal.

Howie, Linda A., Terry Powis and Elizabeth Graham
2005 Can you Judge a Pot by Its Colour? Typological Groupings Under the Microscope at Lamanai, Belize. *Yuk'ab': An Online Journal of Maya Material Culture* 1.
www.units.muohio.edu/mcs/mediaservices/webdev/portal/index.php

Howie-Langs, Linda A.
1999 Ceramic Production and Consumption at Altun Ha, Belize: A Petrographic Study. Unpublished Masters Thesis, Department of Archaeology, University of Sheffield, Sheffield, U.K.

Howry, Jeffrey
1978 Ethnographic Realities of Maya Prehistory. In Cultural *Continuity in Mesoamerica*, edited by David Browman, pp. 239–257. World Anthropology Series. Mouton Publishers, Paris.

Iceland, H.B. and Goldberg, P
1999 Late-Terminal Classic Maya Pottery in Northern Belize: A Petrographic Analysis of Sherd Samples from Colha and Kichpanha. *Journal of Archaeological Science* 26:951–966.

Jones, Lea, D.
1986 *Lowland Maya Pottery: The place of petrological analysis.* BAR International Series 288, Oxford, U.K.

Joyce, Rosemary A. and Julia A. Hendon
2000 Heterarchy, History, and Material Reality: "Communities" in Late Classic Honduras. In The Archaeology of Communities, edited by Marcello A. Canuto and Jason Yaeger, pp. 143-160. Routledge, London.

Kepecs, Susan
1998 Diachronic Ceramic Evidence and its Social Implications in the Chikinchel Region, Northeast Yucatan, Mexico. *Ancient Mesoamerica* 9:121–135.

Kesler, Stephen E., J. Howard Bateson, William L. Josey, George H. Cramer and Weldon A. Simmons
1971 Geological Notes: Mesoscopic Structural Homogeneity of Maya Series and Macal Series, Mountain Pine Ridge, British Honduras. *The American Association of Petroleum Geologists Bulletin* 55(1):97–123.

Kesler, S. E., C. F. Kienle and J. H. Bateson
1974 Tectonic significance of Intrusive Rocks in the Maya Mountains, British Honduras. *Geological Society of America Bulletin* 85:549–552.

Kilikoglou, V.
1994 Scanning Electron Microscopy. In Ceramic Regionalism in Prepalatial Central Crete: The Mesara Imports at EM I to EM IIA Knossos (D.E. Wilson and P.M. Day). *The Annual of the British School at Athens* 89:70–75

Kilikoglou, V., Y. Maniatis and A.P. Grimanis
1988 The Effect of Purification and Firing of Clays on Trace Element Provenance Studies. *Archaeometry* 30(1):37–36.

King, R. B., I. C. Ballie, T. M. B. Abell, J. R. Dunsmore, D. A. Gray, J. H. Pratt, H. R. Versey, W. A.C.S. and S. A. Zisman
1992 *Land Resource Assessment of Northern Belize (Vol.1 and Vol. 2)*. Natural Resource Institute. Bulletin 43.

Lemonnier, Pierre
1980 *Les Salines de l'Ouest: Logique technique, logique sociale*. Editions de la Maison de sciences de l'homme, Presses Universitaires de Lille, Paris, Lille.

1986 The Study of material Culture Today: Toward an Anthropology of Technical Systems. *Journal of Anthropological Archaeology* 5:147–186.

1993 Introduction. In *Technological Choices: Transformation in materials cultures since the Neolithic*, edited by Pierre Lemonnier, pp. 1–35. Routeledge, London.

López Varela, Sandra L., Patricia A. McAnany and Kimberly A. Berry
2001 Ceramics Technology at Late Classic K'axob, Belize. *Journal of Field Anthropology* 28:177–191.

Loten, S.
1985 Lamanai Postclassic. In *The Lowland Maya Postclassic*, edited by Arlen F. Chase and Prudence M. Rice, pp. 85–90. University of Texas Press, Austin.

Lucero, Lisa J.
1992 Problems in Identifying Ceramic Production in the Maya Lowlands: Evidence from the Belize River Area. *Memorias del Primer Congreso Internacional de Mayistas: Mesas Redondas, Arqueologia, Epigrafia* II:143–154.

Mahias, Marie-Claude
1993 Pottery techniques in India: Technical variants and social choice. In *Technological Choices: Transformation in materials cultures since the Neolithic*, edited by Pierre Lemonnier, pp. 157–180. Routeledge, London.

Maniatis, Y and M.S. Tite
1981 Technological examination of the Neolithic-Bronze Age pottery from Central and Southeast Europe and the Near East. *Journal of Archaeological Science* 8: 59–76.

Marcus, Joyce
2000 Toward an Archaeology of Communities. In The Archaeology of Communities, edited by Marcello A. Canuto and Jason Yaeger, pp. 231-242. Routledge, London.

Martin, Simon and Nikolai Grube
2000 *Chronicle of the Maya Kings and Queens: Deciphering the Dynasties of the Ancient Maya.* Thames and Hudson Ltd, London.

Masson, Marilyn A.
2000 *In the Realm of Nachan Kan: Postclassic Maya Archaeology at Laguna de On, Belize*. University Press of Colorado, Boulder, Colorado.

2002 Community Economy and the Mercantile transformation in Postclassic Northeastern Belize. In *Ancient Maya Political Economies,* edited by Marilyn A. Masson and David A. Freidel, pp. 335–364. Altamira Press, New York.

Masson, Marilyn A. and Robert M. Rosenswig
n.d. The Evolution of Postclassic Maya Pottery Traditions in Northern Belize. Submitted to *Latin American Antiquity* in February 2003 (accepted, in revision).

Masson, Marilyn A. and Shirley Boteler Mock
2004 Ceramics and Settlement Patterns at Terminal Classic-Period Lagoon Sites in Northern Belize. In *The Terminal Classic in the Maya Lowlands: Collapse, Transition, and Transformation,* edited by Arthur A. Demarest, Prudence M, Rice and Don S. Rice, pp.367–401. University Press of Colorado, Boulder.

Mazzullo, S.J., and A.M. Reid
1988 Sedimentary textures of Recent Belizean peritidal dolomite. *Journal of Sedimentary Petrology* 58:479–488.

Mazzullo, S.J., W.D. Bischoff, and C.S. Teal
1995 Holocene shallow-subtidal dolomitization by near- normal seawater, northern Belize; *Geology*, 23:341–344.

Mock, Shirley Boteler
1994 Yucatecan Presence in Northern Belize as Seen in the Postclassic Ceramics at Colha. In *Continuing Archaeology at Colha, Belize,* edited by Thomas R. Hester, Harry J. Shafer and Jack D. Eaton, pp. 233–243. Studies in Archaeology 16. Texas Archaeological Research Laboratory, The University of Texas at Austin, Austin.

1998 Prelude. In *The Sowing and the Dawning: Termination, Dedication, and Transformation in the Archaeological and Ethnographic Record of Mesoamerica,* edited by Shirley Boteler Mock, pp. 3–20. University of New Mexico Press, Albuquerque.

McAnany, Patricia A.
1995 Living with the Ancestors: Kinship and Kingship in Ancient Maya Society. University of Texas Press, Austin.

1998 Ancestors and the Classic Maya Built Environment. In *Function and Meaning in Classic Maya Architecture*, edited by Stephen D. Houston, pp. 271-298 . Dumbarton Oaks, Washington, D.C.

Molloy, John P. and William L Rathje
1974 Sexploitation among the Late Classic Maya. In *Mesoamerican archaeology – New Approaches : Proceedings of a Symposium on Mesoamerican Archaeology held by the University of Cambridge Centre of Latin American Studies, August 1972*, edited by Norman Hammond, pp. 431–444. Duckworth, London.

Morley, Sylvanus G. (revised by G. W. Brainerd)
1956 *The Ancient Maya* (3rd Edition). Stanford University Press, Stanford.

Neff, Hector
1989a Origins of Plumbate Pottery Production. In *Ancient Trade and tribute: Economies of the Soconusco Region of Mesoamerica*, edited by B. Voorhies, pp. 175–193. University of Utah Press, Provo.

1989b The Effect of Interregional Distribution of Plumbate Pottery Production. In *Ancient Trade and tribute: Economies of the Soconusco Region of Mesoamerica*, edited by B. Voorhies, pp. 249–267. University of Utah Press, Provo.

2000 Neutron Activation Analysis for Provenance Determination in Archaeology. In *Modern Analytical Methods in Art and Archaeology*, edited by Enrico Ciliberto and Giuseppe Spoto, pp. 81–143. Chemical Analysis Series, Vol. 155. John Wiley & Sons, Inc., New York.

2001 Sources of Raw Material Used in Plumbate Pottery. In *Incidents of Archaeology in Central America and Yucatan*, edited by M. Love, M.P. Hatch and H. Escobedo. University Press of America, Lanham, M.D.

2003 Analysis of Mesoamerican Plumbate Pottery Surfaces by Laser Ablation Inductively-Coupled Plasma-Mass Spectrometry (LA-ICP-MS). *Journal of Archaeological Science* 30:21–35.

Neff, H. and R.L. Bishop
1988 Plumbate Origins and Development. *American Antiquity* 53:505–522.

Neff, Hector, Ronald L. Bishop and Edward V Sayre
1988 A Simulation Approach to the Problem of Tempering in Compositional Studies of Archaeological Ceramics. *Journal of Archaeological Science* 15:159–172.

1989 More Observations on the problem of Tempering in Compositional Studies of Archaeological Ceramics. *Journal of Archaeological Science* 15:57–69.

Neff, Hector and Frederick J. Bove
1999 Mapping Ceramic Compositional Variation and Prehistoric Interaction in Pacific Coastal Guatemala. *Journal of Archaeological Science* 26:1037–1051.

Noll, W.
1978 Material and Techniques of the Minoan Ceramics of Thera and Crete. In *Thera and the Agean World II: papers presented at the Second International Scientific Congress, Santorini, Greece, August 1978,* edited by C. Doumas, pp. 493–505. Thera and the Aegean World,London.

Noll, Walter, Reimer Holm and Liborius Born
1975 Painting of Ancient Ceramics. *Angewandte Chemie* 14(9):602–613.

O'Mansky, Matt and Nicholas P. Dunning.
2004 Settlement and Late Classic Political Disintegration in the Petexbatun Region, Guatemala. In: *The Terminal Classic in the Maya Lowlands: Collapse Transition and Transformation,* edited by Arthur A. Demarest, Prudence M. Rice and Don S. Rice, pp. 83–101. University Press of Colorado, Boulder Colorado.

Ower, Lesley H.
1926 Geology of British Honduras. *Imperial Geologist, Belize City* :494–509.

1927 Features of British Honduras. *Geographical Journal* 70:372–386.

1928a Geology of British Honduras. *Journal of Geology* 36:494–509.

1928b The Silica Lines of British Honduras. *Geological Magazine* 65:507–509.

Peacock, D.P.S.
1970 The Scientific Examination of Ceramics: A Review. *World Archaeology* 1:375–389.

Pendergast , David M.
1975 Lamanai 1975: The Goods Are Oft Interr'd With Their Bones. *ROM Archaeological Newsletter*, New Series, No. 122.

1978 Lamanai 1978, Part II: Crocodiles and Crypts. *ROM Archaeological Newsletter*, New Series, No. 163.

1979 *Excavations at Altun Ha, Belize, 1964-1970.* Archaeological Monograph 1. The Royal Ontario Museum, Toronto, Canada.

1980 A Dusty Tunnel. *ROM Archaeological Newsletter*, New Series, No. 176.

1981a Lamanai, Belize: Summary of Excavation Results, 1974–1980. *Journal of Field Archaeology* 8(1):29–53.

1981b A Regular Three-Ring Circus. ROM *ROM Archaeological Newsletter*, New Series, No. 192.

1981c The 1980 Excavations at Lamanai, Belize. *Mexicon* 2(6):96–99.

1981d Lamanai, Belize: 1981 Excavations. *Mexicon* 3(4):62–63.

1982 Lamanai, Belice, durante el Post-Clásico. *Estudios de Cultura Maya* XIV:19–58. Seminario de Cultura Maya, Universidad Nacional Autónoma de México, Mexico City.

1984 Excavations at Lamanai, Belize, 1983. *Mexicon* 6(1):5-10.

1985 Lamanai, Belize: An Updated View. In *The Lowland Maya Postclassic*, edited by Arlen F. Chase and Prudence M. Rice, pp. 91–103. University of Texas Press, Austin.

1986 Stability through Change: Lamanai, Belize, from the Ninth to the Seventeenth Century. In *Late Lowland Maya Civilization: Classic to Postclassic*, edited by Jeremy A Sabloff and E Wyllys Andrews V, pp. 223–250. School of American Research Advanced Seminar Series. Douglas W Schwartz, general editor, University of New Mexico Press, Albuquerque.

1988 Lamanai Stela 9: The Archaeological Context. *Research Reports on Ancient Maya Writing* 20. Center for Maya Research, Washington, D. C.

1990 Up from the Dust: The Central Lowlands Postclassic as Seen from Lamanai and Marco Gonzalez, Belize. In *Vision and Revision in Maya Studies*, edited by Flora Clancy and Peter D. Harrison, pp. 169–177. University of New Mexico Press, Albuquerque.

1992 Noblesse Oblige: The Elites of Altun Ha and Lamanai, Belize. In *Mesoamerican Elites: An Archaeological Assessment*, edited by Diane Z. Chase and Arlen F. Chase, pp. 61–79. University of Oklahoma Press, Norman.

1998 Intercessions with the Gods: Caches and Their Significance at Altun Ha and Lamanai, Belize. In *The Sowing and the Dawning: Termination, Dedication, and Transformation in the Archaeological and Ethnographic Record of Mesoamerica,* edited by Shirley Boteler Mock, pp. 55–63. University of New Mexico Press, Albuquerque.

Pohl, Mary and John Pohl
1994 Cycles of Conflict: Political Factionalism in the Maya Lowlands. In *Factional Competition and Political Development in the New World*, edited by Elizabeth M. Brumfiel and John W. Fox, pp. 138–157. Cambridge University Press, Cambridge.

Powis, Terry G.
2002 An Integrative Approach to the Analysis of the Late Preclassic Ceramics at Lamanai, Belize. Unpublished Ph.D. dissertation, Department of Anthropology, University of Texas, Austin.

Powis, Terry and Linda Howie
n.d. Compositional and Stylistic Variability of Late Preclassic and Protoclassic Ceramics at Lamanai, Belize. In *Archaeological Investigations in the Eastern Maya Lowlands: Papers of the 2005 Belize Archaeology Symposium*, edited by Jaime Awe, John Morris and Sherilyne Jones. Research Reports in Belizean Archaeology Volume 3.

Powis, T., L.A. Howie and E. Graham
2002 The Style and Petrography of Late Preclassic and Protoclassic Pottery at Lamanai, Belize. Paper presented at the 67th Annual Meeting of The Society for American Archaeology, Denver.

Purdy, E.G., W.C. Pusey III, and K.F. Wantland
1975 Continental Shelf of Belize: regional shelf attributes. In *Belize Shelf-Carbonate Sediments, Clastic Sediments and Ecology*, edited by K.F. Wantland and W.C. Pusey, pp. 1–39. American Association of Petroleum Geologists, Studies in Geology 2.

Pusey, W. C.
1975 Carbonate sedimentation on northern Belize shelf. In *Belize Shelf-Carbonate Sediments, Clastic Sediments and Ecology*, edited by K.F. Wantland and W.C. Pusey, pp. 131–233. American Association of Petroleum Geologists, Studies in Geology 2.

Rainbird, Paul
1999 Entangled Biographies: Western Pacific Ceramics and the Tombs of Pohnpei. *World Archaeology* 31(2):214-224.

Rands, Robert L
1973 The Classic Maya Collapse: Usumacinta Zone and the Northwestern Periphery. In *The Classic Maya Collapse*, edited by T. Patrick Culbert, pp. 165–206. School of American Research Advanced Seminar Series. Douglas W. Schwartz, general editor, University of New Mexico Press, Albuquerque.

1991 The Intellectual Development of Mesoamerican Archaeology and an Assessment of Shepard's Influence. In *The Ceramic Legacy of Anna O. Shepard*, edited by Ronald L. Bishop and Frederick W. Lange, pp. 154–176. University Press of Colorado, Niwot, Colorado.

Rands, Robert and Ronald Bishop

1980 Resource Procurement Zones and Patterns of Ceramic Exchange in the Palenque Region, Mexico. In *Models and Methods in Regional Exchange*, edited by Robert Fry, pp. 19–46. SAA Papers No. 1, Society for American Archaeology, Washington, D.C.

Rands, Robert L., Ronald L. Bishop and Jeremy A. Sabloff

1982 Maya Fine Paste ceramics: An archaeological perspective. In *Excavation at Seibal: Ceramics*, edited by J. A. Sabloff, pp. 315–338. Memoirs of the Peabody Museum of Archaeology and Ethnology, Harvard University, Vol. 13, No. 2, Harvard University, Cambridge, Massachusetts.

Rathje, William L.

1972 Praise the Gods and Pass the Metates: A Hypothesis of the Development of Lowland Rainforest Civilizations in Mesoamerica. In *Contemporary Archaeology: A Guide in Theory and Contributions*, edited by Mark P. Leone, pp. 365–392. Southern Illinois University Press, Carbondale, Illinois.

1973 Classic Maya Development and Denouement: A Research Design. In: *The Classic Maya Collapse*, edited by T. Patrick Culbert, pp. 405–456. School of American Research Advanced Seminar Series. Douglas W. Schwartz, general editor, University of New Mexico Press, Albuquerque.

1975 The Last Tango in Mayapán: A Tentative Trajectory of Production-Distribution Systems. In *Ancient Civilization and Trade,* edited by Jeremy A. Sabloff and C. C. Lamberg-Karlovsky, pp. 409–448. A School of American Research Book. University of New Mexico Press, Albuquerque.

Rathje, William L and Jeffrey A. Sabloff

1973 Ancient Maya Commercial Systems: A Research Design for the Island of Cozumel Mexico. *World Archaeology* 5: 221–231

1975 *A Study of Pre-Columbian Commercial Systems: The 1972–1973 Seasons at Cozumel, Mexico.* Peabody Museum of Archaeology and Ethnology. Harvard University, Cambridge, Massachusetts.

Reents-Budet, Dorie

1988 The Iconography of Lamanai Stela 9, Center for Maya Research, *Research Reports on Ancient Maya Writing* 22. Center for Maya Research, Washington, D. C.

1994 *Painting the Maya Universe: Royal Ceramics of the Classic Period.* Duke University Press, Durham, NC.

Reents-Budet, Dorie and Ronald L. Bishop

2003 More than Methodology: INAA and Classic Maya Painted Ceramics. In *Patterns and Process: A Festschrift in Honor or Dr. Edward V. Sayre*, edited by Lambertus van Zelst, pp. 93–108. Smithsonian Center for Materials Research and Education, Suitland, Maryland.

Reents-Budet, Dorie, Ronald L. Bishop, Jennifer T. Tascheck and Joseph W. Ball

2000 Out of the Palace Dumps: Ceramic production and Use at Buenavista del Cayo. *Ancient Mesoamerica* 11:99–121.

Reid, R. P., I. G. Macintyre and J. E. Post

1992 Micritized Skeletal Grains in Northern Belize Lagoon: A Major Source of Mg-Calcite Mud. Journal of Sedimentary Petrology 62(1):145-156.

Reina, Ruben F. and Robert M. Hill, II

1978 The Traditional Pottery of Guatemala. University of Texas Press, Austin.

Rice, Prudence M.

1976 Rethinking the Ware Concept. *American Antiquity* 41(4):538–543.

1980 Peten Postclassic Pottery Production and Exchange: A View from Macanche. In *Models and Methods in Regional Exchange*, edited by Robert Fry, pp. 67–82. SAA Papers No. 1, Society for American Archaeology, Washington, D.C.

1987a Lowland Maya Pottery Production in the Late Classic Period. In *Maya Ceramics: Papers from the 1985 Maya Ceramic Conference*, edited by P.M. Rice and R.J. Sharer, pp. 525–543. BAR International Series no. 345(ii). Oxford, U.K.

1987b *Macanche Island, El Peten, Guatemala: Excavations, Pottery and Artifacts*. University of Florida Press, Gainesville.

1987c *Pottery Analysis: A Sourcebook*. University of Chicago Press, Chicago.

Rice, Prudence M. and Don S. Rice
2004 Late Classic to Postclassic Transformations in the Petėn Lakes Region, Guatemala. In: *The Terminal Classic in the Maya Lowlands: Collapse Transition and Transformation,* edited by Arthur A. Demarest, Prudence M. Rice and Don S. Rice, pp. 125–139. University Press of Colorado, Boulder Colorado.

Rice, Prudence M., Arthur A. Demarest and Don S. Rice
2004 The Terminal Classic and the "Classic Maya Collapse" in Perspective. In: *The Terminal Classic in the Maya Lowlands: Collapse Transition and Transformation,* edited by Arthur A. Demarest, Prudence M. Rice and Don S. Rice, pp. 1–11. University Press of Colorado, Boulder Colorado.

Rice, Prudence M. and D. Forsyth
2004 The Terminal Classic-Period Lowland Ceramics. In: *The Terminal Classic in the Maya Lowlands: Collapse Transition and Transformation,* edited by Arthur A. Demarest, Prudence M. Rice and Don S. Rice, pp. 28-59. University Press of Colorado, Boulder Colorado.

Ringle, William M., Tomás Gallareta Negrón and George J. Bey III
1998 The Return of Quetzalcoatl: Evidence for the Spread of a World Religion during the Epiclassic Period. *Ancient Mesoamerica* 9:183–232.

Ringle, William M, George A Bey III, Tara Freeman Bond, Craig A. Hanson, Charles W Houck and Gregory J. Smith.
2004 The Decline to the East: The Classic to Postclassic Transition at Ek Balam, Yucatán. In *The Terminal Classic in the Maya Lowlands: Collapse Transition and Transformation,* edited by Arthur A. Demarest, Prudence M. Rice and Don S. Rice, pp. 485–516. University Press of Colorado, Boulder Colorado.

Robinson, G. M.
1983 The agricultural geography of Belize. In *Resources and Development in Belize: An Account of the University of Edinburgh Expedition to Central America, 1981*, edited by G. M. Robinson and P. A. Furley, pp. 21–74. Department of Geography University of Edinburgh, Edinburgh.

Robinson and Furley
1983 *Resources and Development in Belize: An Account of the University of Edinburgh Expedition to Central America, 1981*. Department of Geography University of Edinburgh, Edinburgh.

Rouse, Irving
1960 The Classification of Artifacts in Archaeology. *American Antiquity* 25(3):313-323

Ruz, Alberto L.
1965 Tombs and Funerary Practices in the Maya Lowlands. In *Archaeology of Southern Mesoamerica Part 1,* edited by G. R. Willey, pp.441-461. Handbook of Middle American Indians, Vol. 2, Robert Wauchope, general editor, The University of Texas Press, Austin.

Rye, Owen S.
1976 Keeping Your Temper Under Control: Materials and the Manufacture of Papuan Pottery. *Archaeology and Physical Anthropology in Oceania* 11:106-137.

1981 Pottery Technology: Principles and Reconstruction. Manuals on Archaeology 4. Taraxacum, Washington, D.C.

Sabloff, Jeremy A.
1973 Continuity and Disruption During Terminal Late Classic Times at Seibal: Ceramic and Other Evidence. In: *The Classic Maya Collapse*, edited by T. Patrick Culbert, pp. 107–132. School of American Research Advanced Seminar Series. Douglas W. Schwartz, general editor, University of New Mexico Press, Albuquerque.

1975 *Excavations at Seibal, Department of Peten, Guatemala: The Ceramics*. Memoirs of the Peabody Museum of Archaeology and Ethnology, Vol. 13, No. 2. Harvard University, Cambridge, Massachusetts.

Sabloff, Jeremy A and E. Wyllys Andrews V
1986 Introduction. In: *Late Lowland Maya Civilization: Classic to Postclassic*, edited by Jeremy A. Sabloff and E. Wyllys Andrews V, pp. 3–16. School of American Research Advanced Seminar Series. Douglas W Schwartz, general editor, University of New Mexico Press, Albuquerque.

Sabloff ,Jeremy A. and William L. Rathje
1975 The Rise of a Maya Merchant Class. *Scientific American* 233:72–82.

Sabloff, Jeremy A., William L Rathje, David A. Friedel, Judith G. Connor and Paula LW. Sabloff
1974 Trade and power in Postclassic Yucatan: initial observations. In *Mesoamerican archaeology – New Approaches : Proceedings of a Symposium on Mesoamerican Archaeology held by the University of Cambridge Centre of Latin American Studies, August 1972*, edited by Norman Hammond, pp. 397–416. Duckworth, London.

Sabloff, J.A. and G.R. Willey
1967 The collapse of Maya civilization in the southern lowlands: a consideration of history and process. *Southwestern Journal of Archaeology* 23:311–336.

Sanders, William T.
1973 The Cultural Ecology of the Lowland Maya: A Reevaluation. In: *The Classic Maya Collapse*, edited by T. Patrick Culbert, pp. 325–366. School of American Research Advanced Seminar Series. Douglas W. Schwartz, general editor, University of New Mexico Press, Albuquerque.

Sellett, F.
1993 Chaîne Opératoire: The Concept and Its Applications. *Lithic Technology* 18:106–112.

Sharer, Robert J.
1977 The Maya Collapse Revisited: Internal and External Perspectives. In *Social process in Maya prehistory: studies in honour of Sir Eric Thompson*, edited by Norman Hammond, pp. 531–552, Academic Press, London.

1994 *The Ancient Maya*, 5th edition. Stanford University Press, Stanford.

Shaw, Joseph W., Aleydis Van de Moortel, Peter M. Day and Vassilis Kilikoglou
2001 *A LM IA Ceramic Kiln in South-Central Crete: Function and Pottery Production*. Hesperia Supplement 30. American School of Classical Studies at Athens, Princeton, New Jersey.

Shaw, Justine M.
2003 Climate Change and Deforestation. Implications for the Maya Collapse. Ancient Mesoamerica 14:157-167

Shelby, Thomas Mark
1999 Report of the 1998 and 1999 Investigations on the Archaeology and Iconography of the Polychrome Stucco Façade of Structure N10–28, Lamanai, Belize. FAMSI Report No. 98037. Electronic document, http://www.famsi.org/reports/98037/index.html, accessed December 10, 2005.

Shepard, Anna O.
1939 Technological Notes of the Pottery of San Jose. In *Excavations at San Jose, British Honduras*, edited by John Eric Thompson, pp. 251–277. Carnegie Institution of Washington, Washington.

1948 *Plumbate: A Mesoamerican Trade Ware*. Carnegie Institution of Washington Publication 573. The William Byrd Press, Inc., Richmond, Virginia.

1956 *Ceramics for the Archaeologist*. Carnegie Institution of Washington Publication 609. Washington, DC.

1964 Temper Identification: "Technological Sherd-Splitting" or an Unanswered Challenge. *American Antiquity* 29:518–520.

Sidrys, Raymond V.
1983 *Archaeological Excavations in Northern Belize, Central America*. Monograph XVII. Institute of Archaeology, University of California, Los Angeles.

Simmon, M. and G.F. Brem

1979 The analysis and Distribution of Vocalic Ash-Tempered Pottery in the Lowland Maya Area. *American Antiquity* 44:79–91.

Shotton, F.W. and G.L. Hendry

1979 The Developing Field of Petrology in Archaeology. *Journal of Archaeological Science* 6:75–84.

Smith, Robert E.

1955 *Ceramic Sequence at Uaxactun Guatemala, Vol. 1.* Middle American Research Institute Publication No. 20. Tulane University, New Orleans.

1971 *The Pottery of Mayapan Including Studies of Ceramic Material from Uxmal, Kabah, and Chichen Itza, 2 vols.* Peabody Museum of Archaeology and Ethnology Papers vol. 66. Harvard University, Cambridge, Massachusetts.

Smith, Robert E., Gordon R. Willey and James C. Gifford

1960 The Type–Variety Concept as a Basis for the Analysis of Maya Pottery. *American Antiquity* 25(3):330–340.

Smyth, Michael P., Christopher D. Dore, Hector Neff and Michael D. Glascock

1995 The Origin of Puuc Slate Ware: New Data from Sayil, Yucatan, Mexico. *Ancient Mesoamerica* 6:119–134

Stanchly, N.

2001 Preliminary Observations: The Faunal Assemblage from Structure N10-9 (2000, 2000, 2001 Field Seasons). Ms. On file with the author.

Stuart

1993 Historical Inscriptions and the Maya Collapse. In *Lowland Maya Civilization in the Eighth Century A.D.*, edited by Jeremy A. Sabloff and John S. Henderson, pp. 321–354. Dumbarton Oaks, Washington, D.C.

Suhler, Charles, Traci Arden, David Freidel and Dave Johnstone

2004 The Rise and Fall of Terminal Classic Yaxuna, Yucatán, Mexico. In: *The Terminal Classic in the Maya Lowlands: Collapse Transition and Transformation,* edited by Arthur A. Demarest, Prudence M. Rice and Don S. Rice, pp. 450–484. University Press of Colorado, Boulder Colorado.

Sullivan, Lauren A.

2002 Dynamics of Regional Integration in Northwestern Belize, In *Ancient Maya Political Economies,* edited by Marilyn A Masson and David A. Freidel, pp. 197–222. Altamira Press, New York.

Thompson, John Eric

1939 *Excavations at San Jose, British Honduras.* Carnegie Institution of Washington, Washington.

1954 *The Rise and Fall of Maya Civilization.* University of Oklahoma Press, Oaklahoma.

Tite, M.S.

1995 Firing temperature determinations – how and why? In *The Aim of Laboratory Analysis of Ceramics in Archaeology*, edited by A Lindahl and O. Stilborg, pp. 37–42. Kungl. Vitterhets Historie och Antikvitets Akadamien Konferenser 34, Stockholm.

1999 Pottery Production, Distribution and Consumption – Contribution of the Physical Sciences. *Journal of Archaeological Method and Theory* 6(3):181–233.

Tite, M.S. and Y. Maniatis

1975 Scanning Electron Microscopy of Fired Calareous Clays. *Transactions and Journal of the British Ceramic Society* 74:19–22.

Tite, M.S., I.C. Freestone, N.D. Meeks and M. Bimson

1982 The Use of Scanning Electron Microscopy in the Technological Examination of Ancient Ceramics. In *Archaeological Ceramics*, edited by J.S. Olin and A.D. Franklin, pp. 109–120. Smithsonian Institution Press, Washington, D.C.

Tomkins, Peter, Peter M. Day and Vassilis Kilikoglou

2004 Knossos and the earlier Neolithic landscape of the Herakleion Basin. In Knossos: Palace, City State: Proceedings of the Conference in Herakleion organized by the British School at Athens and the 23rd Ephoreia of Prehistoric and Classical Antiquities of Herakleion, in November 2000, for the Centenary of Sir Arthur Evans's Excavations at Knossos, edited by Gerald Cadogan, Eleni Hatzaki and Adonis Vasilakis. British School at Athens Studies 12, Athens.

van der Leeuw, Sander

1993 Giving the Potter a Choice: Conceptual aspects of pottery techniques. In *Technological Choices: Transformation in materials cultures since the Neolithic*, edited by Pierre Lemonnier, pp. 238–288. Routeledge, London.

van der Leeuw, Sander E., Dick Papousek and Anick Coudart

1991 Technical Traditions and Unquestioned Assumptions: The Case of Pottery in Michoacan. *Techniques et culture* 17–18:145–173.

Valdez Jr., Fred

1987 The Prehistoric Ceramics of Colha, Northern Belize. Unpublished Ph.D. dissertation Department of Anthropology, Harvard University, Cambridge M.A.

1994 The Colha Ceramic Complexes. In *Continuing Archaeology at Colha, Belize,* edited by Thomas R. Hester, Harry J. Shafer and Jack D. Eaton, pp. 9–16. Studies in Archaeology 16. Texas Archaeological Research Laboratory, The University of Texas at Austin, Austin.

Valdez Jr., Fred and Richard.E.W. Adams

1982 The Ceramics of Colha after Three Field Seasons, 1979–1981. In *Archaeology at Colha, Belize: The 1981 Interim Report*, edited by Thomas R. Hester, Harry J. Shafer, and Jack D. Eaton, pp. 21–20. Center for Archaeological Research, The University of Texas at San Antonio, San Antonio, Texas.

Versey, H.R.

1972 Karst in Jamaica. In *Karst: Important Karst Regions of the Northern Hemisphere*, edited by M. Herak & V.T. Springfield, Amsterdam and New York: Elsevier.

Vinson, G. L.

1962 Upper Cretaceous and Tertiary Stratigraphy of Guatemala, *Bulletin of the American Association of Petroleum Geologists* 46:425–456.

Walker, S. H.

1973 *Summary of Climatic Records for Belize.* Land Resource Division, Supplement No.3, Surrey, England.

Walker, Debra Selsor

1990 Cerros Revisited: Ceramic Indicators of Terminal Classic and Postclassic Settlement and Pilgrimage in Northern Belize. Unpublished Ph.D. dissertation, Graduate Faculty of Dedman College, Southern Methodist University, Dallas, Texas.

Webb, Malcolm C.

1973 The Peten Maya Decline Viewed in Perspective in the Perspective of State Formation. In: *The Classic Maya Collapse*, edited by T. Patrick Culbert, pp. 367–404. School of American Research Advanced Seminar Series. Douglas W. Schwartz, general editor, University of New Mexico Press, Albuquerque.

Webster, David L.

1998 Warfare and Status Rivalry: Lowland Maya and Polynesian Comparison. In *Archaic States*, edited by G. M. Feinman and J. Marcus, pp. 311–351. School of American Research Press, Santa Fe, New Mexico.

Webster, David, AnnCorinne Freter and Rebecca Storey.

2004 Dating Copán Culture-History: Implications for the Terminal Classic. In: *The Terminal Classic in the Maya Lowlands: Collapse Transition and Transformation,* edited by Arthur A. Demarest, Prudence M. Rice and Don S. Rice, pp. 231–259. University Press of Colorado, Boulder, Colorado.

Weigand, P.C., G. Harbottle and E.V. Sayre
1977 Turquoise sources and source analysis: Mesoamerica and the Southwestern U.S.A.. In *Exchange Systems in Prehistory*, edited by T.K. Earle and J.E. Ericson, pp. 15–34. Academic Press, New York.

Welsh, W. Bruce M.
1988 *An Analysis of Classic Lowland Maya Burials*. BAR International Series 409. British Archaeological Reports, Oxford.

West, Georgia
2002 Ceramic Exchange in the Late Classic and Postclassic Maya Lowlands: a Diachronic Approach. In *Ancient Maya Political Economies,* edited by Marilyn A. Masson and David A. Freidel, pp. 140–196. Altamira Press, New York.

Wheat, Joe Ben, James C. Gifford, and William W. Wasley
1958 Ceramic Type, Type Cluster, and Ceramic System in Southwestern Pottery Analysis. *American Antiquity* 24(1):34-47.

Whitbread, I.K.
1986 The Characterisation of Argillaceous Inclusions in Ceramic Thin Sections. *Archaeometry* 28:79–88.

1989 A Proposal for the Systematic Description of Thin Sections Towards the Study of Ancient Ceramic Technology. In *Archaeometry: Proceedings of the 25th International Symposium*, edited by Y. Maniatis, pp. 127–138. Elsevier, Amsterdam.

1995 *Greek Transport Amphorae: A Petrological and Archaeological Study*. The British School at Athens Fitch Laboratory Occasional Paper 4. The Short Run Press, Exeter.

1996 Detection and Interpretation of Preferred Orientation in Ceramic Thin Sections. In *Imaging the Past: Electronic Imaging and Computer Graphics in Museums and Archaeology*, edited by Janet Lang, Peter Main , Tony Higgins, pp. 413–425. The British Museum, Department of Scientific Research, Occasional Paper 114, London.

White, Christine D.
1997 Ancient Diet at Lamanai and Pacbitun: Implications for the Ecological Model of Collapse. In *Bones of the Maya: Studies of Ancient Skeletons,* edited by Stephen L. Whittington and David M. Reed, pp. 171–180. Smithsonian Institution, Washington D.C.

Whitelaw, T.M., P.M. Day, E. Kiriatzi, V. Kilikoglou and D.E. Wilson
1997 Ceramic traditions at EM IIB Myrtos, Fournou Korfi. In *TEXNH. Craftsmen, Craftswomen and Craftsmanship in the Aegean Bronze Age (Aegaeum)*, edited by R. Laffineur and P.P. Betancourt, pp. 265–275. Liege and Austin.

Willey, Gordon R.
1974 The Classic Maya hiatus: a 'rehearsal' for the collapse? In *Mesoamerican archaeology – New Approaches : Proceedings of a Symposium on Mesoamerican Archaeology held by the University of Cambridge Centre of Latin American Studies, August 1972*, edited by Norman Hammond, pp. 417–430. Duckworth, London.

Willey, Gordon R., T. Patrick Culbert and E.W. Adams
1967 Lowland Maya Ceramics: A Report from the 1965 Guatemala City Conference. *American Antiquity* 32:289–315.

Willey, Gordon R and Demitir B. Shimkin.
1973 The Maya Collapse a Summary View. In: *The Classic Maya Collapse*, edited by T. Patrick Culbert, pp. 457–502. School of American Research Advanced Seminar Series. Douglas W. Schwartz, general editor, University of New Mexico Press, Albuquerque.

Wilson, D.E. and P.M. Day
1994 Ceramic Regionalism in Prepalatial Central Crete: The Mesara Imports at EM I to EM IIA Knossos. *The Annual of the British School at Athens* 89:1–95.

Wright, A. C. S., D. H. Romney, R. H. Arbuckle and V. E. Vial
1959 *Land in British Honduras: Report of the British Honduras Land Use Survey Team.* Colonial Office, Colonial Research Publications No. 24, D. H. Romney, general editor, Her Majesty's Stationary Office, London.

Yaeger, Jason and Marcello A. Canuto
2000 Introducing an archaeology of communities. In *The Archaeology of Communities,* edited by Marcello A. Canuto and Jason Yaeger, pp. 1–15. Routledge, London.

www.ingramcontent.com/pod-product-compliance
Lightning Source LLC
LaVergne TN
LVHW072251300926
849160LV00018B/1641